*CAMBRIDGE MONOGRAPHS ON PHYSICS*

GENERAL EDITORS

A. HERZENBERG, Ph.D.
*Reader in Theoretical Physics in the University of Manchester*
M. M. WOOLFSON, D.Sc.
*Professor of Theoretical Physics in the University of York*
J. M. ZIMAN, D.Phil, F.R.S.
*Professor of Theoretical Physics in the University of Bristol*

# AN INTRODUCTION TO HAMILTONIAN OPTICS

# AN INTRODUCTION TO HAMILTONIAN OPTICS

BY

H. A. BUCHDAHL

*Professor of Theoretical Physics*
*Australian National University*



CAMBRIDGE
AT THE UNIVERSITY PRESS
1970

Published by the Syndics of the Cambridge University Press
Bentley House, 200 Euston Road, London N.W.1
American Branch: 32 East 57th Street, New York, N.Y.10022



Library of Congress Catalogue Card Number: 69-19372

Standard Book Number: 521 07516 5

Printed in Great Britain
at the University Printing House, Cambridge
(Brooke Crutchley, University Printer)

# CONTENTS

## CHAPTER 5

# The symmetric system (part II)

### A. THE SINE-CONDITION



### B. THE SPHERICAL POINT CHARACTERISTIC $V^{\dagger}$, WAVE-FRONTS AND ABERRATION FUNCTIONS



### C. THE DEPENDENCE OF THE ABERRATIONS ON THE POSITIONS OF THE OBJECT AND OF THE STOP



### D. INVARIANT AND SEMI-INVARIANT ABERRATIONS



## CHAPTER 6

# Symmetric systems with additional symmetries

### A. REVERSIBLE SYSTEMS

### B. CONCENTRIC SYSTEMS



## CHAPTER 7

## The semi-symmetric system



## CHAPTER 8

## Systems with translational symmetry, and plane-symmetric systems



## CHAPTER 9

## Systems with toroidal symmetry



## CHAPTER 10

## Chromatic defects of the image

## CHAPTER 11

## Anisotropy



## CHAPTER 12

## The computation of aberration coefficients

# FOREWORD

Geometrical optics is an outstanding example of the permanence of a basic subject. Though its history extends over many centuries there is today a ground swell of renewed interest in it, for which the probe into space, the laser, and computing machines are perhaps responsible. The large computers, in particular, have required critical examination of all the tried and tested procedures of analysis. Persons involved in laser cavity design, electron optics, crystal optics and focusing of nuclear particles and Cherenkov radiation, have all turned to the basic literature in geometrical optics.

After hundreds of years of use and development, it is not surprising that the literature in the field is extensive. It is difficult for a newcomer to the field to know where to start his learning. The original works are usually excellent sources, but they are often difficult to obtain and to read because of their language and fragmented thought. Most readers today look for a modern treatment, and here lurk severe difficulties.

Many of the modern writings suffer from a variety of basic defects. Thus, the author's interests are too mathematical, or else too casual, and their work does not reveal the kind of feeling for the subject which actual practice demands. Then again several designers of optical instruments have attempted to describe the understanding they acquired through years of experience. These authors are often not trained in the discipline of mathematical writing. The result is that they derive equations clumsily, use poor notation, and usually over-step their understanding by incorrectly generalizing from a few special cases.

This book now provides us with a modern account of the Hamiltonian treatment of geometrical optics which overcomes the criticisms above. Dr Buchdahl has written in various fields of theoretical physics in which he has proven his ability to straighten things out. His interest in geometrical optics is far from casual. He has spent several years writing on aberration theory, and the formulae derived include fifth, seventh and higher orders. His formulation has been coded into machine language and resides on the memory disks of computing machines throughout the world.

Persons with a wide variety of interests can now expand their understanding of this fascinating field. Practising engineers—who have learned optics by doing optics—will find that their vague hunches and concepts can indeed often be formulated with precision. Those who are primarily interested in the more formal aspects of the subject will find the book satisfying and instructive. The beginner may be confident that it will keep him on a straight path; and he will find that Buchdahl's understanding runs so deep that his presentation flows with ease.

It is my hope that this book will serve as an authoritative updating of optical aberration theory, and that future authors in this field will be guided by Buchdahl's notation and organization of concepts.

R. E. HOPKINS

*Rochester*, *April 1968*

# PREFACE

Although the propagation of light is described in all its details by Maxwell's equations, many optical problems can be solved to a sufficient degree of approximation on the basis of the laws of geometrical optics. On this level a thorough understanding of the over-all features of the imagery produced by the types of systems commonly encountered in practice is most easily gained through Hamilton's method. This presentation of it is partly intended to stimulate the teaching of the subject; in at least equal measure it is directed towards those whose task it is actually to design image-forming instruments. For this reason, if no other, the mathematical knowledge required of the reader has been kept to a minimum. For example, although the whole theory arises directly from a variational principle, Euler's equations or formal variational derivatives are never considered explicitly. They are replaced throughout by elementary geometrical constructions. If some of the latter pages look a little bewildering at first glance, this is solely because they concern themselves with the kind of detail which is often quietly swept under the carpet; without which, however, one has merely a general framework, far removed as yet from what one needs to know in everyday practice.

I have resisted the temptations of sophistication: with rare exceptions mathematical niceties have been given short shrift, whilst the many analogies which can be drawn between the methods of Hamiltonian optics and those arising in other branches of physics have been left aside. Pedagogically this is regrettable, bearing the present unpopularity of geometrical optics as a subject in mind. In this context the following remarks may not be out of place.

The power of the Hamiltonian method lies in its ability to yield over-riding results governing the behaviour of types of systems defined solely with respect to the symmetries they possess. In terms of contemporary jargon, we have indeed a typical 'black box theory': the actual constitution of the optical system is left unspecified, that is, nothing is said as to whether it is made up of a finite number of refracting (or reflecting) surfaces, or whether the refractive index varies continuously, or partly both; nor is anything said as to the precise shape of any refracting or reflecting surfaces which may be known to be present. In short, the only information which is regarded as given is that the system as a whole has certain symmetries. Any single one of these is symmetry about either a line or about a plane; but the system may have several such symmetries simultaneously. Thus the kind of system conventionally called merely 'symmetric' has two, namely about a line, and a plane containing this line; whereas the semi-symmetric system has only the first of these. Now, to every symmetry of a system there corresponds an invariance property of its characteristic function; that is to say, there is a linear substitution of the ray-coordinates which leaves its form unaffected. Every symmetry therefore implies a restriction upon the generic form of the characteristic function, or more particularly, of its representation as a power series. Its coefficients are the (characteristic) aberration coefficients; and if we contemplate in the first place the set of $n$th-order coefficients of the most general system, any condition of symmetry subsequently imposed upon the latter implies relations between these coefficients which may, in particular, take the form of necessarily assigning to some of them the value zero. At any rate, these relations impose powerful limitations upon the generic character of the aberrations produced by the system; and we may study them without being encumbered by the need to determine how their values depend upon the details of its constitution.

The situation just outlined is a counterpart to that which prevails in quantum mechanics. To any symmetry of a physical system there corresponds an invariance property (or, as one says, 'a symmetry') of the Hamiltonian, which now takes the place of the characteristic function. The analogy can be pursued in various ways. For example, we might contemplate the wave-function of a general physical system (the number of degrees of freedom alone being prescribed), and suppose it to be expanded in terms of some basis: then any symmetry of the Hamiltonian will imply certain relations between the coefficients of the expansion. Suffice it to say that throughout theoretical physics a great deal of attention is focused upon the exploitation of known 'symmetries', especially in circumstances in which they constitute all the available information; as may be the case in some aspects of scattering theory in which the region of interaction is then simply a 'black box'. Granted that it is desirable to introduce students to such relatively sophisticated notions as early as possible, it is evidently much easier to do so in the context of geometrical optics, since one is then not confronted with the largely irrelevant conceptual difficulties which quantum mechanics presents.

Even when the pedagogic virtues of the theory which have just been discussed are set entirely aside, it is still desirable to deal in orderly progression with classes of systems defined by the various symmetries they possess; and that is just what is done in this monograph. Any redundancy which such a procedure entails is more than counterbalanced by the way in which repetition engenders easy familiarity. With the demands of practice uppermost in my mind I have tended to steer clear of 'general' theory—there is no discussion of the Maxwell fish eye, for example. Instead, I have elaborated in much detail all manner of problems relating to higher-order aberrations; whilst I have tried to clear up the doubts and misunderstandings which one encounters again and again in the context of the sine-relation and sine-condition, and of the so-called $D-d$ method in the theory of chromatic aberrations, to cite but two examples. Further, the detailed treatment of questions related to reversibility, the sine- or cosine-relations, and the like, is not confined to the 'symmetric system' alone. Again, anyone who has ever encountered the (primary) Petzval curvature of field must have wondered whether, or in what sense, there exist higher-order aberrations with analogous properties. This question also is answered in much detail under the heading of 'invariant and semi-invariant aberrations', once the theory of shifts of the object and of the stop has been dealt with at length. At any rate, at this point the reader may do well to read through the list of contents, for the section headings will tell him concisely what he may expect to find. He may be surprised at the presence of a chapter on anisotropic media; but its inclusion is made necessary by the need to show that the explanations of certain earlier, seemingly paradoxical, results are indeed meaningful within the framework of the theory as a whole.

At least by implication I have already expressed the view that the importance of Hamilton's method derives mainly from its ability to yield, in the simplest possible way, general statements about what, in principle, a given type of system can or cannot do.

That is not to say, however, that Hamiltonian optics deals solely with such generalities. On the contrary, it is quite possible to compute the actual numerical values of the coefficients of the characteristic functions of a given system upon the basis of expressions derived from within the theory in hand. Chapter 12 indeed represents a brief introduction to this aspect of the subject. If I have any lingering doubts in this context, they concern merely certain purely practical considerations; namely, whether direct, that is to say Lagrangian, methods are not better suited to the purposes of numerical work. At any rate, this vexed question is discussed at length in Section 116.

It has already been remarked that the mathematical level of this work has deliberately been left as low as possible. That the array of symbols and affixes of all kinds is rather formidable cannot be denied. It is sometimes claimed that the presence of a variety of superscripts, subscripts, asterisks, primes, and so on makes a theory difficult to read. It seems to me that this is true only from the point of view of superficial appearance. The more affixes there are the greater is the amount of explicit information contained within each (composite) symbol, and the interpretation of any given equation becomes correspondingly easier. As more and more affixes are suppressed, more and more sources of ambiguity arise, and ever greater feats of memory are required. At any rate, a list of the principal symbols used is included which may, in particular, help anyone who is not in the habit of reading books systematically from end to end.

Whilst on the subject of notation, it should be remarked that the sign conventions used are, of course, those of analytical geometry. The basic terminology, and the use of the symbols $V$, $T$ and $W$ for the point, angle, and mixed characteristics respectively naturally goes back to Hamilton. However, in one important point I have failed to follow Hamilton's example, albeit very reluctantly: the values of quantities defined both in the object space and in the image space are represented by unprimed and primed symbols respectively. Hamilton adopted the opposite convention, which is much more sensible, since one is concerned most of the time with the details of the disposition of rays in the image space. Alas, the modern convention is so firmly entrenched that it seemed better not to run counter to it.

Problems are provided at the end of each chapter. Since they would be useless for any reader who cannot solve them even after prolonged attempts to do so, their solutions are given in reasonable detail at the end of the book.

It is my earnest hope that no author will feel aggrieved if he finds no acknowledgement of his work here. The truth is that I may not know of it, or else that it has not contributed substantially to the material presented, for I have worked almost everything out from scratch. G. C. Steward alone I must mention explicitly, for the style and content of his writings have been a constant source of inspiration to me. Consistently with what has just been said, the bibliography is very circumscribed, and if it seems to lean towards the mathematical side, this is solely because I have found just the quoted works most useful in securing my own knowledge; and at the same time I nowhere make any claims to priority, though I believe certain parts of this work to be original.

It is a pleasure to express my warmest thanks to the Institute of Optics in the University of Rochester, N.Y., and particularly to its Director, Dr W. L. Hyde, and to Professor R. E. Hopkins for the hospitality I enjoyed whilst I was a visitor there as New York State Professor of Optics during the academic year 1967–8. Funds to support this professorship were provided by the New York State Science and Technology Foundation under a grant from the state legislature. It was during that time that this monograph was written, and I presented the material of the first seven or eight chapters as a one-semester course to graduate students of the Institute. My thanks go to Mrs Dorothy McCarthy for the competent way in which she typed a demanding manuscript. I derived much benefit from the many discussions I had with Dr Peter Sands; and for these I am most grateful to him. To the Cambridge University Press I wish to express my warmest appreciation for their unfailing helpfulness and courtesy; and last, but not least, it is a pleasure to express my appreciation to Miss Jackie Flint for her remarkably careful and patient assistance in reading the proofs.

H. A. BUCHDAHL

*Rochester, March 1968, and*
*Canberra, May 1969*

CHAPTER I

# INTRODUCTION

## 1. Introductory remarks

Light is an electromagnetic disturbance characterized by its rapid oscillation in time and space. Its propagation through a transparent medium is therefore described in all its detail by Maxwell's equations, at any rate on the classical (i.e. non-quantal level). However, the wavelengths with which one is concerned are very small, being of the order of $10^{-5}$ cm; for which reason a good approximation to the laws of propagation may be obtained in many situations by disregarding the finiteness of the wavelength altogether. In other words, one contemplates the formal limit in which the wavelength $\lambda$ is allowed to tend to zero (see Section 115). All questions concerning the phenomena of interference and diffraction (which are of course present in any actual physical situation) then remain unanswered. One often disregards also the polarization of light, as indeed we shall do here. One is left with a basically very simple theory in which the laws of propagation bear an essentially geometrical character; for which reason this theory is called *Geometrical Optics*.

At this stage it is convenient to introduce the restriction that, except in Chapter 11, all optical media to be considered are isotropic. A medium is *isotropic* if at every point its physical properties are independent of direction; in the contrary case it is anisotropic. (Actually, as will be explained in detail in Section 110, almost all of our work remains valid in the presence of internal anisotropy; meaning that isotropy is required only in certain regions surrounding the object and image.) Isotropy has the important consequence that the direction of energy flow is in the direction of the normal to the electromagnetic wavefront. One may therefore contemplate a set of curves, the *rays*, such that the tangent to a ray at any point gives the direction of energy transport at that point, i.e. the direction of propagation of light. This construction remains meaningful in the geometrical-optical limit. Accordingly two alternatives offer themselves: to base the theory either directly on the construction and properties of wavefronts, or else on the consideration of the properties of families of rays. It is a matter of taste which of these alternatives one adopts. In any case the two methods are so closely related that the differences between them are little more than a matter of detail. Given a set, or 'congruence', of rays which originated from a point source, these rays must be the orthogonal trajectories of a set of surfaces, and each such surface is just a wave-surface or *wavefront*. Accordingly, the wave-surfaces can be constructed from the rays, and vice versa. Parenthetically, given any congruence of curves, there need by no means exist a family of surfaces cut orthogonally by the curves. If such a family does exist the congruence is *normal*, otherwise it is *skew*. For the present we assume that in every case through every point of the region of interest there passes one and only one curve.

It may not be out of place here to remark that to some extent a ray can be realized physically by allowing light from a distant source to fall upon a screen provided with a circular aperture of sufficiently small radius $r$. The tube of light which emerges from the screen shrinks to a curve as $r \to 0$, and this curve is a 'ray'. It is of course essential to recall that diffraction effects are being disregarded, since we are pretending that $\lambda = 0$. In actual practice one can construct a 'ray' in this manner only approximately, since the condition $r \gg \lambda$ has to be satisfied.

The stage has been reached to make a decision as to whether to place verbal emphasis on rays or on wave-surfaces; and we choose the former. The reason for this is as follows. Our primary interest centres around discovering the over-all properties of any given generic type of optical system, the general character of its imagery, and the way in which this depends upon the physical symmetries of the system. The most suitable tool for this purpose is probably Hamilton's method, the foundations of which were laid by Sir William Rowan Hamilton in his *Theory of Systems of Rays*, published in the years 1828–37. The very title of his work implies an emphasis on rays. This naturally arises from the fact that the whole theory is based very directly on Fermat's Principle, which is a statement about a general property of rays; as we shall see.

The main virtue of Hamiltonian optics lies in its ability directly to yield general statements about the over-all properties of optical systems, without any need to inquire into the details of their construction. Thus, taking the symmetric system as an example—to be studied in detail later on in Chapters 4, 5 and 6—it is irrelevant whether one has a finite or an infinite number of refracting or reflecting surfaces, and it is likewise irrelevant whether these surfaces are spherical or not. In short, all one needs to know is that the system is symmetric. Analogous conclusions hold whatever symmetries, if any, the system may possess.

## 2. Fermat's Principle

Rather than introduce Fermat's Principle immediately as an axiom we shall motivate it by reference to the elementary laws of refraction. To this end recall that in a homogeneous medium light rays are straight; and that any such medium is characterized by a certain number $N$, its *refractive index*. What part this plays we shall see in a moment.

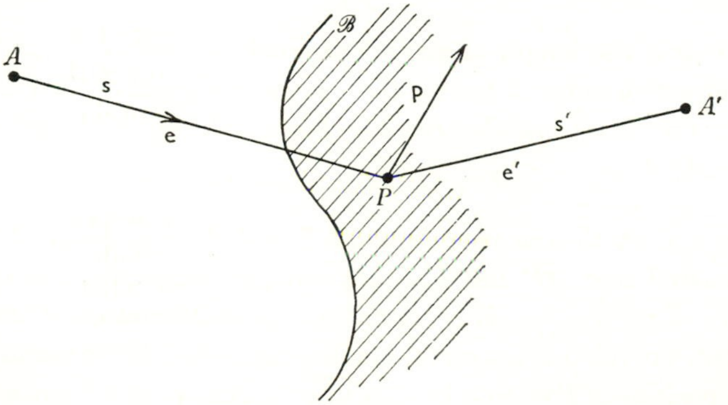


Fig. 1.1

Now suppose one has a pair of homogeneous media whose refractive indices are $N$ and $N'$ respectively, separated by a smooth boundary $\mathscr{B}$. A light ray passes from a point $A$ in the first medium to a point $A'$ in the second, intersecting $\mathscr{B}$ in a point $P$, the *point of incidence*. $AP$ and $PA'$ must of course be straight, according to what has already been said, but the angle between $AP$ and $PA'$ will in general be different from zero. Now let $\mathsf{e}$, $\mathsf{e}'$ be unit vectors in the directions $AP$, $PA'$ respectively, whilst $\mathsf{p}$ is the unit normal to $\mathscr{B}$ at $P$. The laws of refraction can now be summed up in the single equation

$$N'\mathsf{e}' - N\mathsf{e} = (N' \cos I' - N \cos I)\,\mathsf{p}, \tag{2.1}$$

where $I$ and $I'$ (the 'angles of incidence and refraction') are the angles between $\mathsf{e}$ and $\mathsf{p}$, and between $\mathsf{e}'$ and $\mathsf{p}$ respectively. We see at once that the three vectors $\mathsf{e}$, $\mathsf{e}'$ and $\mathsf{p}$ are coplanar, whilst taking the vector product of both members of (2.1) with $\mathsf{p}$ gives Snell's Law:

$$N' \sin I' = N \sin I. \tag{2.2}$$

We note in passing that if $\mathscr{B}$ is a reflecting surface, one can retain (2.1) by adopting the formal device of setting $N' = -N$. At any rate, one can now use (2.1) to trace a ray step by step through any system composed of homogeneous media and reflecting surfaces; but this is just the kind of thing with which we do not want to concern ourselves now.

Given any curve $\mathscr{C}$ joining two points $A$ and $A'$, the *optical length* of $\mathscr{C}$ is defined to be the value of the integral

$$V^* = \int_A^{A'} N\,ds, \tag{2.3}$$

where $ds$ is the length of an infinitesimal element of $\mathscr{C}$, and $N$ is the refractive index of the medium at the midpoint of the element. Under the conditions illustrated by Fig. 1.1 the optical length of the curve $APA'$ is

$$V^*(APA') = N's' + Ns, \tag{2.4}$$

where $s$, $s'$ are the distances between $P$ and $A$, $A'$ respectively.

Granted that $AP$ and $PA'$ are straight segments, the optical length of the curves $APA'$ is defined no matter where $P$ may be located, provided it lies on $\mathscr{B}$. Having calculated $V^*$ for some arbitrary position of $P$ we may inquire into the change of $V^*$ consequent upon $P$ being slightly displaced. Thus

$$\delta V^* = N'\delta s' + N\delta s = N'\mathsf{e}' \, . \, \delta\mathsf{s}' + N\mathsf{e} \, . \, \delta\mathsf{s}.$$

Since $\mathsf{s} + \mathsf{s}'$ is obviously constant it follows that

$$\delta V^* = (N'\mathsf{e}' - N\mathsf{e}) \, . \, \delta\mathsf{s}'. \tag{2.5}$$

However, recall that $P$ is constrained to lie in $\mathscr{B}$, i.e. $\delta\mathsf{s}'$ must be normal to $\mathsf{p}$. This condition can be accommodated by writing

$$\delta\mathsf{s}' = \mathsf{p} \times \delta\mathsf{a},$$

where $\delta\mathsf{a}$ is an arbitrary infinitesimal vector. (2.5) becomes

$$\delta V^* = [(N'\mathsf{e}' - N\mathsf{e}) \times \mathsf{p}] \, . \, \delta\mathsf{a}.$$

Accordingly $\delta V^* = 0$ for all allowed variations of $P$ if and only if

$$(N'\mathsf{e}' - N\mathsf{e}) \times \mathsf{p} = 0,$$

i.e.
$$N'\mathsf{e}' - N\mathsf{e} = \sigma\mathsf{p}, \tag{2.6}$$

where $\sigma$ is some scalar factor. Scalar multiplication of (2.6) throughout by $\mathsf{p}$ then gives (2.1) at once.

We have just shown that the optical length $V^*(APA')$ is stationary for all small displacements of $P$ in $\mathscr{B}$ if $\mathsf{e}$ and $\mathsf{e}'$ satisfy the laws of refraction; and that, conversely, the stationary property $\delta V^* = 0$ implies that the laws of refraction are satisfied. In other words, the condition that $V^*$ be stationary with respect to arbitrary small variations of the position of the point $P$ in $\mathscr{B}$ just singles out that path joining $A$ and $A'$ which is a light ray as it actually occurs in nature. The value of $V^*$ calculated for this particular path will be called the *optical distance* between $A$ and $A'$, to be denoted by $\tilde{V}$.

The crucial point of the preceding analysis of a special case is that it reveals the possibility of stating the laws governing rays in a form in which all explicit reference to planes of incidence or angles of incidence and refraction is avoided. We may expect that this possibility exists also under general circumstances, since one may deal with rays passing through a succession of homogeneous media in much the same way as when there are only two. Indeed, this generalization is almost trivial; and then one can think of an inhomogeneous medium as stratified into an indefinitely large number of homogeneous layers, the mutual boundary between any two of them being smooth. In short, the time has come to state Fermat's Principle, which is, after all, nothing but a generalization of the results obtained in the special situation contemplated above.

In terms of a Cartesian set of coordinates $\bar{x}, \bar{y}, \bar{z}$ let $A(x, y, z)$ and $A'(x', y,' z')$ be any two points in an optical medium. (Note that any optical system $K$ is just an 'optical medium'.) Let $\mathscr{C}$ be a curve joining $A$ and $A'$. Recall that the optical length of $\mathscr{C}$ is

$$V^*(A, A') = \int_A^{A'} N\,ds, \tag{2.7}$$

where $ds$ is an element of arc of $\mathscr{C}$, $N$ is a prescribed function of $\bar{x}, \bar{y}, \bar{z}$ and the integral is of course extended along $\mathscr{C}$. Then we have

*Fermat's Principle: The ray joining any two arbitrary points $A$, $A'$ is determined by the condition that its optical length be stationary as compared with the optical length of arbitrary neighbouring curves joining $A$ and $A'$.* (2.8)

Fermat's Principle is a particular example of a variational principle with fixed end points, for one is comparing a property of different curves all of which are constrained to pass through a pair of fixed points. It should moreover be noted that all comparison curves must lie within a certain small neighbourhood of any one of them. This restriction can be motivated physically by the following example. Let $A$ and $A'$ lie in a homogeneous medium, but let a plane mirror be also present somewhere, so that one has two possible rays, i.e. the straight line $\mathscr{R}$ joining $A$ and $A'$ as well as the ray $\mathscr{R}^*$ reflected at the mirror. Certainly the optical lengths of all curves joining $A$ and $A'$ and lying in a sufficiently small neighbourhood of $\mathscr{R}$ are greater than that of $\mathscr{R}$, and likewise the optical lengths of all curves joining $A$ and $A'$ and lying in a sufficiently small neighbourhood of $\mathscr{R}^*$ are greater than that of $\mathscr{R}^*$; yet the optical length of $\mathscr{R}$ is less than that of $\mathscr{R}^*$. Therefore, were one to allow arbitrarily distant comparison curves in (2.8) the ray $\mathscr{R}^*$ would never appear, in contradiction with physical experience. In short, $\tilde{V}$ is a relative minimum for $\mathscr{R}^*$ and an absolute minimum for $\mathscr{R}$.

The actual value of any quantity when it is stationary with respect to some specified class of variations is called an *extremum*. It will be noticed that statement (2.8) of Fermat's Principle merely implies that the optical length of a ray is in fact an extremum of some kind, but it avoids any reference to its specific character. In particular, it does not require it to be a minimum. One easily sees that the optical length of the path $APA'$ of Fig. 1.1 is in fact a minimum for the ray, just as in the example concerning the rays $\mathscr{R}$ and $\mathscr{R}^*$ above an absolute and a relative minimum were encountered. However, in general the situation is more complex. Take the example of a ray from a point $A$ which is reflected at the point $P$ of a concave mirror and then passes to a point $A'$, $AP$ and $PA'$ being taken as straight, the medium being supposed homogeneous. It is not difficult to arrange the shape of the mirror to be such that $\tilde{V}(APA')$ is a maximum, granted that only variations of the position of the point of incidence $P$ are contemplated. It is, however, obvious that one can find curves joining $A$, $P$ and $A'$ of greater optical length: one need only consider curves for which $AP$ is no longer straight. In other words, $\tilde{V}$ is here a maximum for variations of $P$ alone, but a minimum for variations in which $P$ is left fixed. The extremum is therefore of the kind of a saddle point. Of course, one can apply Fermat's Principle to points lying between $A$ and $P$ or between $P$ and $A'$ and so infer that $AP$ and $PA'$ must be straight. This argument, however, does not affect the conclusion that $\tilde{V}$ is neither a minimum nor a maximum. Evidently one can never have a true maximum of $\tilde{V}$ since any curve can always be so deformed as to increase its over-all optical length. To some extent the term optical 'distance' is therefore conventional, since under everyday circumstances the (geometrical) distance between two points is always the length of the shortest path connecting them.

## Problems

**P.1 (i).** A two-parameter family of rays is given by the equations $x = t^2, y = at,\ z = bt$. Show that the wave-surfaces constitute a family of spheroids.

**P.1 (ii).** A congruence of curves is specified by giving the direction cosines $\alpha$, $\beta$, $\gamma$ of the tangent to a curve at the point $x, y, z$ as functions of $x$, $y$, $z$. Show that if the congruence is normal then

$$\alpha\left(\frac{\partial\gamma}{\partial y}-\frac{\partial\beta}{\partial z}\right)+\beta\left(\frac{\partial\alpha}{\partial z}-\frac{\partial\gamma}{\partial x}\right)+\gamma\left(\frac{\partial\beta}{\partial x}-\frac{\partial\alpha}{\partial y}\right)=0.$$

**P.1 (iii).** (*a*) This problem is to be treated as two-dimensional. Two points, $A$, $A'$, have the coordinates $(a, b)$, $(a, -b)$ respectively $(a > 0)$. A ray through $A$ and $A'$ is reflected at the parabola $x = ky^2$, the point of incidence $P$ being of course at the origin. Show that (with $N$ = constant) the optical length of the path $APA'$ is a minimum or a maximum for the ray according as $k$ is less than or greater than $\frac{1}{2}a/(a^2+b^2)$.

(*b*) Obtain the equation of the reflecting curve which is such that $\tilde{V}$ = constant, and show that your result is consistent with the conclusion above.

**P.1 (iv).** Show that in a medium with continuously varying refractive index equations (2.1) become $d(N\mathbf{e})/ds = \operatorname{grad} N$.

CHAPTER 2

# CHARACTERISTIC FUNCTIONS

## 3. The point characteristic $V$. The idea of the characteristic function

Let $K$ be some optical system, that is to say, an optical medium the refractive index $N$ of which is some prescribed function of the coordinates $\bar{x}, \bar{y}, \bar{z}$. As before $A(x,y,z)$ and $A'(x',y',z')$ shall be two arbitrary points in $K$. Then Fermat's Principle selects, save in exceptional circumstances, a particular curve $\mathscr{R}$ from amongst all possible curves $\mathscr{C}$ joining $A$ and $A'$, as we have seen. In other words, Fermat's Principle thus serves directly to associate an optical distance $\tilde{V}(A, A')$ with $A'$ and $A$, i.e. a certain definite number which depends solely upon the positions of these points. In short, given $A$ and $A'$, $\tilde{V}(A, A')$ is a definite function of the six variables $x', y'$, $z'$, $x$, $y$, $z$ alone, and we write

$$\tilde{V} = V(x', y', z', x, y, z). \tag{3.1}$$

The function $V$ now defined is known as the *point characteristic* of $K$. The reason for this terminology will be made clear shortly.

Our immediate task is to compare the optical distance between $A$ and $A'$ with that between neighbouring points $A_1$ and $A_1'$, the magnitudes of the displacements $\delta s\,(\equiv \delta x, \delta y, \delta z)$ and $\delta s'\,(\equiv \delta x', \delta y', \delta z')$ leading from $A$ to $A_1$, and from $A'$ to $A_1'$ being supposed sufficiently small. $\mathscr{R}$ is the ray between $A$ and $A'$, $\mathscr{R}_1$ that between $A_1$ and $A_1'$. Speaking in terms of Fig. 2.1 for convenience, these rays are of course independent of the values of the refractive index to the right of $A'$ and $A_1'$ on the one hand, and to the left of $A$ and $A_1$ on the other. In other words, we are at liberty to pretend that the refractive index is constant in these regions, its values on the left and right being equal to those at $A$ and $A'$ respectively, $N$ and $N'$, say. If $B$ and $B'$ are two distant points so located that $\mathscr{R}$ is part of the *ray* $\mathscr{S}$ joining them, then $BA$ and $A'B'$ will both be straight. Finally, let $\mathscr{S}_1$ be the *curve* consisting of $\mathscr{R}_1$ together with the two straight lines $BA_1$ and $A_1'B'$.

Observe now that, by construction, $\mathscr{S}_1$ will coalesce smoothly with $\mathscr{S}$ when $ds$ and $ds'$ are allowed to tend to zero. $\mathscr{S}_1$ is thus a curve neighbouring to $\mathscr{S}$ of the kind contemplated in Fermat's Principle. It follows that, since $\mathscr{S}$ is a ray, the optical length of $\mathscr{S}_1$ is (to the first order of small quantities) the same as that of $\mathscr{S}$. One therefore has

$$\tilde{V}(B,A)+\tilde{V}(A,A')+\tilde{V}(A',B') = \tilde{V}(B,A_1)+\tilde{V}(A_1,A_1')+\tilde{V}(A_1',B'). \tag{3.2}$$

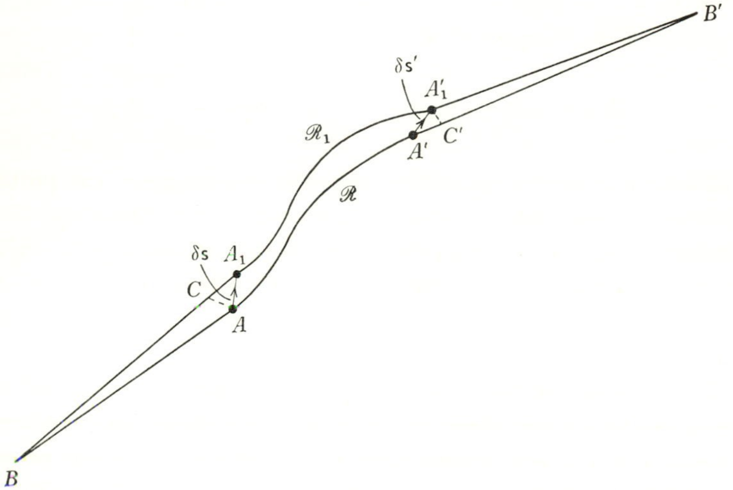


Fig. 2.1

Now, if $\mathsf{e}$, $\mathsf{e}'$ are unit vectors in the direction of $\mathscr{R}$ at $A$ and $A'$ respectively

$$\tilde{V}(B,A_1)-\tilde{V}(B,A) = N\mathsf{e}\,.\,\delta\mathsf{s},$$
$$\tilde{V}(A',B')-\tilde{V}(A_1',B') = N'\mathsf{e}'\,.\,\delta\mathsf{s}',$$

so that (3.2) becomes

$$\tilde{V}(A_1,A_1')-\tilde{V}(A,A') = N'\mathsf{e}'\,.\,\delta\mathsf{s}'-N\mathsf{e}\,.\,\delta\mathsf{s}. \tag{3.3}$$

Both terms of the left-hand member of (3.3) are optical distances, so that, recalling (3.1), one has just the difference

$$V(x'+\delta x', y'+\delta y', z'+\delta z', x+\delta x, y+\delta y, z+\delta z)$$
$$-V(x',y',z',x,y,z).$$

The components of the vectors $\mathsf{e}$ and $\mathsf{e}'$ are respectively the direction cosines $\alpha, \beta, \gamma$ and $\alpha', \beta', \gamma'$ of the tangents to the ray at the points in question. Written out in full, (3·3) becomes

$$\delta V = \frac{\partial V}{\partial x'}\delta x' + \frac{\partial V}{\partial y'}\delta y' + \frac{\partial V}{\partial z'}\delta z' + \frac{\partial V}{\partial x}\delta x + \frac{\partial V}{\partial y}\delta y + \frac{\partial V}{\partial z}\delta z$$

$$= N'(\alpha'\,\delta x' + \beta'\,\delta y' + \gamma'\,\delta z') - N(\alpha\,\delta x + \beta\,\delta y + \gamma\,\delta z). \tag{3·4}$$

Since this must hold identically for all values of the six coordinate differentials $\delta x', \ldots$, we have the *basic equations of Hamiltonian optics*:

$$\left.\begin{aligned} N'\alpha' &= \frac{\partial V}{\partial x'}, \quad N'\beta' = \frac{\partial V}{\partial y'}, \quad N'\gamma' = \frac{\partial V}{\partial z'}, \\ N\alpha &= -\frac{\partial V}{\partial x}, \quad N\beta = -\frac{\partial V}{\partial y}, \quad N\gamma = -\frac{\partial V}{\partial z}. \end{aligned}\right\} \tag{3·5}$$

Hitherto $x, y, z$ and $x', y', z'$ have been the coordinates of two points, referred to the *same* set of Cartesian axes $\bar{x}, \bar{y}, \bar{z}$. However, (3·5) remains valid if $x, y, z$ are referred to one set of axes $\bar{x}, \bar{y}, \bar{z}$, whilst $x', y', z'$ are referred to a different set of Cartesian axes, $\bar{x}', \bar{y}', \bar{z}'$. The direction cosines of course then relate in each case to the appropriate axes.

If $K$ is some optical system, one conventionally speaks of the region in which the object is situated as the *object space*, and unprimed symbols will always refer to this. Primed symbols on the other hand will always refer to the *image space*, i.e. the region in which one inquires into the disposition of rays from the object after their passage through $K$. We shall suppose throughout that the refractive index is constant in the object space and in the image space. This restriction is by no means essential, but it is very convenient, and is appropriate to the majority of cases encountered in practice. It has the advantage that the *initial ray* and the *final ray* (i.e. the parts of any given ray $\mathscr{R}$ which lie in the object space and image space respectively) are straight lines.

Now suppose that the form of the function $V$, appropriate to the system $K$, is known. Then, given that some initial ray passes through the point $A(x, y, z)$, and the final ray through the point $A'(x', y', z')$, the equations (3·5) immediately provide the directions of $\mathscr{R}$ at $A$ and $A'$. One therefore knows exactly all the data $x', y', z'$, $\beta', \gamma'$ of the final ray which correspond to the data $x, y, z, \beta, \gamma$

of the initial ray. In other words, given only the form of the function $V$, the particular correspondence established by $K$ between initial and final rays is available. This correspondence completely characterizes the geometrical–optical behaviour of $K$; so that it is natural to refer to $V$ as a *characteristic function*. The possibility of, as it were, summing up the properties of any given system in a single function has of course very great advantages in optical theory. Even the very existence of a characteristic function, coupled only with assumptions of a general kind (such as assumptions about differentiability) leads to interesting consequences.

As we have seen, $V$ has a simple physical meaning: it is the optical distance between pairs of points, expressed as a function of the coordinates of these points. For this reason it is often called the *point characteristic*, a terminology which at the same time serves to distinguish it from the alternative characteristic functions yet to be considered.

When investigating the properties of optical systems by using the point characteristic one must always bear in mind that (3.5) cannot be applied in situations in which there exist several rays connecting the points $A'(x', y', z')$ and $A(x, y, z)$. For example, if these two points are the foci of an ellipsoid of revolution, all rays from the first pass through the second after reflection at the ellipsoid, granted that the interior medium is homogeneous. The optical distance between the foci (calculated for reflected rays) is in fact constant; and it is quite obvious that in this situation the equations (3.5) become meaningless. Quite generally, the use of the point characteristic is precluded if there exists a *set* of rays the members of which pass through both $A$ and $A'$.

Inspection of (3.5) reveals that $V$ must satisfy both of the differential equations

$$\left(\frac{\partial V}{\partial x'}\right)^2 + \left(\frac{\partial V}{\partial y'}\right)^2 + \left(\frac{\partial V}{\partial z'}\right)^2 = N'^2,$$

$$\left(\frac{\partial V}{\partial x}\right)^2 + \left(\frac{\partial V}{\partial y}\right)^2 + \left(\frac{\partial V}{\partial z}\right)^2 = N^2. \tag{3.6}$$

In particular, these are satisfied, in the case of a homogeneous medium of refractive index $N$, by

$$V = N[(x'-x)^2 + (y'-y)^2 + (z'-z)^2]^{\frac{1}{2}}. \tag{3.7}$$

It is virtually impossible to obtain the exact form of the point characteristic in all but effectively trivial cases, and in practice one has to be satisfied with approximations of one kind or another. This, however, is fortunately of little concern to us here since we are mainly interested in general properties of systems, and these can be discussed without knowing the explicit form of the characteristic function. The following analogy may be informative for the reader who has some acquaintance with quantum mechanics. Thus, a great deal of information can be gained on the basis of known symmetries of the Hamiltonian $H$ of an atomic system, even though the explicit form of $H$ may be unknown. This situation is similar to that which exists in Hamiltonian optics, as we shall see in due course.

## 4. The angle characteristic $T$

In theoretical investigations the use of the point characteristic occasionally leads to irrelevant difficulties, brought about, for example, by the need to observe the restriction to which attention was drawn just prior to equations (3.6). When such a situation arises the use of some alternative characteristic function may be appropriate. Accordingly, recall that according to (3.4) the total differential of the point characteristic is

$$dV = N'(\alpha' dx' + \beta' dy' + \gamma' dz') - N(\alpha dx + \beta dy + \gamma dz). \qquad (4.1)$$

Let $\quad U = N(\alpha x + \beta y + \gamma z), \quad U' = N'(\alpha' x' + \beta' y' + \gamma' z'), \qquad (4.2)$

and recall that $N$ and $N'$ are both constant. Then, using (4.1),

$$\begin{aligned} d(V - U' + U) &= -N'(x' d\alpha' + y' d\beta' + z' d\gamma') + N(x\, d\alpha + y\, d\beta + z d\gamma) \\ &= -N'[(y' - \beta' x'/\alpha')\, d\beta' + (z' - \gamma' x'/\alpha')\, d\gamma'] \\ &\quad + N[(y - \beta x/\alpha)\, d\beta + (z - \gamma x/\alpha)\, d\gamma]. \qquad (4.3) \end{aligned}$$

Let the function whose total differential appears on the left be expressed in terms of the independent variables $\beta'$, $\gamma'$, $\beta$, $\gamma$, and denote it by the symbol $T$. Then, by inspection of (4.3),

$$N'(y' - \beta' x'/\alpha') = -\frac{\partial T}{\partial \beta'}, \quad N'(z' - \gamma' x'/\alpha') = -\frac{\partial T}{\partial \gamma'};$$

$$N(y - \beta x/\alpha) = \frac{\partial T}{\partial \beta}, \qquad N(z - \gamma x/\alpha) = -\frac{\partial T}{\partial \gamma}. \qquad (4.4)$$

These equations are precisely analogous to equations (3.5). In fact, suppose the form of the function $T$ to be known. Then, if one direction in the object space and one direction in the image space be prescribed, equations (4.4) immediately provide the equations of the initial and final parts of that ray which has these directions. $T$ evidently completely characterizes the geometrical-optical properties of $K$, so that it is also a characteristic function. In view of the fact that it must be given as a function of direction cosines, it is known as the *angle characteristic*.

Here again one has to bear one exceptional situation in mind. It arises when a set of mutually parallel initial rays is transformed by $K$ into a set of mutually parallel final rays; for obviously the mere specification of the directions of these cannot lead to the selection

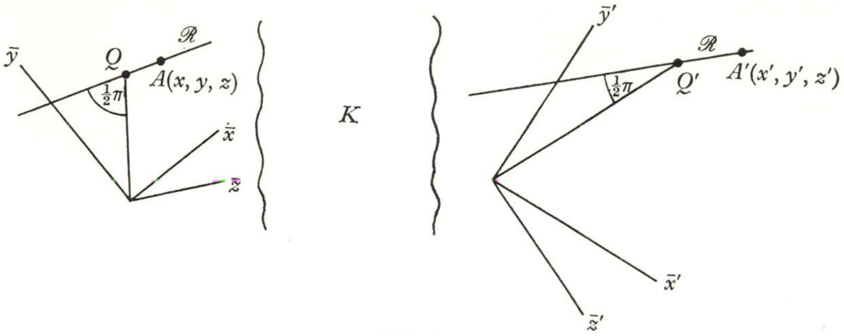


Fig. 2.2

of unique initial and final rays. The kind of system now contemplated is called *afocal* or *telescopic*; so that if we are confronted with such a system the use of the angle characteristic is precluded.

The angle characteristic has a simple physical interpretation. Thus, let the normal to the initial ray drawn from the origin of the axes in the object space meet the ray in the point $Q$; a point $Q'$ being defined analogously in the image space. Then $U/N$ is the distance between $Q$ and $A$, and likewise $U'/N'$ is the distance between $Q'$ and $A'$. Hence $T$ is the optical distance between $Q$ and $Q'$. It should be remarked that when $N$ and $N'$ are not constant one easily convinces oneself that $T$ has to be defined as a function of the components of the 'ray vectors' $N\mathsf{e}$, $N'\mathsf{e}'$ rather than those of $\mathsf{e}$, $\mathsf{e}'$; and the geometrical interpretation of $T$ given above no longer obtains. We shall not, however, concern ourselves with these complications.

The angle characteristic can be obtained in explicit, closed form for certain simple systems. By way of example, consider a system consisting of a single spherical refracting surface $\mathscr{S}$, i.e. two homogeneous media, the boundary between them being of constant curvature $1/r$. The geometrical situation is shown in Fig. 2.3. The origins of axes $B$, $B'$ lie on a diameter of the spherical surface of which $\mathscr{S}$ is a part, at distances $q$ and $q'$ respectively to the right of the pole $A_1$ of $\mathscr{S}$. $C$ is the centre of curvature, so that the line $PC$, of length $r$, is normal to $\mathscr{S}$. Then one can show that

$$T = (N'-N)r\{1-2\kappa[\eta+(1-\xi)^{\frac{1}{2}}(1-\zeta)^{\frac{1}{2}}-1]\}^{\frac{1}{2}} \\ -N'(r-q')(1-\xi)^{\frac{1}{2}}+N(r-q)(1-\zeta)^{\frac{1}{2}}, \quad (4.5)$$

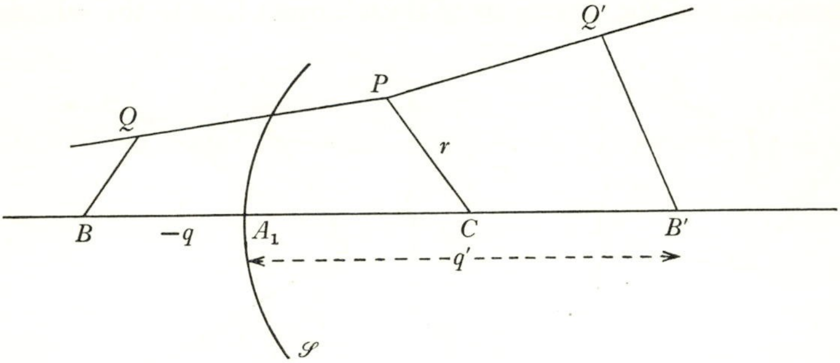


Fig. 2.3

where
$$\left.\begin{aligned} \xi &= \beta'^2+\gamma'^2, & \eta &= \beta\beta'+\gamma\gamma', \\ \zeta &= \beta^2+\gamma^2, & \kappa &= NN'/(N'-N)^2. \end{aligned}\right\} \quad (4.6)$$

Most situations encountered in practice are of course a good deal more complex, and one has again to be content with approximations.

## 5. The mixed characteristics $W_1$ and $W_2$

At times it proves convenient to employ one or other of a pair of characteristic functions which are, as it were, 'intermediate' between $V$ and $T$. Thus, in the notation of Section 4, let

$$W_1 = V+U.$$

Then

$$dW_1 = N'(\alpha' dx'+\beta' dy'+\gamma' dz')+N[(y-\beta x/\alpha)\,d\beta \\ +(z-\gamma x/\alpha)\,d\gamma].$$

Thus, if $W_1$ be regarded as a function of the five independent variables $x'$, $y'$, $z'$, $\beta$, $\gamma$, one has

$$\left.\begin{aligned} N'\alpha' = \frac{\partial W_1}{\partial x'}, \quad N'\beta' = \frac{\partial W_1}{\partial y'}, \quad N'\gamma' = \frac{\partial W_1}{\partial z'}, \\ N(y - \beta x/\alpha) = \frac{\partial W_1}{\partial \beta}, \quad N(z - \gamma x/\alpha) = \frac{\partial W_1}{\partial \gamma}. \end{aligned}\right\} \tag{5.1}$$

Clearly $W_1$ is a characteristic function, and we shall call it the *first mixed characteristic*. It must evidently satisfy the *one* differential equation

$$\left(\frac{\partial W_1}{\partial x'}\right)^2 + \left(\frac{\partial W_1}{\partial y'}\right)^2 + \left(\frac{\partial W_1}{\partial z'}\right)^2 = N'^2.$$

Geometrically it is the optical distance between $Q$ and $A'$. Here again we have one situation in which the use of this particular characteristic function is precluded, namely when there exists a set of rays through $A'$ the members of which are mutually parallel in the object space.

Without further ado we now define the *second mixed characteristic* $W_2$ to be the optical distance between $A$ and $Q'$,

$$W_2 = V - U',$$

regarded as a function of the five variables $\beta'$, $\gamma'$, $x$, $y$, $z$. One has

$$N'(y' - \beta' x'/\alpha') = -\frac{\partial W_2}{\partial \beta'}, \quad N'(z' - \gamma' x'/\alpha') = -\frac{\partial W_2}{\partial \gamma'},$$

$$N\alpha = -\frac{\partial W_2}{\partial x}, \quad N\beta = -\frac{\partial W_2}{\partial y}, \quad N\gamma = -\frac{\partial W_2}{\partial z}. \tag{5.2}$$

In the context of $W_2$ one has to exclude the situation in which there exists a set of rays through $A$ the members of which are mutually parallel in the image space.

Twelve further mixed characteristics may be defined after the fashion above (see towards the end of Section 6). These rarely occur in practice, though we shall have occasion to use one of them in Section 92. In conclusion, attention must be drawn to the fact that, until we come to Chapter 10, all rays through a given system are taken to have a definite, single wavelength.

## 6. Reduced distances. Developments of notation. Further characteristic functions

The ubiquitous appearance of the constants $N$ and $N'$ in the various equations just considered suggests their removal by some simple convention. This may be achieved by the introduction of *reduced* distances. It amounts to this: that if $A$, $B$ are two points in the object space and $s$ is the distance between them, then henceforth we agree to use the same symbol $s$ for the reduced distance between $A$ and $B$, this being defined as $N$ times the (geometrical) distance between them. An analogous convention is to be observed in the image space, $N'$ of course replacing $N$. The refractive indices in equations (3.5) now no longer appear explicitly, since $x'$ replaces $N'x'$, and so on for all other variables which have the character of distances. Similarly $N$ and $N'$ will be absent from equations (4.4), (5.1) and (5.2), and throughout our work $N$ and $N'$ will hardly ever appear explicitly again; see, however, Section 102.

Next, it may have been noticed that $V$ is a function of six variables, $W_1$ and $W_2$ functions of five variables, whilst $T$ is a function of only four variables. However, the variables $x$ and $x'$ are in effect redundant in the sense that in any particular situation their values may be taken as fixed. To see this, notice first that in the context of the point characteristic we have hitherto specified all the coordinates of two points $A$ and $A'$ to select a particular ray. However, choose now in the object space a fixed plane $\mathscr{B}$, called the *anterior base-plane*, and likewise choose in the image space a fixed plane $\mathscr{B}'$ called the *posterior base-plane*. As a matter of convenience let the origins and orientations of the coordinate axes be chosen in such a way that the equations of the base-planes are $\bar{x} = 0$ and $\bar{x}' = 0$ respectively. The point characteristic $V$ is now explicitly a function of only the four variables $y'$, $z'$, $y$, $z$, i.e. we consider only the optical distances between points $A(y, z)$ of the first base-plane and points $A'(y', z')$ of the second. Nevertheless, by giving a set of values of these variables a particular $\mathscr{R}$ will have been selected. The situation is simply that whereas originally the dependence of $V$ on $x$ and $x'$ was known explicitly, it is now contained only implicitly in the form of the function $V(y', z', y, z)$. This is good enough for most purposes, and in any event the 'complete' function $V(x', y', z', x, y, z)$ can be recovered if necessary (cf. Section 35).

As regards $W_1$ and $W_2$, their explicit dependence on $x'$ and $x$ respectively will no longer be given. Finally, with our choice of coordinate axes, wherever the variables $x'$ and $x$ appear in the left-hand members of the equations (4.4), (5.1) and (5.2) they are to be replaced by zero. In short, as a result of the various conventions which have been introduced we have the four sets of equations

$$\beta' = \frac{\partial V}{\partial y'}, \quad \gamma' = \frac{\partial V}{\partial z'}, \quad \beta = -\frac{\partial V}{\partial y}, \quad \gamma = -\frac{\partial V}{\partial z}; \tag{6.1}$$

$$y' = -\frac{\partial T}{\partial \beta'}, \quad z' = -\frac{\partial T}{\partial \gamma'}, \quad y = \frac{\partial T}{\partial \beta}, \quad z = \frac{\partial T}{\partial \gamma}; \tag{6.2}$$

$$\beta' = \frac{\partial W_1}{\partial y'}, \quad \gamma' = \frac{\partial W_1}{\partial z'}, \quad y = \frac{\partial W_1}{\partial \beta}, \quad z = \frac{\partial W_1}{\partial \gamma}; \tag{6.3}$$

$$y' = -\frac{\partial W_2}{\partial \beta'}, \quad z' = -\frac{\partial W_2}{\partial \gamma'}, \quad \beta = -\frac{\partial W_2}{\partial y}, \quad \gamma = -\frac{\partial W_2}{\partial z}. \tag{6.4}$$

At this point we may conveniently insert a remark concerning the further characteristic functions referred to at the end of Section 5. Recall that we arrived at $W_1$, for example, by contemplating the total differential of the function $V + U (= V + \beta y + \gamma z)$. There is, however, nothing to prevent us from defining, say, the function $W_3(y', z', y, \gamma) = V + \gamma z$. Then

$$\beta' = \frac{\partial W_3}{\partial y'}, \quad \gamma' = \frac{\partial W_3}{\partial z'}, \quad -\beta = \frac{\partial W_3}{\partial y}, \quad z = \frac{\partial W_3}{\partial \gamma};$$

and $W_3$ is a characteristic function. One has altogether *sixteen* possibilities: any particular choice corresponds to regarding as independent variables *any* two taken from amongst $y'$, $z'$, $\beta'$, $\gamma'$ together with *any* two taken from amongst $y$, $z$, $\beta$, $\gamma$. We note parenthetically that the list of formal possibilities is still not exhausted; but the discussion of these will be deferred to Sections 36 and 37.

Evidently, if $F$ stands generically for any one of the alternative characteristic functions introduced hitherto, each of them depends upon an appropriate set of *four* independent variables, to be called *ray-coordinates*, and these may be uniformly denoted by $q_i$ $(i = 1, \ldots, 4)$, primed coordinates always preceding the unprimed. The derivatives of $F$ then yield in each case four variables which may be denoted by $p_i$ $(i = 1, \ldots, 4)$; so that

$$p_i = \frac{\partial F}{\partial q_i} \quad (i = 1, \ldots, 4). \tag{6.5}$$

At times $p_i$ will be called the *conjugate* of $q_i$. Note that, for example, the conjugate of $\beta$ is $y$, whereas the conjugate of $y$ is $-\beta$. Since now

$$dF = \sum_{i=1}^{4} p_i dq_i, \tag{6.6}$$

one has the six *conditions of integrability*

$$\frac{\partial p_i}{\partial q_k} - \frac{\partial p_k}{\partial q_i} = 0 \tag{6.7}$$

on the linear differential form on the right of (6.6).

## 7. The idea of the aberration function

In the most general terms, the design of an optical system $K$ which is intended for some specific purpose is so arranged that $K$ will produce, as nearly as possible, an image of a certain desired character of the kind of object being contemplated. The degree of success achieved in this endeavour depends of course upon limitations of both a theoretical and a practical nature. At any rate, one is usually (though not always) confronted with the problem of giving $K$ a structure such that all the rays from any one point $O$ of a set of points in the object space will pass through a corresponding point $O'$ in the image space, the coordinates of $O'$ depending in a preassigned way on those of $O$. By way of example, a very common situation is this: one desires $K$ to produce a sharp image of a plane object and, moreover, this image is to be geometrically similar to the object. As already remarked, such 'perfect' imagery cannot in general be achieved for one reason or another. In other words, if $\mathscr{I}'$ is the plane in which the image is intended to be formed, rays from any point $O$ of the object will, in general, fail to pass through the appropriate point $O'$ of $\mathscr{I}'$. In that case one says that 'the imagery of $K$ is imperfect', or that '$K$ is afflicted with aberrations'. If any particular ray through $O$ in fact intersects $\mathscr{I}'$ in $O_1'$, the displacement $\boldsymbol{\epsilon}'$ from $O'$ to $O_1'$ is a measure of the *aberration* of the ray, and $\boldsymbol{\epsilon}'$, regarded as a function of the variables specifying particular rays, characterizes the *aberrations of* $K$. Of course, the desired imagery of $K$ may be of a different kind, e.g. one might want $K$ to be such that it produces a rectangular image of a square object (*anamorphotic* systems), or one might wish a plane object to have a sharp image lying on some surface other than a plane.

Whatever the actual situation may be in detail, we shall take it for granted that one can specify an *ideal characteristic function* $F_0$, such that *if* $K$ had this characteristic function then its imagery would be precisely that which one wants it to be. The actual characteristic $F$ will in general not be $F_0$, as discussed above, and we write

$$F = F_0 + f. \tag{7.1}$$

Then $f$ is called the *aberration function*; and all aberrations of $K$ will be absent if $f = 0$. (See also the end of Section 36.)

It may be useful to present a specific example illustrating the preceding remarks. Let it be desired that a system $K$ produce a sharp, undistorted, plane image in $\mathscr{I}'$ of a plane object in $\mathscr{I}$. The anterior base-plane $\mathscr{B}$ will be taken to coincide with $\mathscr{I}$, whereas the posterior base-plane $\mathscr{B}'$ is chosen to be parallel to $\mathscr{I}'$, the equation of the latter therefore being $\bar{x}' = d'$, where $d'$ is some constant. A ray through the point $O(0, y, z)$ of $\mathscr{I}$ intersects $\mathscr{I}'$ in the point $O'(d', Y', Z')$. The geometrical similarity of object and image implies that there is a constant $m$ such that

$$Y' = my, \quad Z' = mz \tag{7.2}$$

for all rays through $O$, the orientation of the $\bar{y}'$-axis having been suitably chosen. (The coordinate-system in the image space may possibly have had to be taken as left-handed.) The constant $m$ is the *reduced magnification* associated with $\mathscr{I}$ and $\mathscr{I}'$. The actual magnification refers to the ratio of corresponding unreduced coordinates, and so has the value $Nm/N'$.

Now the optical distance from $O$ to $O'$ is the same for all rays through $O$ and so depends on $y$ and $z$ alone:

$$\tilde{V}(O, O') = g(y, z). \tag{7.3}$$

On the other hand, if a ray through $O$ and $O'$ intersects $\mathscr{B}'$ in the point $D'(y', z')$, then

$$\tilde{V}(O, D') = \tilde{V}(O, O') - \tilde{V}(D', O'),$$

whence

$$V_0(y', z', y, z) = g(y, z) - [d'^2 + (y' - my)^2 + (z' - mz)^2]^{\frac{1}{2}}, \quad (7.4)$$

where (7.2) has been used. Thus, if $K$ has a point characteristic of the generic form (7.4), where $g$ is any function of $y$ and $z$, its imagery will have the desired properties. Whether or not a system of some generic type (to which one might be restricted in practice) can actually have a point characteristic of the form (7.4) is irrelevant.

We have introduced the aberration function in unusually general terms in order clearly to bring out the fact that the form of the aberration function $f$ depends on the form of $F_0$, and the latter is not absolutely prescribed, i.e. to the extent that it depends on the properties of $K$ which one happens to aim at. Thus, in the case of the (axially) symmetric system, to be described in detail in Chapters 4–7, one tends to think of $F_0$ as corresponding to the existence of a plane, undistorted image of a plane object, the image lying in the 'ideal image plane'. On the other hand, in other situations, for example imagery by systems of cylindrical refracting surfaces, such a 'natural' choice of image plane is not at hand. In any event, even in the symmetric case one might want the image of a plane object to lie on some curved surface; see, for example, Section 62. Further, if one considers imagery by light which is not monochromatic, the system will not produce a sharp image lying in a particular image plane for all colours simultaneously, so that $F_0$ will correspond to a choice of position of receiving plane which is inevitably arbitrary to some extent (see Section 100).

## Problems

**P.2 (i).** A system $K$ consists of two homogeneous media in mutual contact, the plane $\bar{x} = a$ being the boundary between them. Choosing the coordinate axes suitably, carry through as far as possible the determination of the point characteristic of $K$. (Note that this involves the solution of an algebraic equation of the fourth degree.)

**P.2 (ii).** Why would you expect the result (4.5) to become meaningless in the limit $r \to \infty$?

**P.2 (iii).** Obtain the result (4.5).

**P.2 (iv).** Show that the first mixed characteristic of the plane refracting surface $\bar{x} = 0$ is

$$W_1 = x'[N'^2 - N^2(\beta^2 + \gamma^2)]^{\frac{1}{2}} + N(\beta y' + \gamma z').$$

**P.2 (v).** Find the angle characteristic $T_0$ which corresponds to the point characteristic $V_0$ given by equation (7.4).

CHAPTER 3

# SYMMETRIES AND REGULARITY

## 8. Symmetries of optical systems and invariance properties of characteristic functions

Many—perhaps most—of the optical systems encountered in practice exhibit *symmetries* of one kind or another. That a system $K$ has a certain symmetry means that there exists an appropriate displacement of $K$, the end result of which is a system indistinguishable from $K$. This general definition may be clarified by means of some specific examples. They will serve at the same time to elucidate the somewhat formal sense which may have to be attached to the phrase 'displacement of $K$'.

Our first example is that of *axial symmetry*. In this context the displacement to be contemplated is a rigid rotation. However, for the sake of verbal uniformity, we imagine the elementary parts into which $K$ may be thought of as subdivided all to be rotated through the same angle $\theta$ about some line $\mathscr{A}$. If $\mathscr{A}$ can be so chosen that the resulting system is indistinguishable from $K$, independently of the value of $\theta$, $K$ is axially symmetric, and $\mathscr{A}$ is its *axis* (of symmetry). If $K$ includes a *diaphragm*, or *stop*, which limits the passage of rays through $K$, it is preferable not to regard it as being a part of $K$ in the present context. Without this convention the inclusion of a non-circular stop, for instance, in an otherwise axially symmetric system would suffice to destroy this symmetry; but it would do so in a somewhat trivial way. It is now evident that if the $\bar{x}$-axis of a Cartesian system is the axis of $K$, then the refractive index depends on $\bar{y}$ and $\bar{z}$ only in the combination $\bar{y}^2+\bar{z}^2$.

As a second example, suppose that $K$ is plane-symmetric, i.e. has a plane of symmetry, which we may here take to be the plane $\bar{z} = 0$. The displacement appropriate to this situation is of a formal kind. One has to imagine all elementary parts of $K$ to be so displaced that after the displacement the part of $K$ lying to one side of the plane of symmetry is the mirror image of itself as it was before the displacement. (Of course, the plane of symmetry is the 'mirror' here.) Plane-symmetry therefore demands that $N(\bar{x},\bar{y},\bar{z})$ be independent of the sign of $\bar{z}$.

These two examples will suffice to illustrate the notion of symmetry. It should be borne in mind that a given system may have several symmetries simultaneously; and various cases of interest will be examined in subsequent chapters.

We come now to a crucial point in the argument. It consists of the observation that to every symmetry of a given system there corresponds a substitution of the ray-coordinates which leaves the point characteristic *invariant*. This means that on replacing $y'$, $z'$, $y$, $z$ by certain functions of these variables, the resulting function is just $V(y',z',y,z)$ again. To see this, let $A(y,z)$, $A'(y',z')$ be two points lying in the base-planes; and let them go over into the points $A^*(y^*,z^*)$ and $A^{*\prime}(y^{*\prime},z^{*\prime})$ as a result of the displacement appropriate to the given symmetry. The base-planes are, in the present context, to be regarded as 'parts of the system' so that they take part in the displacement. Evidently $A^*$ is intended to refer to the object space, as the notation indicates, so that it may correspond to one or other of the undisplaced points $A$, $A'$, depending on the particular symmetry being contemplated (cf. Section 53). At any rate, the point characteristic is the same *function* for the original as for the displaced system, these being indistinguishable from each other. Since $\tilde{V}(A,A')$ is obviously equal to $\tilde{V}(A^*,A^{*\prime})$, the relation

$$V(y^{*\prime},z^{*\prime},y^*,z^*) = V(y',z',y,z) \tag{8.1}$$

must therefore hold *identically*, the coordinates on the left being known functions of those which appear on the right. (8.1) is just the formal expression of the invariance of $V$ which was to be demonstrated. If one considers other characteristic functions the situation is not essentially different, though in the context of the mixed characteristics one may have certain complications, arising from the different geometrical character of the coordinates in the object space on the one hand, and those in the image space on the other. At any rate, we shall deal with problems relating to special cases as we come to them.

As a particularly simple example of (8.1) we return to the case of plane-symmetry. Then $A(y,z)$ evidently goes into $A^*(y,-z)$, and $A'(y', z')$ goes into $A^{*\prime}(y', -z')$. (8.1) therefore reads

$$V(y', -z', y, -z) \equiv V(y', z', y, z). \tag{8.2}$$

In other words, if $K$ is symmetric about the plane $\bar{z}(=\bar{z}') = 0$ its point characteristic must be invariant under the simultaneous reversal of sign of $z$ and $z'$.

An interesting conclusion may be drawn when one has a continuous symmetry group, i.e. when the displacements corresponding to the symmetry in question depend continuously upon a parameter $\omega$. In (8.1) the starred variables are then continuous functions of $\omega$, and we can always arrange the value $\omega = 0$ to correspond to the identity between unstarred and starred variables. For infinitesimal $\omega$ there exist functions $s$, $t$, together with their primed counterparts, such that

$$y^* = y + \omega s, \quad z^* = z + \omega t, \; \ldots. \tag{8.3}$$

Inserting these in (8.1) and expanding the left-hand member in powers of $\omega$, one obtains, upon retaining only terms linear in $\omega$, the identity

$$\left(s'\frac{\partial V}{\partial y'} + t'\frac{\partial V}{\partial z'} + s\frac{\partial V}{\partial y} + t\frac{\partial V}{\partial z}\right)\omega = 0. \tag{8.4}$$

Recalling that $\omega$ can be chosen at will it follows from this, together with (6.1), that

$$\beta' s' + \gamma' t' = \beta s + \gamma t. \tag{8.5}$$

In the case of axial symmetry, for example, one has $\omega = \theta$, $s = z$, $t = -y$, $s' = z'$, $t' = -y'$ (cf. equation (12.1)), so that (8.5) then becomes

$$\beta' z' - \gamma' y' = \beta z - \gamma y. \tag{8.6}$$

The quantity $j = \beta z - \gamma y$ has therefore the same value in the object space as its primed counterpart $j'$ has in the image space. Any quantity which has this property is called an *optical invariant*. In particular, the existence of the optical invariant $j$ is evidently a consequence of axial symmetry alone.

If a system has several symmetries there will be several simultaneous identities of the form (8.1), and we shall have occasion to deal with some of the more interesting possibilities later on. The central point is that any such identity represents a generic restriction on the form of the characteristic function, which immediately brings with it a corresponding characterization of the imagery produced by the system. Herein lies the power of Hamilton's method, especially when a certain generic assumption of regularity is satisfied. It is this, therefore, to which we now turn our attention.

## 9. Regularity

Whichever of the alternative characteristic functions of a given system be chosen, we may be sure that in almost all cases of practical interest it will be an exceedingly complicated function of the variables on which it depends. In these circumstances one is forced to devise approximations of some kind. Expansions in power series immediately spring to mind; and in this context the idea of *regularity* arises.

For convenience let us refer to the pair of sets of coordinate axes $\bar{x}$, $\bar{y}$, $\bar{z}$ and $\bar{x}'$, $\bar{y}'$, $\bar{z}'$ as a *coordinate basis*. Then we have the following definition:

> *A system is called regular with respect to a chosen coordinate basis if F can be written as a power series in the ray-coordinates* $q_i$.

It should be carefully noted that regularity is a joint property of the system and the coordinate basis. By way of illustration, let $K$ consist of a single refracting surface of revolution. Specifically, take it to be the paraboloid whose equation, referred to auxiliary axes $\tilde{x}, \tilde{y}, \tilde{z}$, is $k\tilde{x}^2 = (\tilde{y}^2 + \tilde{z}^2)^{\frac{1}{2}}$ ($k$ = constant). If $\mathscr{R}$ is some ray through $K$, consider now alternative coordinate bases such that (i) the $\bar{x}$-axis lies along the initial ray and the $\bar{x}'$-axis along the final ray; or (ii) these axes lie along the axis of the paraboloid. Then regularity obtains in the first case, but not in the second; yet in both cases one has the same optical system as such. Still, in cases where a definite choice of coordinate basis is understood, one may sometimes speak simply of the 'regularity of the system'.

Even when it is known that the condition of regularity is satisfied in cases of practical interest, the situation is bedevilled by the almost total absence of knowledge concerning the radii of convergence of the series which are encountered. This is an unfortunate state of affairs since the approximation to the actual characteristic function represented by a truncated power series is meaningless if the series diverges, and useless if it converges too slowly. If $F(q_1, \ldots, q_4)$ converges sufficiently rapidly only within a range of values of the $q_i$ which is inadequate, one can help oneself by representing $F$ as a power series in $q_i - \hat{q}_i$, where the $\hat{q}_i$ are appropriately chosen constants. In our language this amounts to going over to a new coordinate basis.

## 10. Parabasal optics in general

Accounts of the theory of axially symmetric systems often begin with a discussion of *paraxial optics*, i.e. of the imagery by rays which lie in a sufficiently small neighbourhood of the axis $\mathscr{A}$. That such rays do in fact exist is an assumption; it is in effect an assumption of the regularity of the system with respect to a coordinate basis such that the $\bar{x}$-axis and the $\bar{x}'$-axis both lie along $\mathscr{A}$. The generalization appropriate to a system $K$ without symmetries is to consider some ray $\mathscr{R}_0$ through $K$ as given, and to investigate imagery by rays which lie in a sufficiently small neighbourhood of the *base-ray* $\mathscr{R}_0$. In this more general context we shall speak of *parabasal optics*. There is, of course, the implicit assumption that $K$ is regular with respect to a coordinate basis such that the $\bar{x}$-axis lies along the initial ray and the $\bar{x}'$-axis along the final ray.

To go into a little more detail, let us choose the point characteristic. Note that because of our various conventions the base-planes are normal to $\mathscr{R}_0$. Regularity implies that

$$V = a_0 + a_1 y' + a_2 z' + a_3 y + a_4 z + O(2), \tag{10.1}$$

where the $a_i$ are constants. Throughout, the symbol $O(n)$ will denote terms which are of degree not less than $n$ in the ray-coordinates. The limit in which $y'$, $z'$, $y$, $z$ go to zero represents coincidence with $\mathscr{R}_0$. Using (6.1) it follows that $a_1, \ldots, a_4$ must be zero. Including now terms of the second degree explicitly, we may write

$$V = a_0 + \tfrac{1}{2}b_1 y'^2 + b_2 y'z' + b_3 y'y + b_4 y'z + \tfrac{1}{2}b_5 z'^2 + b_6 z'y + b_7 z'z + \tfrac{1}{2}b_8 y^2 + b_9 yz + \tfrac{1}{2}b_{10} z^2 + O(3), \tag{10.2}$$

where the numerical factors have been included merely for convenience. As far as parabasal optics is concerned the term $O(3)$ is irrelevant, and may be omitted. Then, from (6.1),

$$\begin{aligned} \beta' &= b_1 y' + b_2 z' + b_3 y + b_4 z, \\ \gamma' &= b_2 y' + b_5 z' + b_6 y + b_7 z, \\ -\beta &= b_3 y' + b_6 z' + b_8 y + b_9 z, \\ -\gamma &= b_4 y' + b_7 z' + b_9 y + b_{10} z. \end{aligned} \tag{10.3}$$

Parabasal optics is thus characterized by the *linearity* of the relations between the ray-coordinates and their conjugates, there being at most ten constants $b_i$ which enter into these relations. These constants, called *parabasal coefficients*, are determined by the constitution of the system in hand (for a given choice of $\mathscr{R}_0$). In the absence of symmetries no restriction in principle is laid upon the values which the parabasal coefficients might take, except in as far as the very use of $V$ implies that the inequality

$$b_3 b_7 - b_4 b_6 \neq 0 \tag{10.4}$$

must hold; for this ensures that to any given initial ray there corresponds just one final ray.

In practice one often traces parabasal rays through a given system by some step-by-step procedure, starting with chosen values of the 'initial variables' $y$, $z$, $\beta$, $\gamma$, say. Granted the linearity of parabasal optics, one thus determines the coefficients of the linear transformation

$$\begin{aligned} y' &= A_1 y + B_1 \beta + E_1 z + F_1 \gamma, \\ z' &= E_2 y + F_2 \beta + A_2 z + B_2 \gamma, \\ \beta' &= C_1 y + D_1 \beta + G_1 z + H_1 \gamma, \\ \gamma' &= G_2 y + H_2 \beta + C_2 z + D_2 \gamma. \end{aligned} \tag{10.5}$$

One will therefore be confronted with *sixteen* constants, and mere inspection will not in general reveal to what extent they are independent of each other. However, if we now draw upon the existence of the point characteristic we know at once that these sixteen constants can all be expressed in terms of the *ten* constants $b_i$, so that in the most general case there must exist *six* identities between the constants defined by (10.5). Probably the easiest way to obtain these identities is to use (10.5) in the expression

$$dV = \beta' dy' + \gamma' dz' - \beta\, dy - \gamma\, dz, \tag{10.6}$$

so regarding $y$, $z$, $\beta$, $\gamma$ now as independent variables. The six conditions of integrability on the resulting linear differential form give just the six required identities. One thus finds

$$\left.\begin{aligned} A_1 D_1 - B_1 C_1 &= 1 + (F_2 G_2 - E_2 H_2), \\ A_2 D_2 - B_2 C_2 &= 1 + (F_1 G_1 - E_1 H_1), \end{aligned}\right\} \qquad (10.7)$$

together with the four homogeneous relations

$$\left.\begin{aligned} A_1 G_1 - C_1 E_1 &= A_2 G_2 - C_2 E_2, \\ A_1 H_1 - C_1 F_1 &= B_2 G_2 - D_2 E_2, \\ B_1 G_1 - D_1 E_1 &= A_2 H_2 - C_2 F_2, \\ B_1 H_1 - D_1 F_1 &= B_2 H_2 - D_2 F_2. \end{aligned}\right\} \qquad (10.8)$$

The existence of these identities is thus a simple illustration of the very direct way in which interesting consequences flow from the general theory developed above. Moreover, no assumptions of symmetry were hitherto made in this section. To illustrate the effects of symmetries of $K$, let us suppose $K$ to be plane-symmetric; see Section 8. (Other cases will be considered later.) The base-ray is of course taken to lie in the plane of symmetry of $K$. Then, recalling (8.2), inspection of (10.2) shows that in this case one must have

$$b_2 = b_4 = b_6 = b_9 = 0. \qquad (10.9)$$

It will be noticed that if at the same time $K$ has a second plane of symmetry normal to the first, no further restrictions on the parabasal terms of $V$ are implied. At any rate, the plane-symmetric system has at most *six* parabasal coefficients. As a result of (10.9) equations (10.3) simplify greatly, for now

$$\left.\begin{aligned} \beta' &= b_1 y' + b_3 y, & \gamma' &= b_5 z' + b_7 z, \\ -\beta &= b_3 y' + b_8 y, & -\gamma &= b_7 z' + b_{10} z. \end{aligned}\right\} \qquad (10.10)$$

If these be substituted in (10.5) one finds at once that only the eight constants $A_i$, $B_i$, $C_i$, $D_i$ $(i = 1, 2)$ can differ from zero; and these must satisfy the two surviving identities (10.7), i.e.

$$A_i D_i - B_i C_i = 1 \quad (i = 1, 2). \qquad (10.11)$$

If one defines two quantities,

$$\lambda_y = \hat{y}\beta - \hat{\beta} y, \quad \lambda_z = \hat{z}\gamma - \hat{\gamma} z, \qquad (10.12)$$

in terms of two arbitrary parabasal rays, of which the second is distinguished by a circumflex, then it is an almost trivial exercise to show that

$$\lambda_y = \lambda'_y, \quad \lambda_z = \lambda'_z. \tag{10.13}$$

In other words $\lambda_y$ and $\lambda_z$ are optical invariants, sometimes called *Lagrange invariants*. Unlike the invariant $j$ of Section 8 they relate only to parabasal rays; and are, moreover, each defined in terms of a *pair* of such rays.

Returning to the general case, let $\mathscr{I}$ denote some object plane and $\mathscr{I}'$ some receiving plane, both of these being supposed normal to the base-ray. The perpendicular distance of $\mathscr{I}'$ from the posterior base-plane $\mathscr{B}'$ is $d'$, so that the equation of $\mathscr{I}'$ is simply $\bar{x}' = d'$; in short, the symbolism is that used in the discussion leading to equation (7.4). A parabasal ray through $O(y, z)$ intersects $\mathscr{I}'$ in the point $O'$ whose coordinates are

$$Y' = y' + d'\beta', \quad Z' = z' + d'\gamma', \tag{10.14}$$

bearing in mind that $\alpha = \alpha' = 1$ in the parabasal limit. It now proves convenient to make use of the freedom one still has to rotate the axes in the image space about $\mathscr{R}_0$. One can choose the angle of rotation so that the resulting value of $b_2$ is zero. By an analogous rotation in the object space one can also always reduce $b_9$ to zero. We henceforth imagine this to have been done. Then, inserting (10.3) into (10.14), with $b_2 = b_9 = 0$, we find that

$$\left.\begin{aligned} Y' &= (1 + d'b_1)y' + d'(b_3 y + b_4 z), \\ Z' &= (1 + d'b_5)z' + d'(b_6 y + b_7 z). \end{aligned}\right\} \tag{10.15}$$

Setting exceptional vaues of the parabasal coefficients aside, we see immediately that all rays through $O$ pass through a straight line in the plane $d' = -1/b_1$, and they also pass through a straight line in the plane $d' = -1/b_5$. One thus has in general two *focal lines*, the distance $|b_1^{-1} - b_5^{-1}|$ between which is known as the *astigmatic focal distance*. The focal lines are mutually perpendicular, but, contrary to what is often stated in the literature, they need not be perpendicular to $\mathscr{R}_0$. As a matter of fact, the angles between the focal lines and $\mathscr{R}_0$ cannot be calculated on the basis of the quadratic terms of $V$ alone.

If a sharp image is to be obtained, the condition

$$b_1 = b_5 \tag{10.16}$$

must be satisfied. Evidently a square grid is then transformed by $K$ into a grid of similar parallelograms. These in general become rectangles when $K$ is plane-symmetric, since then

$$Y' = d'b_3 y, \quad Z' = d'b_7 z. \tag{10.17}$$

One thus has a magnification $m_1 = d'b_3$ in the direction defined by $Y'$ alone increasing and another magnification $m_2 = d'b_7$ in a direction at right angles to this. The image will therefore be geometrically similar to the object, i.e. it will be undistorted, provided

$$b_3 = b_7. \tag{10.18}$$

It may be noted that the quadratic terms of the ideal characteristic function (7.4) indeed satisfy (10.16) and (10.18).

The following remarks may be appropriate at this point. If $K$ is not regular with respect to the coordinate basis none of the preceding equations of this section obtain. This is so, for example, in the case of the paraboloid of Section 9 with the second of the coordinate bases there defined. The reader whose mind is directed towards practical realities might be inclined to argue that regularity could be restored by replacing the actual refracting surface by a small, spherical patch at and near the cusp; and that one can then take a base-ray $\mathscr{R}_0$ along the axis. This much is certainly true. On the other hand this procedure is quite useless for another reason. Parabasal optics is intended to be an approximation—albeit a crude one—in an extended neighbourhood of $\mathscr{R}_0$, that is to say, a neighbourhood not so small as to be of merely academic interest. However, in the present situation the geometry of the surface of interest does not enter into the parabasal optics at all, so that the latter is irrelevant. In short, one has gained nothing.

It should be clear by now how any question concerning parabasal imagery can be answered by elementary arguments based on equations (10.3) and (10.14), if, in the latter, $d'$ be regarded as a parameter whose value can be chosen at will. The subject will therefore not be pursued further, apart from the remark that the preceding work could of course have been based equally well on the use of one of the other characteristic functions. Using the notation introduced at the end of Section 6, the characteristic function, up to terms of the second degree, will be

$$F = a_0 + \tfrac{1}{2} \sum_{k,l=1}^{4} b_{kl}\, q_k q_l, \tag{10.19}$$

whence

$$p_k = \sum_{l=1}^{4} b_{kl}\, q_l; \tag{10.20}$$

and these equations correspond to (10.2) and (10.3) respectively.

It must be carefully observed that, the notation notwithstanding, the meaning of the constants $b_{kl}$ depends of course on the particular choice of $F$. We shall frequently make use of the convention that certain symbols denote quantities of a certain generic *type*, so that the precise significance to be attached to such symbols will depend upon the context in which they occur. Whatever slight difficulties may be encountered as a result of adhering to this convention, it is more than compensated for by the way in which it enables us to avoid an endless and confusing proliferation of symbols.

## 11. Aberration coefficients

Given a system which is regular with respect to a suitable coordinate basis, its aberration function $f$ may be written as a power series:

$$f = \sum_{n=1}^{\infty} f_n. \tag{11.1}$$

Here $f_n$ is a homogeneous polynomial of degree $n+1$ in the ray-coordinates, and it is called the *aberration function of order n*. Its generic form is

$$f_n = \sum_{\lambda=0}^{n+1} \sum_{\mu=0}^{\lambda} \sum_{\nu=0}^{\mu} f_{n\lambda\mu\nu}\, q_1^{n+1-\lambda} q_2^{\lambda-\mu} q_3^{\mu-\nu} q_4^{\nu}. \tag{11.2}$$

Then the $f_{n\lambda\mu\nu}$, which are constants of the system for a given coordinate basis, are called the *aberration coefficients* of order $n$. Occasionally we shall refer to them more precisely as *characteristic aberration coefficients*, to distinguish them from the *effective* aberration coefficients to be considered later on. Since a polynomial of degree $m$ in $s$ variables has $\binom{m+s-1}{m}$ coefficients, the number of aberration coefficients of order $n$ is not greater than

$$n_1 = \tfrac{1}{6}(n+2)(n+3)(n+4). \tag{11.3}$$

This number relates to arbitrary positions of the object surface. If the latter is taken as fixed (as is frequently done in practice) all terms of (11.1) which depend on the object space coordinates $q_3$ and $q_4$ alone need not be counted. There are $n+2$ of these, and so the number $n_1$ is reduced to

$$n_2 = \tfrac{1}{6}(n+1)(n+2)(n+6). \tag{11.4}$$

In *this* sense there are at most 7, 16, 30, 50, 77, ... aberration coefficients of orders 1, 2, 3, 4, 5, .... The problem in practice is to calculate these numbers, given the physical constitution of the system; and, as explained at the end of Section 9, one may have to do this for several choices of coordinate basis. The computational problem is a difficult one, and in practice the sequence (11.1) has to be truncated after a very few terms. Fortunately, the number of independent aberration coefficients of the various orders is reduced by every symmetry the system happens to possess; granted, of course, that the base-ray is chosen appropriately. For example, if $K$ is doubly plane-symmetric (see Section 88), i.e. has just two planes of symmetry, with $\mathscr{R}_0$ taken along their common line, then one has in place of (11.3) the number

$$n_3 = \begin{cases} \tfrac{1}{12}(n+3)(n^2+6n+11) & (n \text{ odd}) \\ 0 & (n \text{ even}). \end{cases} \tag{11.5}$$

This gives $n_3 = 0, 19$ for the coefficients of orders 2 and 3, whereas $n_1 = 20, 35$ according to (11.3). Even this moderate degree of symmetry therefore reduces the computational task very considerably.

Every term of (11.2) is said to represent *an aberration*, so that one can speak of the 'number of aberrations' rather than of the number of aberration coefficients. Now consider the displacement $\boldsymbol{\epsilon}'$ introduced in Section 7. (It suffices to restrict ourselves to the state of affairs contemplated there, since the generalization to more complicated situations, e.g. to curved image surfaces, is easily achieved.) Each of the components of $\boldsymbol{\epsilon}'$ appears as a sequence of homogeneous polynomials of degree 1, 2, ... in the ray-coordinates:

$$\epsilon'_y = \sum_{n=1}^{\infty} \epsilon'_{ny}, \quad \epsilon'_z = \sum_{n=1}^{\infty} \epsilon'_{nz}. \tag{11.6}$$

$\boldsymbol{\epsilon}'_n$ is induced by the aberrations of order $m \leqslant n$. This means that, strictly speaking, one has to distinguish between the 'aberrations of order $s$' and the 'displacement of order $s$', for these terms refer to (11.2) and (11.6) respectively. However, in practice one frequently speaks indiscriminately of the aberrations of order $s$ in either case.

Taken by itself, a particular $n$th-order term of (11.2), $\psi_n$ say, gives rise to a certain $n$th-order displacement $\boldsymbol{\epsilon}_n'(\psi_n)$. The latter depends in a specific way upon the ray-coordinates, and its magnitude depends linearly upon the aberration coefficient which governs $\psi_n$, that is to say, which multiplies it. It is usual to examine the curve generated by the points of intersection with the receiving plane of a suitably selected one-parameter set of rays from a fixed point $O$ of the object. The shape and location of this curve, and the way in which it depends upon parameters (such as the position of $O$) which are still at our disposal, completely characterize the aberration in hand. The purpose of this procedure is to gain insight into the geometrical significance of the various terms of (11.2). More detailed investigation reveals that any particular aberration of order $n$ in a certain sense has as its natural counterpart an aberration in every order exceeding $n$; and that generally speaking various aberrations of a given order can be put into groups, which again have their higher-order counterparts; see, in particular, Sections 21, 22 and 77.

The preceding remarks are intended to indicate in general terms that aberrations, or aberration coefficients, can be usefully divided into various classes or types. The precise details of such a classification naturally depend upon various conventions introduced from time to time. However, rather than embark on such a programme at this stage in the most general terms, it seems desirable to proceed directly to the consideration of systems having given symmetries. The investigation of the aberrations and their significance in these more specialized situations is easier and therefore more readily appreciated; the more so as we shall not hesitate occasionally to repeat in the more specialized circumstances work we have already done in quite general terms. This is particularly true of the lengthy treatment of the symmetric system which is intended to serve as a kind of prototype. Having gone through much the same routine a few times, the reader should experience no overwhelming difficulties with whatever case comes to hand, no matter how general.

## Problems

**P.3 (i).** Obtain the angle characteristic $T$ for the paraboloid described in Section 9, the second of the alternative coordinate bases being chosen. Show especially that $T$ is irregular. (See equations (117.2–15).)

**P.3 (ii).** Give an example of a wavefront $\mathscr{W}$ such that all its rays pass exactly (i.e. without restriction to parabasal optics) through two (non-intersecting) curves. Hence show that for an arbitrary $\mathscr{R}_0$ the parabasal focal lines are not necessarily normal to $\mathscr{R}_0$.

**P.3 (iii).** Show that $\lambda_y + \lambda_z$ is an optical invariant even when $K$ has no symmetries.

**P.3 (iv).** A circle in the object plane and with centre on $\mathscr{R}_0$ has as its parabasal image a certain curve $\mathscr{C}$. Determine $\mathscr{C}$ and examine the condition that it must not shrink to a point.

**P.3 (v).** In the parabasal region, for fixed $x, y, z$, $V$ will be a function of $x', y', z'$ which may be written in the form (10.2), with the term $O(3)$ omitted. Obtain differential equations which the functions $a_0(x'), b_1(x'), \ldots, b_{10}(x')$ must satisfy. ($b_2$ and $b_9$ may be taken to be zero, but their derivatives cannot, of course, be supposed to vanish at the same time.) Solve some of the equations.

**P.3 (vi).** An optical system consists of a homogeneous glass cylinder of arbitrary cross-section. What can be said about the generic form of $V$?

**P.3 (vii).** Does the system of the preceding problem in general have non-vanishing second-order aberrations? If so, how many coefficients govern them?

CHAPTER 4

# THE SYMMETRIC SYSTEM (PART I)

## 12. Definition of the symmetric system

A system is called *symmetric* (without qualification) if it has (i) an axis of symmetry which has points in common with both the object space and the image space (cf. Section 93), *and* (ii) a plane of symmetry which contains the axis. Of course, when both conditions are satisfied every plane containing the axis is a plane of symmetry. On the other hand, mere axial symmetry does not imply plane-symmetry. In this context it may be helpful to think about a turbine with non-radial blades; see also Chapter 7.

The displacements corresponding to the present symmetries are just those considered explicitly in Section 8. With regard to axial symmetry, a rotation of $K$ through the angle $\theta$ takes $A$, $A'$ into $A^*$, $A^{*\prime}$, where now

$$\left.\begin{aligned} y^{*\prime} &= y'\cos\theta - z'\sin\theta, \quad & z^{*\prime} &= y'\sin\theta + z'\cos\theta, \\ y^* &= y\cos\theta - z\sin\theta, \quad & z^* &= y\sin\theta + z\cos\theta. \end{aligned}\right\} \qquad (12.1)$$

A coordinate basis has been adopted such that the $\bar{x}$-axis and the $\bar{x}'$-axis both lie along the axis $\mathscr{A}$ of $K$, whilst the $\bar{y}$-axis and $\bar{y}'$-axis are mutually parallel. The plane containing these axes will be called the *meridional* plane (or also *tangential* plane), whilst the *sagittal* plane is normal to this and contains $\mathscr{A}$. In the context of the symmetric system the particular coordinate basis just chosen shall be understood throughout, so that later on reference to the 'regularity of $K$' will be intended to mean regularity with respect to this coordinate basis.

As for the second symmetry condition, we may, without loss of generality, choose the meridional plane to be the plane of symmetry, so that in this case

$$y^{*\prime} = y', \quad z^{*\prime} = -z', \quad y^* = y, \quad z^* = -z. \qquad (12.2)$$

The identity (8.1), i.e.

$$V(y^{*\prime}, z^{*\prime}, y^*, z^*) \equiv V(y', z', y, z), \qquad (12.3)$$

must now hold for all values of $y'$, $z'$, $y$, $z$ *and* $\theta$, for both (12.1) and (12.2) taken separately. Precisely analogous equations to these will obtain in the context of the other characteristics $T$, $W_1$ and $W_2$, since under rotations and reflections direction cosines transform exactly like the corresponding Cartesian coordinates; whilst the ray-coordinates in the object space on the one hand, and those in the image space on the other, merely transform amongst themselves, i.e. do not get mixed up with each other.

In the light of types of symmetries to be considered later, the traditional terminology, to which we have adhered here, is somewhat unfortunate. It is not very sensible to single out a class of systems defined by a *particular* set of symmetries and merely to call the members of this class 'symmetric'. In short, to avoid needless ambiguity we can be specific by speaking, in the present context, of '$r$-symmetry' rather than of 'symmetry'; but we shall do so only on occasions where the usual terminology seems altogether too inadequate.

## 13. Form of the characteristic function. Rotational invariants

Equations (12.1) and (12.3) together express the condition that $V$ must be invariant under rotations. To investigate the explicit consequences of this condition, as regards the generic form of $V$, we may proceed in the following elementary way. First, introduce four new independent variables $\xi, \eta, \zeta, \sigma$ in place of $y'$, $z'$, $y$, $z$, of which the first three are

$$\xi = y'^2 + z'^2, \quad \eta = yy' + zz', \quad \zeta = y^2 + z^2, \tag{13.1}$$

whilst $\sigma$ can be taken arbitrarily, subject, of course, to the requirement that it cannot be written as a function of $\xi$, $\eta$, $\zeta$ alone. We now observe that $\xi$, $\eta$, $\zeta$ are all separately invariant under rotations, i.e. $y^{*2} + z^{*2} \equiv y^2 + z^2$ for all $\theta$, and so on. For this reason these three quantities are often called (*elementary*) *rotational invariants.* Next suppose that $\sigma$ could be defined in such a way that it also is invariant under rotations. Then, since any function of $y'$, $z'$, $y$, $z$ can equally well be written as a function of $\xi$, $\eta$, $\zeta$, $\sigma$, one would be able to conclude that every function of $y'$, $z'$, $y$, $z$ is invariant under rotations, and this is certainly not the case. It follows that no $\sigma$ which is functionally independent of $\xi$, $\eta$, $\zeta$ can be rotationally invariant. $V$ therefore cannot be a function of $\sigma$. In short, we have shown that

*$V$ is invariant under rotations if and only if it depends on $y'$, $z'$, $y$, $z$ through the combinations $\xi, \eta, \zeta$ alone.*

It remains to take the condition of plane-symmetry into account. At first sight it looks as if this were automatically satisfied, since $\xi$, $\eta$, $\zeta$ are invariant under the substitution $z' \to -z'$, $z \to -z$. Clearly, we must be mistaken since, as already remarked, axial symmetry does not imply plane-symmetry; or, in other words, $K$ may have a built-in screw-sense. The error lies in our inadvertently thinking only of single-valued functions of $\xi$, $\eta$, $\zeta$. This point is clearly illustrated by the function

$$\tau = yz' - zy'. \tag{13.2}$$

It is easily shown to be invariant under rotations, and so must be a function of $\xi$, $\eta$, $\zeta$; in fact, $\tau = \pm(\xi\zeta - \eta^2)^{\frac{1}{2}}$. (Incidentally, $\tau$ is not to be included under the heading of elementary rotational invariants.) Now when $z$ and $z'$ reverse sign then so does $\tau$, and no ambiguity arises. When, however, $\tau$ is regarded as a function of $\xi$, $\eta$, $\zeta$ its double-valuedness must be explicitly taken into account. Moreover, if, for example, $V$ contained a term which depended linearly on $\tau$ then it would not be invariant under reflections. In short, the condition of plane-symmetry imposes definite limitations upon the form of the dependence of $V$ on $\xi$, $\eta$, $\zeta$.

These limitations can be stated explicitly when the condition of regularity is satisfied; and we now assume that this is the case. This means of course that $V$ can be written as a power series in the ray-coordinates. We recognize that already because of rotational invariance this series cannot contain terms of odd degree, since a reversal of sign of all the ray-coordinates is equivalent to a rotation through 180°. At this stage of the argument it is convenient temporarily to introduce polar coordinates in both base-planes:

$$y' = \rho\cos\theta, \quad z' = \rho\sin\theta, \quad y = \chi\cos\phi, \quad z = \chi\sin\phi. \tag{13.3}$$

Then $V$ becomes a power series in $\rho$ and $\chi$, a typical term of which, $X$, say, has the form $w\rho^\mu\chi^\nu$ ($\mu+\nu$ even), where $w$ is a sum of products of the trigonometric functions in (13.3). However, $\theta$ and $\phi$ can, in effect, occur in $w$ only in the combination $\psi = \theta - \phi$, since under a rotation $\theta$ and $\phi$ change by the same amount. In short, $w$ is generically a sum of products of $\cos\psi$ and $\sin\psi$. Symmetry about the meridional plane now requires the absence of *odd* powers of $\sin\psi$, since a reversal of sign of $z'$ and $z$ is equivalent to a reversal of sign of $\psi$. This means that $w$ can always be written simply as a polynomial in $\cos\psi$ alone. Now recall that $X$ arose from making the substitutions (13.3) in a series whose terms contained only integral powers of the ray-coordinates, so that the same must be true of $X$. Since $\xi = \rho^2$, $\eta = \rho\chi\cos\psi$, $\zeta = \chi^2$ it follows that $X$ must be a sum of products of $\xi$, $\eta$ and $\zeta$. Consequently we have shown that

> *the characteristic function of a regular symmetric system can be written as a power series in the three elementary rotational invariants.* (13.4)

We have stated this important result in a form which no longer explicitly refers to any particular characteristic function $F$, since it is clearly valid for any choice of the latter, bearing in mind the concluding remarks of the preceding section. The elementary rotational invariants must of course be appropriate in each case to the particular choice of $F$. In terms of the generic notation of Section 6 the invariants

$$\xi = q_1^2+q_2^2, \quad \eta = q_1q_3+q_2q_4, \quad \zeta = q_3^2+q_4^2 \tag{13.5}$$

are admissible. Nevertheless these are not the only possibilities, for any three linearly independent linear combinations of $\xi, \eta, \zeta$ will do as well. Such combinations will then usually be denoted again by $\xi$, $\eta$, $\zeta$. In short, we have here an excellent example of the convention regarding the repeated use of the same set of symbols discussed at the end of Section 6.

Finally, it should be remarked that the result (13.4) is by no means trivial, though it is often taken for granted. Unfortunately some discussions seem to imply that it is valid for *any* axially symmetric regular system. This, however, is clearly not the case (see also Section 72).

## 14. Paraxial optics

The work of this section will to some extent be a repetition of that of Section 10, but will naturally be much simpler in its details. The base-ray $\mathscr{R}_0$ now coincides with the axis $\mathscr{A}$ of $K$, so that we speak of *paraxial*, rather than of parabasal, optics. We have the freedom to use whatever characteristic function we please, and for the time being we continue to use $V$. Retaining only terms of the second degree (in the ray-coordinates) we now have

$$V = a_0 + \tfrac{1}{2}k_1\xi + k_2\eta + \tfrac{1}{2}k_3\zeta, \tag{14.1}$$

where $a_0$, $k_1$, $k_2$, $k_3$ are constants of the system. These of course still depend upon the exact location of the base-planes, which are for the present arbitrary normal planes. (A *normal plane* is any plane normal to $\mathscr{A}$.) It will be recalled that the anterior base-plane $\mathscr{B}$ has the equation $\bar{x} = 0$, the posterior base-plane $\mathscr{B}'$ the equation $\bar{x}' = 0$, in view of the arrangement of the coordinate basis. The axial points of $\mathscr{B}$ and $\mathscr{B}'$ will be denoted by $B$, $B'$ respectively.

Comparing (14.1) with (10.2) we have

$$b_1 = b_5, \quad b_3 = b_7, \quad b_8 = b_{10}, \quad b_2 = b_4 = b_6 = b_9 = 0. \tag{14.2}$$

This is a situation of great simplicity, paraxial imagery being governed by only three constants. Now, using (6.1), or directly from (10.3),

$$\left.\begin{aligned} \beta' &= k_1 y' + k_2 y, & \gamma' &= k_1 z' + k_2 z, \\ -\beta &= k_2 y' + k_3 y, & -\gamma &= k_2 z' + k_3 z. \end{aligned}\right\} \tag{14.3}$$

We have here the first example of sets of equations which naturally group themselves into pairs, and it is convenient to adopt an appropriate notation. For this purpose we take (14.3) as a typical example. Thus the first and second pair of equations will be written as

$$\boldsymbol{\beta}' = k_1\mathbf{y}' + k_2\mathbf{y}, \quad -\boldsymbol{\beta} = k_2\mathbf{y}' + k_3\mathbf{y} \tag{14.4}$$

respectively. In effect, we are using a two-vector notation which is largely self-explanatory. Any symbol in bold-face type stands for two 'components', e.g. $\boldsymbol{\beta}$ for $\beta$ and $\gamma$, as above, or $\boldsymbol{\epsilon}'$ for $\epsilon_y'$ and $\epsilon_z'$; and so on. If desired one may use the usual notation for the scalar product, so that, for example, $\boldsymbol{\beta}.\boldsymbol{\beta} = \beta^2 + \gamma^2$.

The basic equations of 'Gaussian optics' are just the equations (10.5) when specialized to the symmetric case:

$$\mathbf{y}' = A\mathbf{y} + B\boldsymbol{\beta}, \quad \boldsymbol{\beta}' = C\mathbf{y} + D\boldsymbol{\beta}. \tag{14.5}$$

Inserting (14.4) into (14.5) one finds, with $c = 1/(k_2^2 - k_1k_3)$, that

$$A = -k_3/k_2, \quad B = -1/k_2, \quad C = 1/ck_2, \quad D = -k_1/k_2. \tag{14.6}$$

From these the identity $AD - BC = 1$ (14.7)
follows at once.

Let $\mathscr{D}'$ be the normal plane $\bar{x}' = d'$ in the image space. A ray $\mathscr{R}$ intersects it in the point

$$\mathbf{Y}' = \mathbf{y}' + d'\boldsymbol{\beta}' = (1 + d'k_1)\,\mathbf{y}' + d'k_2\mathbf{y}, \tag{14.8}$$

cf. (10.15). If $\mathscr{R}$ is parallel to the axis in the object space, then

$$k_2\mathbf{Y}' = (d'/c - k_3)\,\mathbf{y}.$$

Evidently the ray intersects $\mathscr{D}'$ in its axial point $F'$ if $d'$ has the value

$$d_0' = ck_3. \tag{14.9}$$

The particular point $F'$ so defined is the posterior *focal point* of $K$. It suffices to suppose that $\mathscr{R}$ is meridional, in view of the symmetry of $K$, i.e. to take $z = z' = 0$. Its actual (unreduced) distance from $\mathscr{A}$ in the object space is then $y/N$. Let $\mathscr{R}$ intersect $\mathscr{D}'$ in a point whose actual normal distance from $\mathscr{A}$ is just $y/N$ again, so that

$$Y'/N' = y/N.$$

Then $d'$ must be chosen to have the value

$$d_1' = c(N'k_2/N + k_3). \tag{14.10}$$

The distance $d_0' - d_1'$ is, by definition, the (reduced) posterior *focal length* $\hat{f}'$ of $K$, so that

$$\hat{f}' = -N'ck_2/N. \tag{14.11}$$

The anterior focal length $\hat{f}$ is defined analogously in terms of a ray which is parallel to the axis in the image space, and it turns out that

$$\hat{f}'/N'^2 = \hat{f}/N^2. \tag{14.12}$$

The explicit appearance of the refractive indices in (14.11) and (14.12) is a nuisance, since if they are not removed forthwith by some means or other they will occur time and again later on. Accordingly we define the *mean focal length* $f$ as the geometric mean of $\hat{f}$ and $\hat{f}'$. Thus

$$f = (\hat{f}\hat{f}')^{\frac{1}{2}}; \tag{14.13}$$

but the qualification 'mean' will henceforth be omitted. Now

$$\hat{f}' = N'f/N, \quad \hat{f} = Nf/N', \tag{14.14}$$

whilst the common ratio which appears in (14.12) is just $f/NN'$.

In terms of a terminology in common usage, this ratio is the reciprocal of the *power* of $K$, and $f$ is the reciprocal of the *modified power*. Note that (14.11) now reads simply

$$f = -ck_2. \tag{14.15}$$

All rays from a point $O(\mathbf{y})$ of $\mathscr{B}$ intersect $\mathscr{D}'$ in the same point $O'(\mathbf{Y}')$ if one chooses

$$d' = -1/k_1, \tag{14.16}$$

for then (14.8) reduces to

$$\mathbf{Y}' = d'k_2\mathbf{y}. \tag{14.17}$$

This shows that, as far as paraxial imagery is concerned, a plane object has a sharp, undistorted plane image. Moreover, the (reduced) magnification associated with $\mathscr{B}$ and $\mathscr{D}'$ is given by

$$m = d'k_2. \tag{14.18}$$

When the object point is at infinity the initial rays are characterized by the constancy of $\boldsymbol{\beta}$. Then, from (14.8) and the second member of (14.4),

$$\mathbf{Y}' = (1 - d'/ck_3)\,\mathbf{y}' - (d'k_2/k_3)\,\boldsymbol{\beta}. \tag{14.19}$$

The image is of course formed in the plane which has $d' = d'_0$; and, using (14.9) and (14.15), the *image height* is

$$\mathbf{Y}' = f\boldsymbol{\beta}. \tag{14.20}$$

We now write $\mathscr{I}$ in place of $\mathscr{B}$ to emphasize that it is to be regarded as the object plane, whilst we write $\mathscr{I}'$ in place of $\mathscr{D}'$ to indicate that this plane is *conjugate* to $\mathscr{I}$, i.e. that every point of $\mathscr{I}$ is transformed by $K$ into one point of $\mathscr{I}'$, their respective coordinates bearing a fixed ratio to each other in the sense of equation (14.17). $\mathscr{I}'$ is also called the *ideal image plane* corresponding to $\mathscr{I}$, for obvious reasons. Similarly, if the point $O'$ in this plane is conjugate to the point $O$ of $\mathscr{I}$, then $O'$ is called the *ideal image point* (of $O$).

The system will generally contain a *stop* somewhere, that is to say, some plane screen, normal to the axis, which is provided with an aperture of some shape. It is this stop which we suppose to limit the bundles of rays capable of passing through $K$, though in practice this limitation is often imposed, at least in part, by other obstacles, such as lens rims, in which case one speaks of *vignetting*. At any rate, we suppose the stop (i.e. its aperture) to be circular, concentric with the axis. The paraxial images formed of the axial point of the stop by the parts of $K$ respectively preceding and following it of course lie on the axis, and they will be denoted by $E$ and $E'$. The normal planes through $E$ and $E'$ are called the planes of the (*paraxial*) *entrance and exit pupil* respectively, the pupils being the generally somewhat ill-defined images of the stop which lie approximately in the planes just defined. A family of rays just grazing the rim of the stop will generally intersect the plane of the paraxial exit pupil in a curve which approximates a circle concentric with the axis. It should, however, be borne in mind that cases arise in practice when what has just been said is completely false, for example, in the case of some wide-angle photographic objectives.

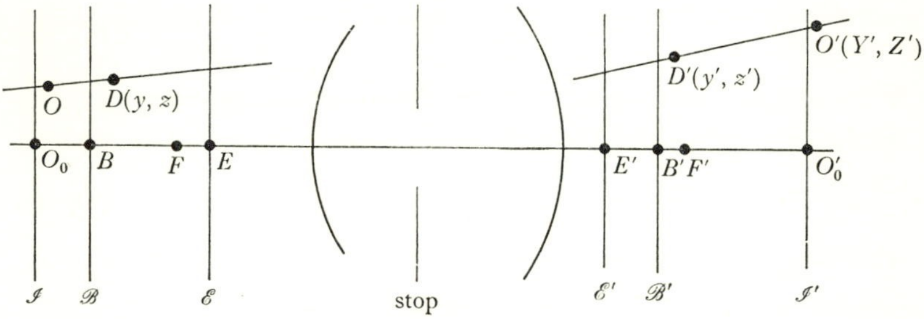


Fig. 4.1

To recapitulate, we have at hand the following planes: the object plane $\mathscr{I}$ and its conjugate ideal image plane $\mathscr{I}'$, the anterior and posterior base-planes $\mathscr{B}$ and $\mathscr{B}'$, the planes $\mathscr{E}$ and $\mathscr{E}'$ of the paraxial entrance and exit pupils; the axial points of these planes, taken in order, being $O_0$, $O_0'$, $B$, $B'$, $E$ and $E'$, whilst $F$ and $F'$ are the focal points; see Fig. 4.1. In any particular situation some of these planes may be mutually coincident, and indeed, will usually be so by choice, depending upon the problem in hand, and on the particular characteristic function used to deal with it. Thus, in the present circumstances $\mathscr{B}$ coincides with $\mathscr{I}$, and we now choose $\mathscr{B}'$ to coincide with $\mathscr{E}'$, so that $d'$ is the fixed distance between $E'$ and $O_0'$.

We may now slightly extend our previous work. To this end let $\bar{x} = q$ be a plane in the object space, and $\bar{x}' = q'$ a plane in the image space. A ray intersects these in the points

$$\mathbf{Y}_1 = \mathbf{y} + q\boldsymbol{\beta}, \quad \mathbf{Y}_1' = \mathbf{y}' + q'\boldsymbol{\beta}'$$

respectively. Thus

$$\mathbf{Y}_1 = -qk_2\mathbf{y}' + (1 - qk_3)\,\mathbf{y}, \;\; \mathbf{Y}_1' = (1 + q'k_1)\,\mathbf{y}' + q'k_2\mathbf{y}. \qquad (14.21)$$

The planes in question are mutually conjugate if $\mathbf{Y}_1'$ and $\mathbf{Y}_1$ stand in a fixed ratio to each other for all $\mathbf{y}'$, $\mathbf{y}$, which will be the case if the discriminant of the equations (14.21) vanishes, i.e. if

$$c^{-1}qq' + k_1q' - k_3q + 1 = 0. \qquad (14.22)$$

The (reduced) magnification $\hat{s}$ associated with these planes is then given by

$$\hat{s} = -(1 + k_1q')/k_2q = k_2q'/(1 - k_3q), \qquad (14.23)$$

provided neither $q$ nor $q'$ is infinite. Take the pupil planes as an example. Then the distance from $O_0$ to $E$ is the analogue of $d'$, and we therefore write $d$ for it. Setting $q' = 0$ in (14.22), we find at once that

$$d = 1/k_3; \qquad (14.24)$$

whilst, if $s$ is the magnification associated with the pupil planes, (14.23) gives

$$s = -k_3/k_2. \qquad (14.25)$$

This may be compared with (14.18), i.e.

$$m = -k_2/k_1. \qquad (14.26)$$

From these results one infers easily that

$$d' = (s - m)f \qquad (14.27)$$

and

$$d = \left(\frac{1}{s} - \frac{1}{m}\right)f. \qquad (14.28)$$

Recalling (14.16) one can therefore write

$$k_1 = -1/d', \quad k_2 = m/d', \quad k_3 = -sm/d', \qquad (14.29)$$

with $d'$ given by (14.27). In this way the paraxial coefficients are expressed in terms of $s$, $m$ and $f$ alone. (14.1) may now be written as

$$V = (a_0 - d') - \frac{1}{2d'}(\xi - 2m\eta + m^2\zeta) - \frac{m}{2f}\zeta. \qquad (14.30)$$

We note in passing that the pairs of conjugate planes defined by the value unity of the actual and of the reduced magnification are known as *principal planes* and *nodal planes* respectively.

It should be clear by now how any question concerning paraxial imagery can be answered by appealing to (14.4) and (14.21). Moreover, though we proceeded above on the basis of the point characteristic, any other characteristic function would have served as well. It is hardly necessary to demonstrate this in detail. Indeed, it will suffice briefly to consider the angle characteristic; doing so chiefly to bring out clearly a certain feature of the notation, which, if not explicitly remarked on, might be a source of confusion for the unwary reader.

Accordingly we now begin with the equation

$$T = a_0 + \tfrac{1}{2}k_1\xi + k_2\eta + \tfrac{1}{2}k_3\zeta. \qquad (14.31)$$

Here the end of Section 13 must be recalled: $\xi$, $\eta$, $\zeta$ now stand for $\beta'^2+\gamma'^2$, $\beta\beta'+\gamma\gamma'$, $\beta^2+\gamma^2$ respectively, and likewise the significance of the constants $k_1$, $k_2$, $k_3$ is different from that which they had in (14.1). By (6.2),

$$-\mathbf{y}' = k_1\boldsymbol{\beta}' + k_2\boldsymbol{\beta}, \quad \mathbf{y} = k_2\boldsymbol{\beta}' + k_3\boldsymbol{\beta}. \qquad (14.32)$$

These equations relate to an arbitrary choice of base-planes; and they are therefore entirely equivalent to (14.4). We could now proceed exactly as before; but the work would differ only trivially from that following equations (14.4).

However, note that in the context of the angle characteristic the use of conjugate base-planes is permitted. If, therefore, we now choose $\mathscr{B}$ and $\mathscr{B}'$ to be $\mathscr{I}$ and $\mathscr{I}'$ respectively, the points $O(\mathbf{y})$ and $O'(\mathbf{y}')$ must be conjugate to each other, i.e. $\mathbf{y}'$ must stand in a fixed ratio to $\mathbf{y}$, this ratio being of course just the magnification $m$. It follows at once that

$$k_1 = -mk_2 = m^2k_3. \qquad (14.33)$$

The posterior focal length is $\hat{f}' = -N'y/N\beta'$, calculated for a meridional ray which is parallel to the axis in the object space. Drawing upon (14.14) and (14.32) with $\beta = 0$, it follows that

$$f = -k_2. \qquad (14.34)$$

Therefore $$T = a_0 + (f/2m)(m^2\xi - 2m\eta + \zeta). \qquad (14.35)$$

At this point a word needs to be said about *telescopic systems*, which have $f = \infty$. (14.35) shows at once that the theory breaks down in this case; as we already know, of course, from Section 4.

It is, however, not necessary to go again through the preceding work in detail. Thus, inspection of equations (14.27) and (14.28) shows that we must, of necessity, now have $s = m$; which means that the magnification associated with *all* pairs of conjugate planes is the same, say $m$. (14.22) simplifies to

$$k_1 q' - k_3 q + 1 = 0, \tag{14.36}$$

i.e. $q'$ and $q$ are linearly related to each other. With the usual choice of base-planes, equations (14.29) apply as before (with $s = m$), whilst (14.30) becomes

$$V = \text{const.} - 1/(2d')(\xi - 2m\eta + m^2\zeta). \tag{14.37}$$

One also has the simple relation

$$d' = -m^2 d. \tag{14.38}$$

When $\mathscr{I}$ is at infinity, and therefore $\mathscr{I}'$ also, one has a still more specialized situation, to which (14.37) and (14.38) of course do not apply. One then has to choose any convenient pair of finitely situated (non-conjugate) base-points $B$, $B'$. $V$, as given by (14.37), may then be taken to refer to these, if $m$ be interpreted as the magnification associated with $\mathscr{B}$ and the plane conjugate to it, with a corresponding interpretation of $d'$.

We return to the point of notation which was alluded to just before equation (14.31). It is this: in the context of $V$ the image point $O'$ had coordinates denoted by $Y'$, $Z'$, whereas in the context of $T$ these were denoted by $y'$, $z'$. The reason for this state of affairs is that we have agreed once and for all to use lower-case symbols for all ray-coordinates and their conjugates. Granted the very convenient choice of base-planes made hitherto, the conjugates $\mathbf{y}'$ of $-\boldsymbol{\beta}'$ simply *are* the coordinates of points in the ideal image plane. In the case of $V$ on the other hand the ray-coordinates $\mathbf{y}'$ relate to points in $\mathscr{B}'$ and this plane must be distinct from $\mathscr{I}'$; so that new symbols are required for the coordinates of points in $\mathscr{I}'$. No genuine ambiguity is inherent in this notation. Once again, the meaning of a particular symbol depends upon the context in which it occurs; a situation which reflects the wide freedom one has in the choice of characteristic functions, and of base-planes, once a particular characteristic function has been adopted.

## 15. Ideal characteristic functions

To specify a particular ideal characteristic function $F_0$ we must first decide under which circumstances the behaviour of a given system $K$ is to be regarded as 'ideal'; as was emphasized in the course of the discussion of Section 7. Accordingly, bearing in mind what is perhaps the most common situation in practice, we desire to form a sharp, undistorted, plane image of an object lying in a fixed normal plane $\mathscr{I}$. Our immediate task is to determine the characteristic function $F_0$ which $K$ would have to have to exhibit the desired behaviour. Although later on we shall investigate the properties of systems with particular symmetries with the aid of whatever characteristic function appears to be the most convenient, we do not want to make any particular choice for the present and so determine $V_0$, $T_0$, $W_{10}$ and $W_{20}$ in turn.

$V_0$ has of course already been considered in a more general setting in Section 7. Nevertheless it will do no harm to recapitulate the work in the present context. To begin with, we note that since all rays from a point $O$ of $\mathscr{I}$ are to pass through one point $O'$ of $\mathscr{I}'$ this must be true for paraxial rays in particular. $\mathscr{I}'$ is therefore the ideal image plane and $O'$ the ideal image point $I'$ conjugate to $O$. As base-planes we choose $\mathscr{I}$ and $\mathscr{E}'$. Then, if a ray through $O$ and $O'$ intersects $\mathscr{E}'$ in $D'$,

$$\bar{V}(O,D') = \bar{V}(O,O') - \bar{V}(D',O'),$$

in the usual notation. Now $\bar{V}(O,O')$ is a constant for all rays through $O$, and so depends on $y$ and $z$ only. Since these variables can occur only in the combination $\zeta$, we have

$$\bar{V}(O,O') = \bar{g}(\zeta), \tag{15.1}$$

where the function $\bar{g}(\zeta)$ is not further determined here. Bearing in mind that $\mathbf{Y}' = m\mathbf{y}$, the required result

$$V_0 = \bar{g}(\zeta) - (d'^2 + \xi - 2m\eta + m^2\zeta)^{\frac{1}{2}} \tag{15.2}$$

follows at once. Note that if we write

$$\bar{g}(\zeta) = \bar{g}_0 + \tfrac{1}{2}\bar{g}_1\zeta + O(4), \tag{15.3}$$

(15.2) gives

$$V_0 = (\bar{g}_0 - d') - 1/(2d')(\xi - 2m\eta + m^2\zeta) + \tfrac{1}{2}\bar{g}_1\zeta, \tag{15.4}$$

in agreement with (14.30); and $\bar{g}_1$ is seen to have the value $-m/f$.

When the object is at infinity the choice of $\mathscr{I}$ as anterior base-plane is clearly no longer permitted. (Physically: all optical distances between points of the base-planes become infinite.) Any other choice of anterior base-plane is, however, very inconvenient in the context of the point characteristic. One can help oneself with a limiting process, that is to say, one first takes $\mathscr{I}$ to be at a finite distance from $K$ (so that $m \neq 0$) and eventually allows $\mathscr{I}$ to recede to infinity ($d \to \infty$). To avoid formal complications we therefore introduce independent variables $\mathbf{y}_1$ in place of $\mathbf{y}$, where

$$\mathbf{y}_1 = m\mathbf{y}, \tag{15.5}$$

so that $y_1, z_1$ are simply the coordinates of the ideal image point. $\eta$ and $\zeta$ then refer to $\mathbf{y}_1$ rather than to $\mathbf{y}$, e.g. $\zeta = y_1^2 + z_1^2$. Of course, one now has $\boldsymbol{\beta} = -m\,\partial V/\partial \mathbf{y}_1$. In place of (15.2),

$$V_0 = g(\zeta) - (d'^2 + u)^{\frac{1}{2}}, \tag{15.6}$$

where
$$u = \xi - 2\eta + \zeta, \tag{15.7}$$

and $g$ is the same function of the new $\zeta$ as $\bar{g}$ was of the old. When $m \to 0$ infinities now only occur in $g(\zeta)$, but this function does not contribute to the displacement. In short, the formal device just introduced allows one to discuss aberrations on the basis of $V$ even when the object is at infinity; so that this case, which commonly occurs in practice, need not be treated separately.

We next turn our attention to $T_0$. Here we have to proceed quite differently. As base-planes we now take $\mathscr{I}$ and $\mathscr{I}'$, so that $\mathbf{y}' = m\mathbf{y}$ for all rays. This means that $T_0$ must satisfy each of the simple differential equations

$$\frac{\partial T_0}{\partial \beta'} + m\frac{\partial T_0}{\partial \beta} = 0, \quad \frac{\partial T_0}{\partial \gamma'} + m\frac{\partial T_0}{\partial \gamma} = 0. \tag{15.8}$$

The first of these shows that $\beta$ and $\beta'$ can occur in $T_0$ only in the combination $m\beta' - \beta$, whilst the second likewise shows that it can depend on $\gamma$ and $\gamma'$ only through the combination $m\gamma' - \gamma$. Since, however, $T_0$ must be a function of rotational invariants, it follows that

$$T_0 = g\{(m\beta' - \beta)^2 + (m\gamma' - \gamma)^2\}, \tag{15.9}$$

where $g$ is a function of one argument. The form of (15.9) suggests the introduction of new rotational invariants in place of $\beta'^2 + \gamma'^2$, etc.

A symmetrical choice is

$$\left.\begin{aligned} \xi &= (s\beta'-\beta)^2+(s\gamma'-\gamma)^2, \\ \eta &= (s\beta'-\beta)(m\beta'-\beta)+(s\gamma'-\gamma)(m\gamma'-\gamma), \\ \zeta &= (m\beta'-\beta)^2+(m\gamma'-\gamma)^2. \end{aligned}\right\} \quad (15.10)$$

Here $s$ is still a disposable constant ($s \neq m$), but it will be convenient to take it as the magnification associated with the pupil planes. Note that the relations between the old and the new rotational invariants are linear. Indeed, if we now exceptionally write $\bar{\xi}$, $\bar{\eta}$, $\bar{\zeta}$ for the former,

$$\left.\begin{aligned} \xi &= s^2\bar{\xi}-2s\bar{\eta}+\bar{\zeta}, \\ \eta &= sm\bar{\xi}-(s+m)\bar{\eta}+\bar{\zeta}, \\ \zeta &= m^2\bar{\xi}-2m\bar{\eta}+\bar{\zeta}, \end{aligned}\right\} \quad (15.11)$$

or conversely,

$$\left.\begin{aligned} (s-m)^2\bar{\xi} &= \xi-2\eta+\zeta, \\ (s-m)^2\bar{\eta} &= m\xi-(s+m)\eta+s\zeta, \\ (s-m)^2\bar{\zeta} &= m^2\xi-2sm\eta+s^2\zeta. \end{aligned}\right\} \quad (15.12)$$

The ideal angle characteristic now takes the simple form

$$T_0 = g(\zeta), \quad (15.13)$$

in generic agreement with the paraxial limit (14.35).

When $\mathscr{I}$ recedes to infinity, the only resulting infinities occur in the function $g(\zeta)$; but this is irrelevant to the displacement. Alternatively, one may go over to a new anterior base-plane, taking this to be $\mathscr{E}$ now. Ideal imagery requires that

$$\mathbf{y}' = f\boldsymbol{\beta}/\alpha, \quad (15.14)$$

the value of the constant of proportionality on the right being fixed by the paraxial limit. We therefore have

$$\frac{\partial T_0}{\partial \beta'} = -\frac{f\beta}{\alpha}, \quad \frac{\partial T_0}{\partial \gamma'} = -\frac{f\gamma}{\alpha}, \quad (15.15)$$

and these may immediately be integrated with respect to $\beta'$ and $\gamma'$ respectively. Taking rotational invariance into account it follows that

$$T_0 = g(\bar{\zeta})-f\bar{\eta}(1-\bar{\zeta})^{-\frac{1}{2}}, \quad (15.16)$$

where $g$ as usual denotes some otherwise undetermined function of its argument. In terms of $\xi$, $\eta$, $\zeta$, with $m = 0$, (15.16) reads

$$T_0 = g(\zeta)+s^{-1}f\eta(1-\zeta)^{-\frac{1}{2}}, \quad (15.17)$$

where certain terms which depend on $\zeta$ alone have been absorbed in $g(\zeta)$.

Next we come to $W_1(y', z', \beta, \gamma)$. As in the case of $T_0$ we choose $\mathscr{I}$ and $\mathscr{I}'$ as base-planes. Perfect imagery requires that

$$\mathbf{y}' = m\mathbf{y} = m\,\partial W_{10}/\partial\boldsymbol{\beta}. \tag{15.18}$$

In terms of the appropriate invariants (cf. (13.5)) this implies the result

$$W_{10} = g(\xi) + m^{-1}\eta. \tag{15.19}$$

When the object recedes to infinity this breaks down completely; as was to be expected, since one then has the exceptional situation in which the use of $W_1$ is forbidden in principle. If we therefore go over to the new base-planes $\mathscr{E}$ and $\mathscr{E}'$ we deduce, by an argument similar to that which led to (15.2), that (when $m = 0$)

$$W_{10} = g(\zeta) - \{d'^2 + \xi - 2f\eta(1-\zeta)^{-\frac{1}{2}} + f^2\zeta(1-\zeta)^{-1}\}^{\frac{1}{2}}. \tag{15.20}$$

Finally, in the context of $W_2(\beta', \gamma', y, z)$ we again take $\mathscr{I}$ and $\mathscr{I}'$ as base-planes. Therefore, in place of (15.18),

$$m\mathbf{y} = \mathbf{y}' = -\,\partial W_{20}/\partial\boldsymbol{\beta}',$$

whence

$$W_{20} = g(\zeta) - m\eta. \tag{15.21}$$

As in the case of $V$, it is best to introduce $\mathbf{y}_1$ in place of $\mathbf{y}$ when the object happens to be at infinity.

To end this discussion of ideal characteristic functions we inquire briefly into the kind of condition one might impose upon $K$ in order to fix the form of the function $g$ which appears repeatedly above. Some limitation upon the aberrations associated with the pupil planes clearly suggests itself. Accordingly we seek to determine the function $g(\zeta)$ which occurs in (15.6) so that all rays through $E$ pass through $E'$.

Let a ray through $E$ and $E'$ intersect $\mathscr{I}$ in $O$. Then

$$\tilde{V}(O, E') = \tilde{V}(O, E) + \tilde{V}(E, E'). \tag{15.22}$$

In view of the assumed condition, $\tilde{V}(E, E') = \text{const.} = e$, say. Further

$$\tilde{V}(O, E') = V(0, 0, y, z) = g(\zeta) - (d'^2 + \zeta)^{\frac{1}{2}}, \tag{15.23}$$

from (15.6), whilst

$$\tilde{V}(O, E) = (d^2 + \zeta/m^2)^{\frac{1}{2}}. \tag{15.24}$$

(15.22) now gives

$$g(\zeta) = (d'^2 + \zeta)^{\frac{1}{2}} + (d^2 + \zeta/m^2)^{\frac{1}{2}} + e, \tag{15.25}$$

so that $g(\zeta)$ is indeed fully determined. One could in turn define the 'ideal form' $g_0$ of $g$ to be that given by (15.25), and the extent to which any actual $g(\zeta)$ differs from $g_0$ will be reflected in the extent to which rays through $E$ will fail to pass also through $E'$.

## 16. Aberration functions

In this section we reconsider, in the special context of the symmetric system, much of the general work of Sections 7 and 11, and suitably adapt the notation. The actual characteristic function $F$ of $K$ will in general differ from the desired ideal $F_0$, and we write

$$F = F_0 + f, \tag{16.1}$$

as before (see also the end of Section 36); an equation which defines $f$. Now recall that $f$ can be expanded in ascending powers of the rotational invariants $\xi$, $\eta$, $\zeta$. This means that in (11.1) only terms of even degree can appear, i.e. $f_1, f_3, f_5, \ldots = 0$. We therefore write

$$f = \sum_{n=2}^{\infty} f^{(n)} \tag{16.2}$$

in place of (11.1), where $f^{(n)}$ is a homogeneous polynomial of degree $n$ in $\xi$, $\eta$, $\zeta$. Evidently $f^{(n)}$ is the aberration function of order $2n-1$, whilst there are now no aberration functions of even order. As a matter of convenience, one therefore sometimes refers to the aberrations of order 3, 5, 7, 9, ... alternatively as primary, secondary, tertiary, quarternary, ... aberrations. This is not a good terminology, since the term 'primary' surely connotes dominance, so that in a general system the primary aberrations will be of the *second* order (cf. Section 88, in particular equation (88.16)). Note that in (16.2) there is no term $f^{(1)}$ quadratic in the ray-coordinates. Such a term would relate to paraxial defects of the image, but we already know that when $K$ is symmetric every object plane $\mathscr{I}$ has a definite conjugate image plane $\mathscr{I}'$; and the paraxial imagery associated with $\mathscr{I}$ and $\mathscr{I}'$ is perfect. A term $f^{(1)}$ can therefore arise only if, at some stage, one decides to consider a receiving plane other than the ideal image plane.

Since formal expansions are now in *three* rather than in four variables, we set in place of (11.2)

$$f^{(n)} = \sum_{\mu=0}^{n} \sum_{\nu=0}^{\mu} f_{\mu\nu}^{(n)} \xi^{n-\mu} \eta^{\mu-\nu} \zeta^{\nu}. \tag{16.3}$$

The number of constant coefficients in $f^{(n)}$ is $\frac{1}{2}(n+1)(n+2)$. One of these is, however, essentially redundant. This may be seen by contemplating $V$, for example. The displacement does not depend on the arbitrary function $g(\zeta)$ which appears in (15.6), so that the terms of $v$ which depend upon $\zeta$ alone are irrelevant; indeed, one may think of them as absorbed in $g(\zeta)$. The number of essential constants is therefore reduced to $\frac{1}{2}n(n+3)$. (See, however, the remarks at the end of Section 17.) We thus see that the number $\bar{n}$ of aberration coefficients of *order n* is given by

$$\bar{n} = \begin{cases} \frac{1}{8}(n+1)(n+7) & (n \text{ odd}) \\ 0 & (n \text{ even}). \end{cases} \qquad (16.4)$$

Accordingly one has 5, 9, 14, 20, ... aberrations of orders 3, 5, 7, 9, ... respectively, and none of even order. Of course, should it happen that $K$ has some additional symmetry, these numbers will be further reduced. Also, (16.4) refers to a fixed position of $\mathscr{I}$. If the imagery associated with arbitrarily situated (normal) conjugate planes is to be described, one has to deal with *one* additional coefficient in each odd order. Indeed, it is just that coefficient which was above regarded as redundant, specifically $v_{nn}^{(n)}$ in the case of $V$.

## 17. The displacement

Geometrically the quantity of principal interest is the displacement $\boldsymbol{\epsilon}'$. In the context of $V$ this is, in the usual notation,

$$\boldsymbol{\epsilon}' = \mathbf{Y}' - m\mathbf{y} = \mathbf{y}' - \mathbf{y}_1 + d'\boldsymbol{\beta}'/\alpha'. \qquad (17.1)$$

Proceeding from this as it stands, the constant $d'$ will occur later a very large number of times, and this is a nuisance. Accordingly we choose, in the present context, a new unit of length such that the numerical measure of $d'$ becomes unity. This convention is rather unorthodox when $d'$ happens to be negative, and then the phrase 'unit of length' is somewhat metaphorical. At any rate, every length is to be thought of as expressed as a multiple of $d'$, and an expression such as $(d'^2+u)^{\frac{1}{2}}$ is intended to mean $d'(1+u/d'^2)^{\frac{1}{2}}$, the positive square root being understood. With this prescription one just does not worry about the sign of $d'$ in the course of formal work: one restores $d'$ explicitly by dimensional arguments at the very end of it, and *then* inserts its actual numerical value; see, for example, the remarks following equations (23.3) and (23.5).

Now, recalling (15.6),

$$V = g(\zeta) - (1+u)^{\frac{1}{2}} + v. \tag{17.2}$$

Here one does well to bear in mind that, strictly speaking, (17.2) operates as a definition of $g+v$, rather than of $v$ alone; for the previous definition of $g$ has now become meaningless, since the conjugate planes in question are *in fact* not perfect. Restoration of uniqueness requires a prescription as to how the terms of (17.2) which depend on $\zeta$ alone are to be distributed amongst $g$ and $v$. One might require, for instance, that $v(0, 0, \zeta) = 0$, or else that $g(\zeta)$ depend linearly on $\zeta$.

From (17.2),

$$\boldsymbol{\beta}' = \partial V/\partial \mathbf{y}' = -(1+u)^{-\frac{1}{2}}(\mathbf{y}' - \mathbf{y}_1) + (2\mathbf{y}' v_\xi + \mathbf{y}_1 v_\eta), \tag{17.3}$$

where differentiations with respect to $\xi$, $\eta$, $\zeta$ are indicated by the appropriate subscripts; e.g. $v_\xi = \partial v/\partial \xi$. After some manipulation (17.1) then becomes

$$\boldsymbol{\epsilon}' = (\mathbf{y}' - \mathbf{y}_1)(1 - D) + (1+u)^{\frac{1}{2}}(2v_\xi \mathbf{y}' + v_\eta \mathbf{y}_1) D, \tag{17.4}$$

where

$$D = \{1 + 2(1+u)^{\frac{1}{2}}[2(\xi - \eta) v_\xi + (\eta - \zeta) v_\eta] - (1+u)(4\xi v_\xi^2 + 4\eta v_\xi v_\eta + \zeta v_\eta^2)\}^{-\frac{1}{2}}. \tag{17.5}$$

The general relationship between $\boldsymbol{\epsilon}'$ and $v$ is therefore quite complex.

With regard to $\boldsymbol{\epsilon}'$ we retain the notation of Section 11, and write

$$\boldsymbol{\epsilon}' = \sum_n \boldsymbol{\epsilon}'_n \quad (n = 3, 5, 7, \ldots). \tag{17.6}$$

*If* $v$ has no terms of order less than $2n-1$, then

$$\boldsymbol{\epsilon}'_{2n-1} = 2v_\xi^{(n)} \mathbf{y}' + v_\eta^{(n)} \mathbf{y}_1, \tag{17.7}$$

so that these are the terms induced by the aberration function of order $2n-1$ in the displacement of the *same* order, even when terms of lower orders are present.

Let us contrast the results just obtained with those one gets if one uses $T$ in place of $V$. In view of (15.13) we now have

$$T = g(\zeta) + t, \tag{17.8}$$

the rotational invariants being those defined by (15.12). Now

$$\boldsymbol{\epsilon}' = \mathbf{y}' - m\mathbf{y} = -\frac{\partial T}{\partial \boldsymbol{\beta}'} - m\frac{\partial T}{\partial \boldsymbol{\beta}}. \tag{17.9}$$

Using (17.8), all terms involving $g(\zeta)$ of course disappear and one is left simply with

$$\boldsymbol{\epsilon}' = (s-m)\,[2(\boldsymbol{\beta}-s\boldsymbol{\beta}')\,t_\xi + (\boldsymbol{\beta}-m\boldsymbol{\beta}')\,t_\eta]. \tag{17.10}$$

This equation looks very much simpler than its counterpart (17.4). Alas, this simplicity is somewhat illusory. The reason for this situation is as follows. One is almost always interested in families of rays from points of the object. Any such point is defined by constant values of $y$ and $z$ (granted of course that $\mathscr{I}$ and $\mathscr{B}$ coincide), i.e. the four coordinates which occur in (17.10) are constrained by the conditions

$$\partial T/\partial \boldsymbol{\beta} = \mathbf{y} = \text{const.} \tag{17.11}$$

This means in effect that after the displacement has been written down according to (17.10), two of the four coordinates, or two suitable linear combinations of them, have to be eliminated in favour of **y**. This is, in general, a very tedious task; the difficulties being compounded by the need in practice to introduce coordinates in the exit pupil, say, at the same time (see also Section 23). In the case of $V$ on the other hand, the tedious part of the work consists in expanding the factors multiplying $\mathbf{y}'$ and $\mathbf{y}_1$ in ascending powers of $\xi$, $\eta$, $\zeta$; but in this instance $y$ and $z$ are already amongst the ray-coordinates, and no further complications arise. The sad story in this particular context is that in changing over from one characteristic function to another one may gain simplicity in one place, only to lose it in another. In investigations of a general kind, however, elegance of the theory will often hinge on the appropriate choice of $F$.

One final remark needs to be made concerning an incidental effect of the constraint (17.11). It is obvious that the left-hand member of this equation will contain the coefficients of those terms of $T$ which multiply powers of $\zeta$ alone. This means, in effect, that after the elimination described above has been carried out, the coefficient of order $2n-1$ which was rejected on the grounds of redundancy in Section 16, will in fact appear in the displacement, though only in terms of order exceeding $2n-1$ (see, for example, equations (24.13)). The result (16.4) must therefore be appropriately interpreted. Bearing in mind the work of the end of Section 15, the process of elimination evidently brings in implicitly the aberrations associated with the pupil planes. Note that this feature does not arise in the context of the point characteristic. One would, however, face an analogous situation there, if one eventually decided to introduce coordinates in $\mathscr{E}$ in place of those in $\mathscr{E}'$ as is sometimes done in practice. The effects of such a change of coordinates may be far from trivial, for example in the case of photographic objectives covering very wide fields.

## 18. Out-of-focus image planes

Occasionally it is desirable to consider the image not in $\mathscr{I}'$ but in some normal plane $\mathscr{I}'_\chi$ the axial distance from the first to the second being $\chi d'$, that is to say $\chi$, since we have arranged $d'$ to have the value unity. To define a sensible 'displacement' $\tilde{\boldsymbol{\epsilon}}'$ in $\mathscr{I}'_\chi$ we first have to define some point $I'_\chi$ in this to serve as a counterpart to $I'$. A simple choice appears to be the point whose coordinates are $(1+\chi)\,\mathbf{y}_1$. If a ray through $O$ intersects $\mathscr{I}'_\chi$ in the point $\tilde{\mathbf{Y}}'$ it is natural to define $\tilde{\boldsymbol{\epsilon}}'$ as

$$\tilde{\boldsymbol{\epsilon}}' = \tilde{\mathbf{Y}}' - (1+\chi)\,\mathbf{y}_1. \tag{18.1}$$

Then, referring to (17.1),

$$\tilde{\boldsymbol{\epsilon}}' - \boldsymbol{\epsilon}' = \chi(\boldsymbol{\beta}'/\alpha' - \mathbf{y}_1) = \chi(\boldsymbol{\epsilon}' - \mathbf{y}'),$$

whence

$$\tilde{\boldsymbol{\epsilon}}' = (1+\chi)\,\boldsymbol{\epsilon}' - \chi\mathbf{y}'. \tag{18.2}$$

This result is remarkably simple, for the out-of focus displacement, reckoned relative to $I'_\chi$, is simply the sum of the usual displacement, re-scaled by the factor $1+\chi$, and the linear term $-\chi\mathbf{y}'$. Moreover, when a non-zero order can be ascribed to $\chi$, as is usually the case, the scale factor $1+\chi$ is to be ignored in considering the displacement of a given order generated by the aberration function of the same order.

## 19. The third-order displacement

We proceed to investigate the geometrical significance of the aberrations of the various orders, beginning with the terms of lowest non-vanishing order, i.e. the third. Formal considerations aside, it does not matter which particular characteristic function we use for this purpose, as has been stressed repeatedly. On the whole we shall base our treatment on the point characteristic, except for occasional references to other characteristic functions.

We write the third-order aberration function in the form

$$v^{(2)} = p_1\xi^2 + p_2\xi\eta + p_3\xi\zeta + p_4\eta^2 + p_5\eta\zeta. \tag{19.1}$$

The term in $\zeta^2$ alone has been omitted since it does not enter into the displacement, or else may be thought of as having been absorbed in the function $g(\zeta)$ of (15.6). The aberration coefficients $p_1, \ldots, p_5$ are constants of the system for the given positions of the various reference planes. They are, of course, the constants $v^{(2)}_{\mu\nu}$ of (16.3)—with $f = v$—in a simplified notation, i.e.

$$p_1 = v^{(2)}_{00}, \quad p_2 = v^{(2)}_{10}, \quad \ldots, \quad p_5 = v^{(2)}_{21}.$$

The third-order, or primary, displacement is now immediately obtained from (17.7), with $n = 2$. Thus

$$\boldsymbol{\epsilon}'_3 = (4p_1\xi + 2p_2\eta + 2p_3\zeta)\,\mathbf{y}' + (p_2\xi + 2p_4\eta + p_5\zeta)\,\mathbf{y}_1. \tag{19.2}$$

When investigating rays from a fixed point $O$ of $\mathscr{I}$ one may take $z_1 = 0$ without loss of generality, in view of the symmetry of $K$. At the same time write $y_1 = h'$, and introduce polar coordinates in $\mathscr{E}'$, i.e.

$$y' = \rho\cos\theta, \quad z' = \rho\sin\theta. \tag{19.3}$$

Evidently the family of rays grazing the rim of the stop are defined, at least approximately, by the constancy of $\rho$. At any rate, we now have

$$\xi = \rho^2, \quad \eta = \rho h'\cos\theta, \quad \zeta = h'^2. \tag{19.4}$$

(19.2) then reads

$$\left.\begin{aligned} \epsilon'_{3y} &= 4p_1\rho^3\cos\theta + p_2\rho^2h'(2+\cos 2\theta) + (2p_3+2p_4)\rho h'^2\cos\theta + p_5 h'^3, \\ \epsilon'_{3z} &= 4p_1\rho^3\sin\theta + p_2\rho^2h'\sin 2\theta + 2p_3\rho h'^2\sin\theta. \end{aligned}\right\} \tag{19.5}$$

For historical reasons one often writes this in terms of five other constants $\sigma_1, \ldots, \sigma_5$ (the so-called 'Seidel coefficients'), defined by

$$\sigma_1 = 4p_1, \quad \sigma_2 = p_2, \quad \sigma_3 = p_4, \quad \sigma_4 = 2p_3 - p_4, \quad \sigma_5 = p_5 \tag{19.6}$$

(cf. the remarks following shortly after equations (41.3) and (45.13)).

Then

$$\left.\begin{aligned} \epsilon'_{3y} &= \sigma_1\rho^3\cos\theta+\sigma_2\rho^2h'(2+\cos 2\theta)+(3\sigma_3+\sigma_4)\rho h'^2\cos\theta+\sigma_5 h'^3, \\ \epsilon'_{3z} &= \sigma_1\rho^3\sin\theta+\sigma_2\rho^2h'\sin 2\theta+(\sigma_3+\sigma_4)\rho h'^2\sin\theta. \end{aligned}\right\} \tag{19.7}$$

The total displacement is made up of four distinct *partial displacements*, characterized by their dependence upon the various powers of $\rho$ and $h'$; whilst one of them is governed jointly by two coefficients. It is usual to speak of each of the partial displacements as being an aberration of a certain type, and we proceed to examine them in turn.

(i) *Spherical aberration*

The terms of (19.7) governed by $\sigma_1$ give rise to the partial displacement

$$\hat{y} = \sigma_1\rho^3\cos\theta, \quad \hat{z} = \sigma_1\rho^3\sin\theta. \tag{19.8}$$

To avoid the constant repetition of a large number of affixes we are using the symbols $\hat{y}$ and $\hat{z}$ for the components of partial displacements in the equations such as (19.8), whatever the receiving plane may happen to be. Indeed we may regard $\hat{x}$, $\hat{y}$, $\hat{z}$ as relating to a set of Cartesian axes having the usual orientation but with origin at $I'$ or $I'_\chi$, as the case may be. Evidently a family of rays for which $\rho =$ constant (henceforth called a *zonal* family) intersects $\mathscr{I}'$ in a circle of radius $\sigma_1\rho^3$, concentric with the ideal image point $I'$. Varying now $\rho$ from 0 to its largest value, $\rho_0$ say, we see at once that the image of $O$ is a circular patch of light, centred on $I'$, and of radius $\sigma_1\rho_0^3$. This aberration is called (primary) *spherical aberration*. Its name is somewhat unfortunate, in as far as it suggests some connection with the presence of specifically spherical refracting or reflecting surfaces in the system. However, the suggestion obviously has no foundation in fact, and the terminology survives only by tradition.

It is of advantage to consider an out-of-focus receiving plane $\mathscr{I}'_\chi$. If $\chi$ is taken to be of the second order—an assumption yet to be justified—we have, in view of (18.2),

$$\hat{y} = (\sigma_1\rho^3-\chi\rho)\cos\theta, \quad \hat{z} = (\sigma_1\rho^3-\chi\rho)\sin\theta.$$

It suffices to take $h' = 0$ since spherical aberration is independent of $h'$, and at the same time to restrict oneself to meridional rays,

$\theta = 0$ or $\pi$, in view of the symmetry of $K$. The smallest illuminated disk, the so-called *disk of least confusion*, will be found in that plane for which

$$\hat{y}(\rho_0) = \hat{y}(\bar{\rho}), \tag{19.9}$$

where $\bar{\rho}$ is that value of $\rho$ which maximizes $|\hat{y}(\rho)|$, i.e. $\bar{\rho}^2 = \chi/3\sigma_1$. One then finds easily that

$$|\bar{\rho}| = \tfrac{1}{2}\rho_0, \quad \chi = \tfrac{3}{4}\sigma_1\rho_0^2, \quad \hat{y} = \tfrac{1}{4}\sigma_1\rho_0^3. \tag{19.10}$$

Notice that $\chi$ is indeed of the second order, consistently with our previous assumption. The radius of the disk of least confusion is

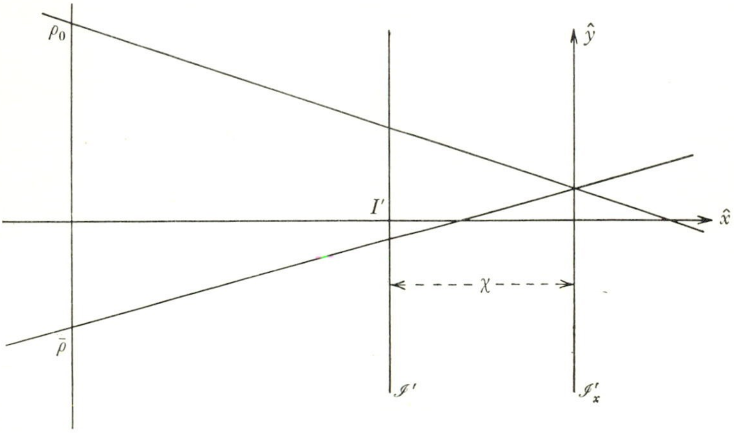


Fig. 4.2

therefore only one-quarter of the radius of the image patch in $\mathscr{I}'$; and the mutual distance between these disks is three-quarters of the distance between $I'$ and the marginal focus, i.e. the point in which rays having $\rho = \rho_0$ intersect the axis. We also note in passing that all rays touch the *caustic* surface

$$4\hat{x}^3 = 27\sigma_1(\hat{y}^2 + \hat{z}^2). \tag{19.11}$$

It will not come amiss at this stage to issue the general reminder that the validity of all results relating to displacements depends of course on the extent to which the geometrical-optical limit does in fact represent any 'reasonable' sort of approximation at all. In practice this means that reliable quantitative information can be expected only when the aberrations are in some sense large. One need only reflect that in the extreme case of the absence of all aberrations $K$ should produce a point image $I'$. In actual fact, however, it produces a small circular patch of light (the *Airy disk*) surrounded by concentric rings. (It is assumed that $h'$ is not too large.) Similarly, in the presence of a very small amount of primary spherical aberration alone, (19.10) does not in fact give the best position of the receiving plane.

(ii) *Coma*

The partial displacement of (19.7) which varies as $\rho^2 h'$ is

$$\hat{y} = \mathfrak{a}(2+\cos 2\theta), \quad \hat{z} = \mathfrak{a}\sin 2\theta, \tag{19.12}$$

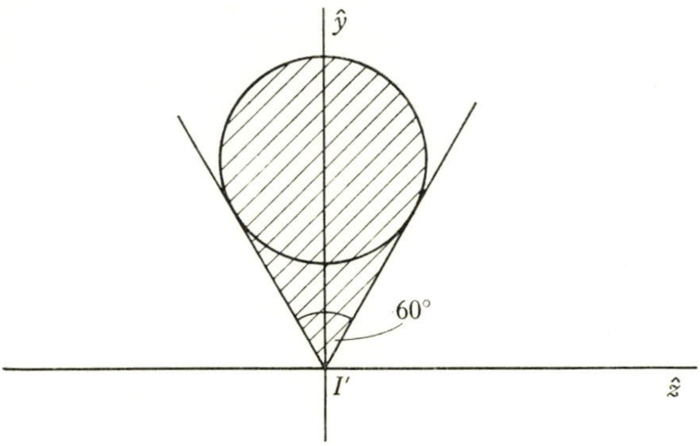


Fig. 4.3

where $\mathfrak{a} = \sigma_2 \rho^2 h'$. A zonal family of rays again intersects $\mathscr{I}'$ in a circle, its radius being $\mathfrak{a}$. This circle has its centre at the point $(2\mathfrak{a}, 0)$, i.e. it is not concentric with $I'$. Moreover it is described twice as the radius vector in the plane of the exit pupil sweeps out its aperture once. The family of circles obtained by varying $\rho$ touches two straight lines inclined at an angle $\psi_3 = 60°$ to each other. The image is therefore a flare-shaped patch of light bounded by these lines and by the arc of a circle of radius $\sigma_2 \rho_0^2 h'$ (Fig. 4.3). This particular aberration is called (primary) *coma*, more exactly *circular*, or *linear*, coma, to distinguish it from other types which occur amongst the aberrations of higher order. The presence of coma is particularly objectionable on account of the asymmetric appearance of the image, and in practice one will aim to remove it as far as possible. It cannot be improved by a different choice of receiving plane.

For any fixed value of $\rho$ *tangential rays* ($\theta = 0$ or $\pi$) and *sagittal rays* ($\theta = \pm\frac{1}{2}\pi$) meet in certain points along the $\hat{y}$-axis. Their respective displacements relative to $I'$ are called *tangential* and *sagittal coma* respectively, and they will be denoted by $\kappa_t$ and $\kappa_s$. Here $\kappa_t = 3\mathfrak{a}$, $\kappa_s = \mathfrak{a}$, so that, upon indicating the order explicitly,

$$\kappa_{t3}/\kappa_{s3} = 3. \tag{19.13}$$

(iii) *Curvature of field*

The terms of (19.7) governed jointly by $\sigma_3$ and $\sigma_4$ are

$$\hat{y} = (3\sigma_3+\sigma_4-k)\rho h'^2\cos\theta, \quad \hat{z} = (\sigma_3+\sigma_4-k)\rho h'^2\sin\theta, \tag{19.14}$$

this being the displacement in an out-of-focus plane $\mathscr{I}'_k$ shifted relative to $\mathscr{I}'$ by a second-order amount $\chi = kh'^2$, where $k$ is some number. For general values of the constants involved in (19.14) a zonal family of rays intersects $\mathscr{I}'_k$ in an ellipse. This degenerates into a straight line for each of the values

$$k_t = 3\sigma_3+\sigma_4 \quad \text{and} \quad k_s = \sigma_3+\sigma_4$$

of $k$. These lines are called *tangential* and *sagittal focal lines* respectively, their mutual separation being $2\sigma_3\rho h'^2$. The ellipse becomes a circle of radius $\sigma_3\rho h'^2$ in the plane defined by $k = 2\sigma_3+\sigma_4$. The focal lines are mutually at right angles: they are in fact nothing but the focal lines whose existence was already inferred in Section 10. In the present context the earlier discussion amounts to taking not the conventional coordinate basis, but one whose $\bar{x}$-axis and $\bar{x}'$-axis point along a *principal ray*, i.e. a ray through $O$ and $E'$. (This definition is used throughout our work. Not infrequently principal rays are defined to be rays through $O$ and the axial point of the stop; and one can run into apparent contradictions if one fails to bear in mind that in general these alternative definitions are inequivalent for all but paraxial rays.)

The shape of the complete image, obtained by varying $\rho$ from 0 to $\rho_0$, is elliptical, circular, or linear, as the case may be, and is in every case disposed symmetrically about $I'_k$. When $\sigma_3 = 0$ a plane object has a sharp image in a surface of revolution whose polar curvature is

$$c = 2\sigma_4. \tag{19.15}$$

Accordingly the aberration governed by $\sigma_4$ is known as (primary) *curvature of field*, more specifically as *Petzval curvature*. Further, according to (19.14) all meridional rays pass through one point of the surface of revolution whose polar curvature is

$$c_t = 2(3\sigma_3 + \sigma_4), \tag{19.16}$$

whilst all sagittal rays pass through one point of a surface of revolution whose polar curvature is

$$c_s = 2(\sigma_3 + \sigma_4); \tag{19.17}$$

so that, incidentally, $$c_t - c = 3(c_s - c). \tag{19.18}$$

The three ancillary surfaces just defined of course touch $\mathscr{I}'$ at $O'_0$. The first of these is often referred to as the *Petzval surface*; whilst the other two are the *tangential and sagittal image surfaces* respectively. The existence of these as *distinct* surfaces is guaranteed if only the coefficient $\sigma_3$ does not vanish, and the defect of the imagery governed by $\sigma_3$ is therefore known as (primary) *astigmatic curvature of field*, or simply *astigmatism*.

(iv) *Distortion*

The last of the partial displacements in (19.7) is simply

$$\hat{y} = \sigma_5 h'^3, \quad \hat{z} = 0. \tag{19.19}$$

$K$ therefore produces a sharp image in $\mathscr{I}'$, but it is distorted unless $\sigma_5 = 0$. Accordingly the defect of the image governed by $\sigma_5$ is known as (primary) *distortion*. Correctly to the present order, a segment of the straight line $y = \text{const.} = k/m$, say, in $\mathscr{I}$ is transformed into an arc of the parabola

$$Y' = k[1 + \sigma_5(k^2 + Z'^2)] \tag{19.20}$$

in $\mathscr{I}'$. Cursory examination of the image of a square grid will reveal the reason for speaking of *pin-cushion* distortion and *barrel* distortion when $\sigma_5 > 0$ and $\sigma_5 < 0$ respectively. Note that this type of aberration is asymmetric, for the image lies to one side of $I'$.

When all four kinds of displacements are present together the curve generated by the points of intersection of a zonal family of rays with $\mathscr{I}'$ will be quite complicated. Nevertheless, certain features of the image can be isolated without much difficulty. For instance, the asymmetry of the image relative to $I'$ is still governed by $\sigma_2$ and $\sigma_5$. However, the distortion is really irrelevant to the extent that it does not contribute to the asymmetry of the illuminated patch as a whole. One may therefore define the *tangential asymmetry*

$$K_t = \tfrac{1}{2}[\hat{y}(\rho, 0) + \hat{y}(\rho, \pi)] - \hat{y}(0, \theta) \tag{19.21}$$

and the *sagittal asymmetry*

$$K_s = \hat{y}(\rho, \pi/2) - \hat{y}(0, \theta), \tag{19.22}$$

as appropriate partial measures of the asymmetry of the image patch. In the present circumstances one finds of course that

$$K_t = \kappa_{t3}, \quad K_s = \kappa_{s3}.$$

More usefully one might define a *mean asymmetry*

$$K_m = \tfrac{1}{2}(K_t + K_s). \tag{19.23}$$

Then the vanishing of $K_m$ is a necessary, though not a sufficient, condition for the symmetry of the image patch.

We conclude this section by showing briefly how the same results are arrived at when the work is based on the use of $T$ rather than of $V$. We write generically

$$t^{(2)} = (s-m)^{-1}(p_1\xi^2 + p_2\xi\eta + p_3\xi\zeta + p_4\eta^2 + p_5\eta\zeta), \tag{19.24}$$

where the factor $(s-m)^{-1}$ has been inserted for convenience. The rotational invariants are those given by (15.10). (17.10) gives straight away

$$\boldsymbol{\epsilon}_3' = (4p_1\xi + 2p_2\eta + 2p_3\zeta)(\boldsymbol{\beta} - s\boldsymbol{\beta}') \\ + (p_2\xi + 2p_4\eta + p_5\zeta)(\boldsymbol{\beta} - m\boldsymbol{\beta}'). \tag{19.25}$$

Now in view of (14.35) we have here

$$T = a_0 + (f/2m)\zeta + O(4), \tag{19.26}$$

whence, by differentiation,

$$\mathbf{y}_1 = f(\boldsymbol{\beta} - m\boldsymbol{\beta}') + O(3). \tag{19.27}$$

(14.35) can be thought of as relating to the pupil planes, provided $m$ be replaced by $s$, so that

$$\mathbf{y}_e' = f(\boldsymbol{\beta} - s\boldsymbol{\beta}') + O(3). \tag{19.28}$$

Just as it was previously convenient to prevent the continued appearance of the constant $d'$ by choosing a suitable unit of length, so we can now choose a unit of length such that $f = 1$, recalling that $f$ cannot be zero here. Correctly to the present order we therefore have to make the substitutions

$$\left.\begin{aligned} \beta - m\beta' &= h', & \gamma - m\gamma' &= 0, \\ \beta - s\beta' &= \rho\cos\theta, & \gamma - s\gamma' &= \rho\sin\theta \end{aligned}\right\} \tag{19.29}$$

in (19.25). Since (19.4) formally also applies here now, (19.25) will at once be seen to yield (19.5) exactly; and thereafter everything follows as before. In practice, however, one must be careful to bear in mind the conventions regarding the different choices of units of length which have hitherto been made.

## 20. The fifth-order displacement in the absence of third-order aberrations

The considerations of the preceding section will now be extended to aberrations of order five, supposing those of the third order to be absent. Equivalently, we wish to study that part of the fifth-order displacement which is generated by the fifth-order aberration function $v^{(3)}$. In analogy with (19.1) we write

$$v^{(3)} = s_1\xi^3 + s_2\xi^2\eta + s_3\xi^2\zeta + s_4\xi\eta^2 + s_5\xi\eta\zeta + s_6\xi\zeta^2 + s_7\eta^3 + s_8\eta^2\zeta + s_9\eta\zeta^2. \tag{20.1}$$

The quantity of interest is the displacement $\boldsymbol{\epsilon}_5'(v^{(3)})$, the notation corresponding to that introduced shortly after equation (11.6). However, we naturally write simply $\hat{\mathbf{y}}$ for it, since we are, after all, not considering the *total* displacement. Thus, again from (17.7),

$$\hat{\mathbf{y}} = (6s_1\xi^2 + 4s_2\xi\eta + 4s_3\xi\zeta + 2s_4\eta^2 + 2s_5\eta\zeta + 2s_6\zeta^2)\,\mathbf{y}' + (s_2\xi^2 + 2s_4\xi\eta + s_5\xi\zeta + 3s_7\eta^2 + 2s_8\eta\zeta + s_9\zeta^2)\,\mathbf{y}_1, \tag{20.2}$$

from which one gets the fifth-order equations corresponding to (19.5):

$$\begin{aligned} \hat{y} = {} & 6s_1\rho^5\cos\theta + 2s_2\rho^4h'(\tfrac{3}{2} + \cos 2\theta) + 2\rho^3h'^2[(2s_3 + s_4) + s_4\cos^2\theta] \\ & \times \cos\theta + \rho^2h'^3[(2s_5 + \tfrac{3}{2}s_7) + (s_5 + \tfrac{3}{2}s_7)\cos 2\theta] \\ & + 2(s_6 + s_8)\rho h'^4\cos\theta + s_9h'^5, \\ \hat{z} = {} & 6s_1\rho^5\sin\theta + 2s_2\rho^4h'\sin 2\theta + 2\rho^3h'^2(2s_3 + s_4\cos^2\theta)\sin\theta \\ & + s_5\rho^2h'^3\sin 2\theta + 2s_6\rho h'^4\sin\theta. \end{aligned} \tag{20.3}$$

Of the six types of partial displacement, three are now each governed jointly by two distinct aberration coefficients. We proceed to investigate the various types in turn. For this purpose it is appropriate to pretend that the primary aberrations are in fact absent.

(i) *Spherical aberration*

The first partial displacement

$$\hat{y} = 6s_1\rho^5\cos\theta, \quad \hat{z} = 6s_1\rho^5\sin\theta \tag{20.4}$$

is an obvious fifth-order counterpart to (19.3), and it is accordingly called secondary spherical aberration. The image path is circular, and centred on $I'$. There is again a disk of least confusion, located in a certain plane between $\mathscr{I}'$ and the plane containing the marginal focus. See also Section 21.

(ii) *Circular coma*

The terms of (20.3) governed by $s_2$ are

$$\hat{y} = \mathfrak{a}(\tfrac{3}{2}+\cos 2\theta), \quad \hat{z} = \mathfrak{a}\sin 2\theta, \tag{20.5}$$

where now $\mathfrak{a} = 2s_2\rho^4h'$. The resemblance to (19.12) is obvious, and this aberration is accordingly called *secondary circular coma*. The circle of intersection points of a zonal family of rays has radius $\mathfrak{a}$, whilst its centre is at $(\frac{3}{2}\mathfrak{a}, 0)$. The whole family of such circles touches a pair of straight lines, their mutual inclination being $\psi_5 = 2\arcsin\frac{2}{3} \approx 84^\circ$. The general appearance of the image thus closely resembles that of Fig. 4.3, if due allowance be made for the increased angle between the asymptotes. Also, one now has

$$\kappa_{t5} = \tfrac{5}{2}\mathfrak{a}, \quad \kappa_{s5} = \tfrac{1}{2}\mathfrak{a},$$

so that

$$\kappa_{t5}/\kappa_{s5} = 5. \tag{20.6}$$

(iii) *Oblique spherical aberration*

The partial displacement varying as $\rho^3h'^2$ is conveniently written as

$$\hat{y} = \mathfrak{a}(k+1+\cos^2\theta)\cos\theta, \quad \hat{z} = \mathfrak{a}(k+\cos^2\theta)\sin\theta, \tag{20.7}$$

with $\mathfrak{a} = 2s_4\rho^3h'^2$ and $k = 2s_3/s_4$. It has no third-order counterpart, and it represents (secondary) *oblique spherical aberration*. This terminology presumably arose from the fact that this aberration varies with $\rho$ in the same way as does primary spherical aberration, but does not vanish on the axis. Also, when $s_4$ happens to be zero, one does have an effect resembling primary spherical aberration for any given value of $h'$.

The usual curves are drawn in Fig. 4.4 for a few typical values of $k$, $\mathfrak{a}$ having been given the nominal value unity. The scale used is

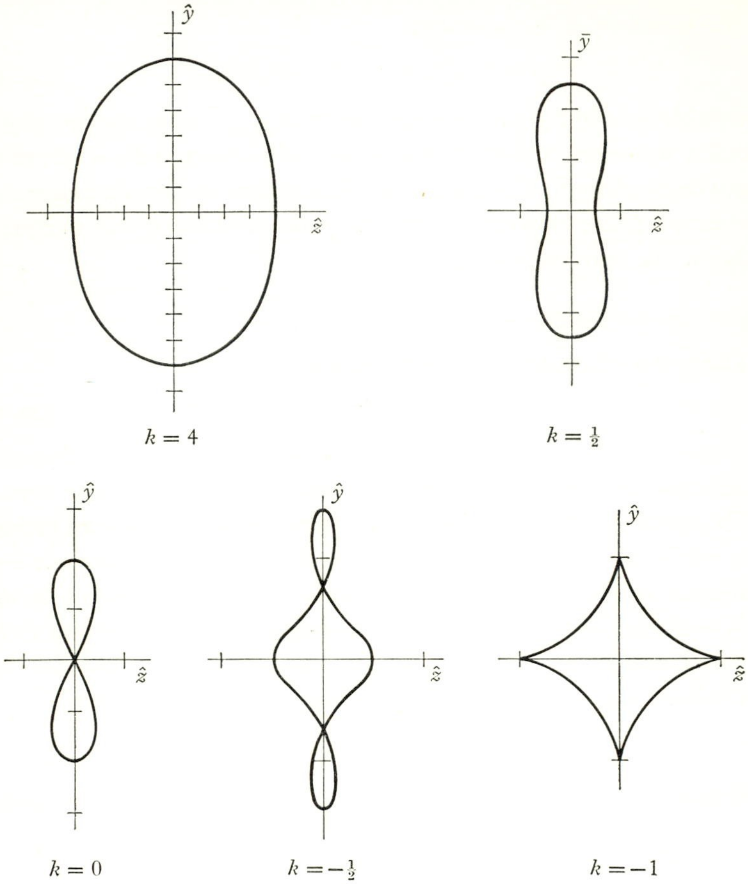


Fig. 4.4

indicated in each case along the axes. In general terms, one has an oval for sufficiently large values of $k$. As $k$ decreases this oval becomes more and more elongated along the $\hat{y}$-axis, until (for $0 < k < 2$) it presents a pinched-in appearance. When $-1 < k \leqslant 0$ the curve intersects itself at points of the $\hat{y}$-axis. The only curve which is symmetric about both axes arises when $k = -1$, and then

$$\hat{y}^{\frac{2}{3}} + \hat{z}^{\frac{2}{3}} = 1. \tag{20.8}$$

Finally, for values of $k$ less than $-1$ one essentially goes through the same sequence of curves again, in the sense that the curve belonging to the value $-k$ is the same as the curve belonging to the value $k-2$, but turned through 90° about the origin.

The full image patch results from letting $\rho$ vary through its full range. In all cases it is symmetric with respect to $I'$, its general appearance being that of the generating curves, except when $-1 < k < 0$. In the exceptional case $s_4 = 0$ the image is exactly circular.

(iv) *Elliptical coma*

We come now to another new type of aberration, the partial displacement varying as $\rho^2 h'^3$. Write

$$\hat{y} = \mathfrak{a}(k+1+k\cos 2\theta), \quad \hat{z} = \mathfrak{a}\sin 2\theta, \tag{20.9}$$

where $\mathfrak{a} = s_5 \rho^2 h'^3$ and $k = 1 + 3s_7/2s_5$. (20.9) is reminiscent of (20.5), though here we have to deal with an additional parameter $k$. The usual curve generated by a set of zonal rays is an ellipse:

$$[\hat{y} - (k+1)\,\mathfrak{a}]^2 + k^2\hat{z}^2 = k^2\mathfrak{a}^2, \tag{20.10}$$

and this is described twice as $\theta$ goes from 0 to $2\pi$. Upon varying $\rho$, all these ellipses will be found to touch a pair of straight lines through $I'$, provided the condition $k > -\frac{1}{2}$ is satisfied. The angle between the tangents is then $2\,\mathrm{arccot}\,(2k+1)^{\frac{1}{2}}$. Under these circumstances the image patch will be bounded by these tangents and the arc of an ellipse, and will therefore have an appearance rather similar to that which results from the presence of circular coma. For this reason the aberration under discussion is known as (secondary) *elliptical coma*. When $k > -\frac{1}{2}$ the image lies wholly to one side of $I'$. In particular, when $k = 0$, $\hat{y}$ no longer depends upon $\theta$, so that the image patch will then be triangular in shape, with a right angle at $I'$, whilst when $s_5$ vanishes it reduces to a part of the $\hat{y}$-axis. The image is always situated asymmetrically with respect to $I'$, save in the exceptional case $k = -1$; for then it is exactly a circle, centred on $I'$. Consistently with this result, (19.23) just gives $K_m = 0$ when $k = -1$.

(v) *Curvature of field*

The aberration governed jointly by $s_6$ and $s_8$ is best considered in an out-of-focus plane $\mathscr{I}'_\chi$. We now take $\chi = kh'^4$, where $k$ is some constant, and then

$$\hat{y} = [2(s_6+s_8)-k]\rho h'^4 \cos\theta, \quad \hat{z} = (2s_6-k)\rho h'^4 \sin\theta. \quad (20.11)$$

This is entirely analogous to (19.14), for which reason this defect of the imagery is called (secondary) *curvature of field*. The tangential and sagittal focal lines will be found in the planes defined by

$$k_t = 2(s_6+s_8), \quad k_s = 2s_6, \quad (20.12)$$

respectively. When $s_8 = 0$ these lines coalesce into a single point, so that an object in $\mathscr{I}$ then has a sharp image lying in a surface of revolution which, to the present order, has the equation

$$\bar{x}' = 1+2s_6(\bar{y}'^2+\bar{z}'^2)^2. \quad (20.13)$$

(Recall that the origin is at $E'$.) Evidently $s_8$ is properly called the coefficient of (secondary) *astigmatism*; and provided $s_8 = 0$, $s_6$ is the secondary analogue of the coefficient of Petzval curvature. (See also the end of Section 51.)

(vi) *Distortion*

The sole remaining terms of (20.3) are represented by

$$\hat{y} = s_9 h'^5, \quad \hat{z} = 0. \quad (20.14)$$

This is entirely analogous to (19.19), and this defect is called (secondary) *distortion*. It is not necessary to go into detail, as any discussion would be little more than a repetition of what was already said in the context of primary distortion.

We could now go on to discuss the tertiary displacement induced by the tertiary aberration function along exactly similar lines. However, little is to be gained by restricting ourselves to this particular order, and we proceed directly to the aberrations of order $2n-1$ (any $n$).

## 21. The displacement of any order in the absence of lower orders

Assume that $v^{(s)} = 0$ for $s = 1, 2, \ldots, n-1$, so that we can examine the displacement $\boldsymbol{\epsilon}'_{2n-1}$ generated by $v^{(n)}$. Recall from Section 16 that

$$v^{(n)} = \sum_{\mu=0}^{n} \sum_{\nu=0}^{\mu} v^{(n)}_{\mu\nu} \xi^{n-\mu} \eta^{\mu-\nu} \zeta^{\nu} \quad (\nu \neq n), \tag{21.1}$$

the term $\nu = n$ having been omitted since, as we know, it does not contribute anything to $\boldsymbol{\epsilon}'$, as far as the present conjugate planes are concerned. It is very convenient to make the substitutions (19.3) and (19.4) already at this stage. Since we need not differentiate with respect to $\mathbf{y}_1$, $h'$ is just a fixed constant, whilst $\rho$ and $\theta$ take the place of $\mathbf{y}'$. Then (21.1) becomes at once

$$v^{(n)} = \sum_{\lambda=0}^{2n-1} \phi^{(n)}_{\lambda} \rho^{2n-\lambda} h'^{\lambda} = \sum_{\lambda=0}^{2n-1} v^{(n)}_{\lambda}, \tag{21.2}$$

say, where

$$\phi^{(n)}_{\lambda} = \sum_{\nu} v^{(n)}_{\lambda-\nu,\nu} \cos^{\lambda-2\nu}\theta. \tag{21.3}$$

The sum over $\nu$ goes from 0 or $\lambda-n$, according as $\lambda-n<0$ or $\lambda-n \geqslant 0$, to $\frac{1}{2}$ if $\lambda$ is even or $\frac{1}{2}(\lambda-1)$ if it is odd. (21.2) exhibits $v^{(n)}$ as the sum of $2n$ terms, each of which just corresponds to one of the $2n$ types of aberrations of order $2n-1$. The partial displacement (of order $2n-1$) generated by $v^{(n)}_{\lambda}$ is given by

$$\hat{y} = \left(\cos\theta \frac{\partial}{\partial\rho} - \frac{\sin\theta}{\rho}\frac{\partial}{\partial\theta}\right) v^{(n)}_{\lambda}, \quad \hat{z} = \left(\sin\theta \frac{\partial}{\partial\rho} + \frac{\cos\theta}{\rho}\frac{\partial}{\partial\theta}\right) v^{(n)}_{\lambda}; \tag{21.4}$$

and we proceed to examine in this section various cases, corresponding to selected values of $\lambda$, in turn.

(i) $\lambda = 0$

We have

$$v^{(n)}_{0} = v^{(n)}_{00} \rho^{2n}, \tag{21.5}$$

so that, since this aberration is independent of $h'$, it suffices to restrict oneself to meridional rays. Then

$$\hat{y} = 2n v^{(n)}_{00} \rho^{2n-1} - \chi\rho \tag{21.6}$$

in an out-of-focus plane $\mathscr{I}'_{\chi}$. The displacement (21.6) is clearly the generalization to all orders of that given by (19.8) and (20.4), and so we have here *spherical aberration of order* $2n-1$. (21.6) may be used to obtain the position and radius of the disk of least confusion. The work involved is elementary, and since the problem is rather academic we merely quote the result. The radius of the disk of least confusion is smaller than that of the image in $\mathscr{I}'$ by a fraction $2(n-1)c^{2n-1}$, where $c$ is the real root of the equation

$$\sum_{s=0}^{2n-3} (-1)^s (s+1) c^s = 0. \tag{21.7}$$

Note that the image is of course disposed symmetrically about $I'$. It should perhaps be remarked that even when image patches corresponding to different values of $n$ happen to have the same radii, the actual distribution of light within them will, of course, depend upon $n$.

(ii) $\lambda = 1$

Again we have only a single term, i.e.

$$v_1^{(n)} = v_{10}^{(n)} \rho^{2n-1} h' \cos\theta. \tag{21.8}$$

The partial displacement obtained from this by means of (21.4) is

$$\hat{y} = \mathfrak{a}[n/(n-1) + \cos 2\theta], \quad \hat{z} = \mathfrak{a} \sin 2\theta, \tag{21.9}$$

with $\mathfrak{a} = (n-1) v_{10}^{(n)} \rho^{2n-2} h'$. This is clearly a generalization of (19.12) and (20.5), so that (21.9) represents *circular coma of order* $2n-1$. One has the familiar family of circles, touched by two straight lines through $I'$, their mutual inclination being $\psi_{2n-1} = 2 \arcsin(1 - 1/n)$. Also

$$\kappa_{t,2n-1}/\kappa_{s,2n-1} = 2n-1, \tag{21.10}$$

so that the value of this ratio is just the order of the aberration being considered. The image lies entirely to one side of $I'$, as before. Note that when spherical aberration has been removed, circular coma is the dominant aberration for systems of small field angle, such as microscope objectives or astronomical telescopes.

(iii) $\lambda = 2$

We now have to deal with two terms, i.e.

$$v_2^{(n)} = \rho^{2n-2} h'^2 (v_{20}^{(n)} \cos^2\theta + v_{11}^{(n)}). \tag{21.11}$$

The corresponding displacement is

$$\hat{y} = \mathfrak{a}[k + 1/(n-2) + \cos^2\theta] \cos\theta, \quad \hat{z} = \mathfrak{a}(k + \cos^2\theta) \sin\theta, \tag{21.12}$$

where

$$\mathfrak{a} = 2(n-2)\, v_{20}^{(n)} \rho^{2n-3} h'^2, \quad k = (n-1)\, v_{11}^{(n)}/(n-2)\, v_{20}^{(n)}.$$

This evidently generalizes (20.7). That it also generalizes (19.14) is fortuitous, in as far as it hinges on the vanishing of $\mathfrak{a}$ when $n$ happens to have the value 2. (21.12) thus represents *oblique spherical aberration of order* $2n-1$. (Note that this terminology now appears even less well motivated than before, as the displacement does not even vary as the cube of $\rho$ when $n > 3$.) The general trend of the usual curves generated by a zonal family of rays is not essentially dissimilar from that already encountered when $n = 3$. Here, whatever the curve for a certain value of $k$ may be, the same curve, but turned through 90°, obtains for $-[k+(n-1)/(n-2)]$. The previous critical value $k = -1$ therefore now becomes

$$k = -\tfrac{1}{2}(n-1)/(n-2);$$

but the critical curve itself is somewhat more complicated than before.

(iv) $\lambda = 2n-3$

Again we have two terms, namely

$$v_{2n-3}^{(n)} = \rho^3 h'^{2n-3}(v_{n,\,n-3}^{(n)} \cos^3\theta + v_{n-1,\,n-2}^{(n)} \cos\theta). \quad (21.13)$$

The displacement is given exactly by (20.9), provided one now takes $\mathfrak{a}$ and $k$ to be given by

$$\mathfrak{a} = v_{n-1,\,n-2}^{(n)} \rho^2 h'^{2n-3}, \quad k = 1 + 3v_{n,\,n-3}^{(n)}/2v_{n-1,\,n-2}^{(n)}.$$

This aberration is therefore *elliptical coma of order* $2n-1$; and its description is fully covered by that given previously in the special case $n = 3$.

(v) $\lambda = 2n-2$

Now
$$v_{2n-2}^{(n)} = \rho^2 h'^{2n-2}(v_{n,\,n-2}^{(n)} \cos^2\theta + v_{n-1,\,n-1}^{(n)}), \quad (21.14)$$

whence, in an out-of-focus plane given by $\chi = kh'^{2n-2}$,

$$\hat{y} = \rho h'^{2n-2}[2(v_{n,\,n-2}^{(n)} + v_{n-1,\,n-1}^{(n)}) - k] \cos\theta,$$

$$\hat{z} = \rho h'^{2n-2}(2v_{n-1,\,n-1}^{(n)} - k) \sin\theta. \quad (21.15)$$

Here we are evidently dealing with *curvature of field of order* $2n-1$. The condition for the formation of a sharp image lying in a certain surface of revolution is that $v^{(n)}_{n,n-2}$ should vanish. When it does not, one again has two focal lines, so that this coefficient governs *astigmatism of order* $2n-1$.

(vi) $\lambda = 2n-1$

Here we have simply

$$v^{(n)}_{2n-1} = \rho h'^{2n-1} v^{(n)}_{n,n-1} \cos\theta, \tag{21.16}$$

which implies that $\hat{y} = v^{(n)}_{n,n-1} h'^{2n-1}, \quad \hat{z} = 0.$ (21.17)

We are evidently dealing with *distortion of order* $2n-1$, any discussion of which is surely superfluous.

All third- and fifth-order aberrations previously encountered have now been generalized to order $2n-1$, and we could go on to consider in detail the new types which arise as we go on to the seventh and higher orders. Such a specific discussion, however, becomes increasingly academic, and we shall therefore deal explicitly with only one further aberration. As we go from fifth to seventh order two new types appear, of which one is governed jointly by two coefficients, the other by three. We therefore contemplate the former, which, upon generalizing to any order, is

$$v^{(n)}_3 = \rho^{2n-3} h'^3 (v^{(n)}_{30} \cos^3\theta + v^{(n)}_{21} \cos\theta). \tag{21.18}$$

Then

$$\hat{y} = \mathfrak{a}\left[\left(\frac{3(n-1)}{n-3} + \frac{n-1}{n-2}k\right) + \left(\frac{2(2n-3)}{n-3} + k\right)\cos 2\theta + \cos 4\theta\right],$$

$$\hat{z} = \mathfrak{a}[(2+k)\sin 2\theta + \sin 4\theta], \tag{21.19}$$

where $\mathfrak{a} = \frac{1}{4}(n-3)\, v^{(n)}_{30}\, \rho^{2n-4} h'^3$

and $k = 4(n-2)\, v^{(n)}_{21} / (n-3)\, v^{(n)}_{30}.$

The usual zonal curves may have a great variety of shapes, and it is hardly worth while enumerating them all. In certain cases one again gets a comatic flare closely resembling that familiar from elliptical coma. It is also possible to get a set of figures-of-eight, such that one set of loops is touched by a pair of straight lines through $I'$. The image is, however, evidently not wholly contained within these tangents; nor does it lie entirely to one side of $I'$.

With this example we conclude our enumeration of particular aberrations, and go on to a general classification and terminology.

## 22. General classification of aberrations

For the purpose of discussing the various aberrations which contribute to the displacement of order $2n-1$, we continue to suppose that the aberrations of order less than $2n-1$ are absent. From (21.2) we know that

$$v_\lambda^{(n)} = \rho^{2n-\lambda} h'^\lambda \Theta, \tag{22.1}$$

where $\Theta$ is a polynomial in $\cos\theta$, containing only even and odd powers of this, according as $\lambda$ is even or odd. We note in passing that in proceeding from order $2n-1$ to order $2n+1$ *two* new types of aberrations appear, and in view of (16.4) these are governed by $n+2$ coefficients altogether. This result alone shows that the general zonal curves corresponding to the various types become increasingly complicated, and for sufficiently large $n$ it is neither feasible nor useful to enumerate them in detail; so that a characterization of a general kind must suffice (see also the end of Section 40).

Consider therefore the displacement generated by (22.1). It has the generic form

$$\hat{y} = q\cos\theta\ \Theta_y, \quad \hat{z} = q\sin\theta\ \Theta_z \quad (q = \rho^{2n-\lambda-1}h'^\lambda), \tag{22.2}$$

where $\Theta_y$ and $\Theta_z$ are polynomials in $\cos\theta$, again containing only even and odd powers of this according as $\lambda$ is even or odd. When, and only when, $\lambda$ is odd (22.2) can be written in the form

$$\hat{y} = q\sum_{s=0} a_s \cos 2s\theta, \quad \hat{z} = q\sum_{s=1} b_s \sin 2s\theta, \tag{22.3}$$

where the $a_s$ and $b_s$ are certain constants. In this case the usual zonal curves are described twice every time the radius vector in the exit pupil sweeps the latter out once. As $\rho$ increases the various curves remain geometrically similar to one another; but, ignoring the over-all change of dimension, the curves suffer an increasing bodily shift, on account of the presence of the term $qa_0$ in $\hat{y}$. If the various curves do not contain $I'$ there will be a pair of fixed tangents touching all of them, and such tangents may also exist when the curves have loops which do not contain $I'$. The image is in general asymmetric, i.e. it has no line of symmetry parallel to the $\hat{z}$-axis.

It is clear that circular coma is essentially characteristic of the aberrations which have just been described, and for this reason they are called *comatic*. Every comatic displacement varies as an odd power of $h'$ and an even power of $\rho$.

When $\lambda$ is even, $\hat{y}$ and $\hat{z}$ are also periodic in $\theta$, but the period is now $2\pi$. Moreover, every zonal curve is now symmetric about the $z$-axis and is centred on $I'$. (Of course *all* zonal curves are always symmetric about the $\hat{y}$-axis.) In any order, astigmatism is amongst these symmetrical aberrations, and, at least as a matter of convenience, these may therefore be called *astigmatic*. Every astigmatic displacement varies as an even power of $h'$ and an odd power of $\rho$.

As one proceeds to higher and higher orders it becomes increasingly futile to invent ever new names for the new aberrations which make their appearance. It is preferable to use a terminology which directly reflects the results embodied in the present discussion. Accordingly, we call the aberration (i.e. the displacement) varying as $\rho^{2n-\lambda-1}h'^{\lambda}$ *coma of degree* $\lambda$ *and order* $2n-1$ when $\lambda$ is odd, and *astigmatism of degree* $2n-\lambda-1$ *and order* $2n-1$ when $\lambda$ is even. Then, for example, secondary oblique spherical aberration becomes secondary cubic astigmatism; circular coma (of any order) becomes linear coma; whilst, when $n = 4$ (21.19) represents tertiary cubic coma. Of course, in special cases one will retain the traditional names, and will continue to refer to spherical aberration rather than to astigmatism of degree $2n-1$, granted that the order is $2n-1$ in each case.

It should be borne in mind that the preceding classification is by no means only of academic interest, as might appear to be the case at first sight. The point is that although we initially supposed that aberrations of order less than $2n-1$ were absent, the generic result (22.3) remains valid when this restriction is removed. The presence of lower orders merely affects the values of the constants $a_s$ and $b_s$, but these were not further specified in any case. At any rate, the time has plainly come to consider the combined effects of aberrations of several orders, present simultaneously.

## 23. Effective aberration coefficients ($v_{\mu\nu}^{(n)}$ given)

Given the aberration function $v$, the displacement $\boldsymbol{\epsilon}'$ is given by two series, each of which is a sequence of homogeneous polynomials of degree 3, 5, 7, ... in the ray-coordinates $\mathbf{y}'$, $\mathbf{y}_1$. The polynomials of degree $2n-1$ have coefficients which are certain combinations of the characteristic aberration coefficients which enter into $v^{(2)}, \ldots, v^{(n)}$.

Write $$\epsilon'_{2n-1} = \sum_{\mu=0}^{n-1} \sum_{\nu=0}^{\mu} (u_{\mu\nu}^{(n)} \mathbf{y}' + \bar{u}_{\mu\nu}^{(n)} \mathbf{y}_1) \xi^{n-1-\mu} \eta^{\mu-\nu} \zeta^{\nu}. \qquad (23.1)$$

Then the $u_{\mu\nu}^{(n)}$, $\bar{u}_{\mu\nu}^{(n)}$ are the new coefficients under discussion, and we call them *effective aberration coefficients*. (See also the discussion following shortly after equation (69.5).) They are defined by (23.1), so that even if one started with the angle characteristic, for instance, one must end up with the *same* series (23.1), provided that at some stage one reverts to the variables $\mathbf{y}'$, $\mathbf{y}_1$. Of course, one could define different effective coefficients through series exactly like (23.1) except that the variables $\mathbf{y}'$ are replaced by coordinates in the entrance pupil or elsewhere; see for example Section 37, after equation (37.7). At any rate, we shall adopt (23.1). Our task is to relate the $u_{\mu\nu}^{(n)}, \bar{u}_{\mu\nu}^{(n)}$ to the characteristic coefficients $v_{\mu\nu}^{(n)}$, or $t_{\mu\nu}^{(n)}, \ldots$, as the case may be.

The desired end may be achieved straightforwardly enough, but the task becomes exceedingly laborious for orders greater than 5, certainly for orders greater than 7. It may be made a good deal easier in the following way. One first represents certain groups of terms in (17.5) by some collective symbol, e.g. one might write $A$ and $B$ for the factors of $2(1+u)^{\frac{1}{2}}$ and $(1+u)$, and $C$ and $E$, say, for $v_\xi$ and $v_\eta$. Then $A = A_2 + A_3 + A_4 + \ldots$, where $A_s$ is a homogeneous polynomial of degree $s$ in the variables $\xi, \eta, \zeta$; and likewise $B$, $C$ and $E$ are series beginning with terms $B_3$, $C_1$ and $E_1$ respectively, whilst $D = 1 + D_2 + \ldots$. Proceeding in this way, it is reasonably easy to keep an eye on the terms of any specific degree, and one is less likely to miss some of them inadvertently. In particular, the quadratic terms of the factors multiplying $\mathbf{y}'$ and $\mathbf{y}_1$ in (17.4) are simply

$$2C_2 + A_2 + uC_1 \quad \text{and} \quad E_2 - A_2 + \tfrac{1}{2}uE_1$$

respectively. Similarly, the cubic terms are

$$2C_3 + \tfrac{1}{2}uA_2 - \tfrac{1}{2}B_3 - 2A_2C_1 + A_3 + uC_2 - \tfrac{1}{4}u^2C_1$$

and $$E_3 - \tfrac{1}{2}uA_2 + \tfrac{1}{2}B_3 - A_2E_1 - A_3 + \tfrac{1}{2}uE_2 - \tfrac{1}{8}u^2C_1,$$

respectively, and these are the terms giving the total seventh-order displacement. From the results just given it is a routine matter to obtain the coefficients $u_{\mu\nu}^{(n)}, \bar{u}_{\mu\nu}^{(n)}$ for $n = 2, 3, 4, \ldots$, the case $n = 2$ being somewhat trivial. We shall content ourselves with the explicit results for orders 3, 5 and 7. In an obvious notation we write in these three cases $p^*_\mu, \bar{p}^*_\mu$ $(\mu = 1, 2, 3)$, $s^*_\mu, \bar{s}^*_\mu$ $(\mu = 1, ..., 6)$ and $t^*_\mu, \bar{t}^*_\mu$ $(\mu = 1, ..., 10)$ respectively for the effective coefficients. Then, when $n = 2$,

$$\left.\begin{aligned} p_1^* &= 4p_1, & \bar{p}_1^* &= p_2, \\ p_2^* &= 2p_2, & \bar{p}_2^* &= 2p_4, \\ p_3^* &= 2p_3, & \bar{p}_3^* &= p_5. \end{aligned}\right\} \tag{23.2}$$

When $n = 3$

$$\left.\begin{aligned} s_1^* &= 6s_1 + 6p_1, & \bar{s}_1^* &= s_2 - 4p_1 + \tfrac{1}{2}p_2, \\ s_2^* &= 4s_2 - 8p_1 + 4p_2, & \bar{s}_2^* &= 2s_4 + 4p_1 - 4p_2 + p_4, \\ s_3^* &= 4s_3 + 2p_1 - p_2 + 3p_3, & \bar{s}_3^* &= s_5 + \tfrac{3}{2}p_2 - 2p_3 + \tfrac{1}{2}p_5, \\ s_4^* &= 2s_4 - 4p_2 + 2p_4, & \bar{s}_4^* &= 3s_7 + 2p_2 - 4p_4, \\ s_5^* &= 2s_5 + p_2 - 4p_3 - 2p_4 + p_5, & \bar{s}_5^* &= 2s_8 + 2p_3 + 3p_4 - 2p_5, \\ s_6^* &= 2s_6 + p_3 - p_5, & \bar{s}_6^* &= s_9 + \tfrac{3}{2}p_5. \end{aligned}\right\} \tag{23.3}$$

Before going to $n = 4$ a word must be said about the effects of dropping the normalization condition $d' = 1$. The effective and the characteristic aberration coefficients of order $2n-1$ have dimensions $(\text{length})^{-2n+2}$ and $(\text{length})^{-2n+1}$ respectively; granted that $v$ is defined as hitherto, i.e. without an additional scale factor. Evidently, therefore, every primary characteristic coefficient takes a factor $d'$ in (23.2), but a factor $1/d'$ in (23.3); whilst in the latter $s_\alpha$ $(\alpha = 1, ..., 9)$ takes a factor $d'$. In short, removal of the condition $d' = 1$ entails modifications of (23.2) and (23.3) which may be represented schematically by

$$p^* = d'p, \quad s^* = d's + p/d'. \tag{23.4}$$

When $n = 4$

$$t_1^* = 8t_1 + 9s_1 + \tfrac{3}{2}p_1 - 24p_1^2,$$

$$t_2^* = 6t_2 - 12s_1 + 7s_2 - 4p_1 + \tfrac{5}{4}p_2 + 16p_1^2 - 32p_1p_2,$$

$$t_3^* = 6t_3 + 3s_1 - s_2 + 6s_3 + p_1 - \tfrac{1}{2}p_2 + \tfrac{3}{4}p_3 + 4p_1p_2 - 24p_1p_3 - \tfrac{1}{2}p_2^2,$$

$$t_4^* = 4t_4 - 8s_2 + 5s_4 - 2p_1 - 3p_2 + p_4 + 16p_1p_2 - 16p_1p_4 - 10p_2^2,$$

$$t_5^* = 4t_5+2s_2-8s_3-2s_4+4s_5+2p_2-2p_3-p_4+\tfrac{1}{2}p_5+16p_1p_3 +8p_1p_4-8p_1p_5+2p_2^2-16p_2p_3-2p_2p_4,$$

$$t_6^* = 4t_6+2s_3-s_5+3s_6-\tfrac{1}{2}p_1-\tfrac{1}{2}p_2+\tfrac{1}{2}p_3-\tfrac{1}{2}p_5+4p_1p_5+2p_2p_3 -p_2p_5-6p_3^2,$$

$$t_7^* = 2t_7-4s_4+3s_7+p_2-2p_4+4p_2^2-8p_2p_4,$$

$$t_8^* = 2t_8+s_4-4s_5-3s_7+2s_8+p_3+3p_4-p_5+8p_2p_3+4p_2p_4 -4p_2p_5-8p_3p_4-2p_4^2,$$

$$t_9^* = 2t_9+s_5-4s_6-2s_8+s_9-\tfrac{1}{4}p_2-p_4+\tfrac{3}{2}p_5+2p_2p_5+4p_3^2+4p_3p_4 -4p_3p_5-2p_4p_5,$$

$$t_{10}^* = 2t_{10}+s_6-s_9-\tfrac{1}{4}p_3-\tfrac{1}{2}p_5+2p_3p_5-\tfrac{1}{2}p_5^2,$$

$$\bar{t}_1^* = t_2-6s_1+\tfrac{1}{2}s_2-2p_1-\tfrac{1}{8}p_2+8p_1^2-4p_1p_2,$$

$$\bar{t}_2^* = 2t_4+6s_1-6s_2+s_4+6p_1-p_2-\tfrac{1}{4}p_4+16p_1p_2-8p_1p_4-3p_2^2,$$

$$\bar{t}_3^* = t_5+\tfrac{3}{2}s_2-4s_3+\tfrac{1}{2}s_5-2p_1+\tfrac{1}{4}p_2-p_3-\tfrac{1}{8}p_5+8p_1p_3-2p_2p_3 -4p_1p_5+\tfrac{3}{2}p_2^2,$$

$$\bar{t}_4^* = 3t_7+4s_2-6s_4+\tfrac{3}{2}s_7-4p_1+\tfrac{7}{2}p_2+8p_1p_4+6p_2^2-16p_2p_4,$$

$$\bar{t}_5^* = 2t_8-4s_2+4s_3+3s_4-2s_5+s_8+2p_1-2p_2+3p_3+\tfrac{1}{2}p_4+8p_1p_5 +8p_2p_3+6p_2p_4-4p_2p_5-4p_3p_4,$$

$$\bar{t}_6^* = t_9+\tfrac{3}{2}s_5-2s_6+\tfrac{1}{2}s_9+\tfrac{3}{8}p_2-p_3+\tfrac{1}{4}p_5-3p_2p_5+2p_3^2-2p_3p_5,$$

$$\bar{t}_7^* = 4t_{11}+2s_4-6s_7-2p_2+p_4+8p_2p_4-4p_4^2,$$

$$\bar{t}_8^* = 3t_{12}+2s_5+\tfrac{9}{2}s_7-4s_8+p_2-2p_3-2p_4+\tfrac{1}{2}p_5+4p_2p_5+8p_3p_4 -4p_4p_5+6p_4^2,$$

$$\bar{t}_9^* = 2t_{13}+2s_6+3s_8-2s_9+p_3+\tfrac{3}{4}p_4-p_5+4p_3p_5+6p_4p_5-p_5^2,$$

$$\bar{t}_{10}^* = t_{14}+\tfrac{3}{2}s_9+\tfrac{3}{8}p_5+\tfrac{3}{2}p_5^2. \qquad (23.5)$$

The analogue of (23.4) is here

$$t^* = d't+s/d'+p/d'^3+pp. \qquad (23.6)$$

The labour involved in obtaining the secondary, and even the tertiary, effective coefficients has after all been quite moderate. It might be remarked that one can also use an iterative method which allows one to proceed even more easily than by that used above.

For this purpose one writes $1/\alpha'$ in a form which contains $\boldsymbol{\epsilon}'$ explicitly (see (35.6)) and inserts this in (17.1). However, the effective coefficients of any order then require one to know only those of lower order, so that the various equations above can be obtained iteratively; cf. the discussion preceding equation (35.12). At any rate, in the case of the angle characteristic the situation is worse, a state of affairs which was already alluded to after equation (17.10). In fact, it is so much worse that one is undoubtedly better off with $V$ than with $T$ as long as one wants to consider the *exact* displacements of the various orders.

We conclude this section with the somewhat obvious remark that between the $n(n+1)$ effective coefficients of order $2n-1$ there must exist $\frac{1}{2}n(n-1)$ *identities*, in the sense that there are $\frac{1}{2}n(n-1)$ distinct linear combinations of them which express themselves entirely in terms of the coefficients of order less than $2n-1$. These identities merely express the requirement that the integrability condition on $\beta' dy' + \gamma' dz'$ must be satisfied (cf. equation (6.7)). In principle they may therefore be obtained directly from the equation

$$\partial\beta'/\partial z' = \partial\gamma'/\partial y', \tag{23.7}$$

where $\boldsymbol{\beta}'$ is to be obtained from (17.1); see (35.6) and (35.7). In the primary domain (23.7) reduces to

$$\partial\epsilon'_y/\partial z' = \partial\epsilon'_z/\partial y', \tag{23.8}$$

and this immediately gives $2\bar{p}_1^* - p_2^* = 0$.

## 24. Effective coefficients ($t_{\mu\nu}^{(n)}$ given)

As an alternative to (23.3) the secondary effective coefficients are now to be related to the characteristic coefficients $t_{\mu\nu}^{(2)}$ and $t_{\mu\nu}^{(3)}$. We may proceed as follows. As in (19.24) we write

$$t = (s-m)^{-1}[(p_1\xi^2 + \ldots + p_6\zeta^2) + (s_1\xi^3 + \ldots + s_{10}\zeta^3) + \ldots]. \tag{24.1}$$

The terms depending on $\zeta$ alone must here be included, so that $t$ is the angle characteristic after the paraxial term has been removed. We now use variables $\sigma_y, \sigma_z (\equiv \boldsymbol{\sigma})$ and $\mu_y, \mu_z (\equiv \boldsymbol{\mu})$ defined by

$$\boldsymbol{\beta} - s\boldsymbol{\beta}' = \boldsymbol{\sigma}, \quad \boldsymbol{\beta} - m\boldsymbol{\beta}' = \boldsymbol{\mu}, \tag{24.2}$$

so that $\xi = \sigma_y^2 + \sigma_z^2$, etc. Then

$$\mathbf{y}' = (2sT_\xi + mT_\eta)\,\boldsymbol{\sigma} + (sT_\eta + 2mT_\zeta)\,\boldsymbol{\mu},$$

$$\mathbf{y}_1 = (2mT_\xi + mT_\eta)\,\boldsymbol{\sigma} + (mT_\eta + 2mT_\zeta)\,\boldsymbol{\mu}, \qquad (24.3)$$

whilst (17.10) becomes

$$\boldsymbol{\epsilon}' = (s-m)(2t_\xi\,\boldsymbol{\sigma} + t_\eta\,\boldsymbol{\mu}). \qquad (24.4)$$

Also, let $\mathbf{y}'_e$ be coordinates in the exit pupil, i.e.

$$\mathbf{y}'_e = (\mathbf{y}' - d'\boldsymbol{\beta}')/\alpha'. \qquad (24.5)$$

(Note that since we have arranged $f$ to have the value unity, we have to carry $d' = s-m$ along here.) Our task is to eliminate $\boldsymbol{\sigma}$ and $\boldsymbol{\mu}$ in favour of $\mathbf{y}'_e$ and $\mathbf{y}_1$.

Now, using the first member of (15.12),

$$1/\alpha' = [1-(s-m)^{-2}(\xi - 2\eta + \zeta)]^{-\frac{1}{2}}$$

$$= 1 + \tfrac{1}{2}(s-m)^{-2}(\xi - 2\eta + \zeta) + O(4).$$

Also, from (24.2), $\boldsymbol{\beta}' = (s-m)^{-1}(\boldsymbol{\mu} - \boldsymbol{\sigma})$, whence

$$-d'\boldsymbol{\beta}'/\alpha' = (\boldsymbol{\sigma} - \boldsymbol{\mu})\,[1 + \tfrac{1}{2}(s-m)^{-2}(\xi - 2\eta + \zeta)] + O(5). \qquad (24.6)$$

One therefore has, on rejecting terms of degree higher than 3,

$$\mathbf{y}'_e = \{1 + [2st_\xi^{(2)} + mt_\eta^{(2)} + \tfrac{1}{2}(s-m)^{-2}(\xi - 2\eta + \zeta)]\}\,\boldsymbol{\sigma}$$
$$+ [st_\eta^{(2)} + 2mt_\zeta^{(2)} - \tfrac{1}{2}(s-m)^{-2}(\xi - 2\eta + \zeta)]\,\boldsymbol{\mu},$$

$$\mathbf{y}_1 = (2mt_\xi^{(2)} + mt_\eta^{(2)})\,\boldsymbol{\sigma} + [1 + (mt_\eta^{(2)} + 2mt_\zeta^{(2)})]\,\boldsymbol{\mu}. \qquad (24.7)$$

We write these temporarily in the abbreviated form

$$\mathbf{y}'_e = (1+P)\,\boldsymbol{\sigma} + Q\boldsymbol{\mu}, \quad \mathbf{y}_1 = R\boldsymbol{\sigma} + (1+S)\,\boldsymbol{\mu}. \qquad (24.8)$$

Bearing in mind that we intend to proceed only to the fifth order, we need to include only terms up to the third order in the series which are the inversion of (24.8). Since $\boldsymbol{\sigma} = \mathbf{y}'_e + O(3)$, $\boldsymbol{\mu} = \mathbf{y}_1 + O(3)$, the quantities $\xi$, $\eta$, $\zeta$ which occur in $P$, $Q$, $R$ and $S$ are, to the required order, simply the same as the invariants $\xi$, $\eta$, $\zeta$ which appeared in the context of the point characteristic; and so here

$$\boldsymbol{\sigma} = (1-P)\,\mathbf{y}'_e - Q\mathbf{y}_1, \quad \boldsymbol{\mu} = -R\mathbf{y}'_e + (1-S)\,\mathbf{y}_1. \qquad (24.9)$$

However, from (24.1) and (24.4),

$$\boldsymbol{\epsilon}' = [(4p_1\xi+2p_2\eta+2p_3\zeta)+(6s_1\xi^2+4s_2\xi\eta+4s_3\xi\zeta+2s_4\eta^2+2s_5\eta\zeta+2s_6\zeta^2)]\,\boldsymbol{\sigma}+[(p_2\xi+2p_4\eta+p_5\zeta)+(s_2\xi^2+2s_4\xi\eta+s_5\xi\zeta+3s_7\eta^2+2s_8\eta\zeta+s_9\zeta^2)]\,\boldsymbol{\mu}. \tag{24.10}$$

Continuing to reject terms of degree greater than 5, the substitutions (24.9) will leave the quintic terms, taken by themselves, unaffected, but they will produce a large number of additional terms arising from the cubic members. Denoting these additional terms by $\Delta\boldsymbol{\epsilon}'$ one thus has

$$\begin{aligned}-\Delta\boldsymbol{\epsilon}' = \{&4p_1(3P\xi+2Q\eta)+p_2[3R\xi+2(2P+S)\eta+2Q\zeta]\\ &+2p_3[2R\eta+(P+2S)\zeta]+2p_4R\eta+p_5R\zeta\}\,\mathbf{y}'\\ &+\{4p_1Q\xi+p_2[(2P+S)\xi+4Q\eta]+2p_3Q\zeta\\ &+2p_4[R\xi+(P+2S)\eta+Q\zeta]+p_5(2R\eta+3S\zeta)\}\,\mathbf{y}_1.\end{aligned} \tag{24.11}$$

Here, in turn,

$$\left.\begin{aligned}(s-m)P &= (4sp_1+mp_2+a)\xi+2(sp_2+mp_4-a)\eta\\ &\quad+(2sp_3+mp_5+a)\zeta,\\ (s-m)Q &= (sp_2+2mp_3-a)\xi+2(sp_4+mp_5+a)\\ &\quad+(sp_5+4mp_6-a)\zeta,\\ (s-m)R &= m(4p_1+p_2)\xi+2m(p_2+p_4)\eta+m(2p_3+p_5)\zeta,\\ (s-m)S &= m(p_2+2p_3)\xi+2m(p_4+p_5)\eta+m(p_5+4p_6)\zeta,\end{aligned}\right\} \tag{24.12}$$

where $a = \frac{1}{2}(s-m)^{-1}$. (24.12) must now be substituted in (24.11), and the subsequent expression suitably rearranged. This is a most tedious process, the final outcome of which is embodied in the following set of equations giving the secondary effective in terms of the characteristic coefficients. (The primary relations (23.2) apply here also.)

$$s_1^* = 6s_1-2a(48sp_1^2+24mp_1p_2+3mp_2^2+12ap_1),$$

$$\begin{aligned}s_2^* = 4s_2-8a[&12sp_1p_2+8mp_1p_3+8mp_1p_4+3mp_2^2+2mp_2p_3\\ &+2mp_2p_4-a(8p_1-p_2)],\end{aligned}$$

$$\begin{aligned}s_3^* = 4s_3-2a[&32sp_1p_3+16mp_1p_5+2sp_2^2+10mp_2p_3+mp_2p_5\\ &+8mp_3^2+a(12p_1-2p_2+2p_3)],\end{aligned}$$

$$s_4^* = 2s_4 - 2a[16sp_1p_4 + 16mp_1p_5 + 8sp_2^2 + 14mp_2p_3 + 16mp_2p_4 + 7mp_2p_5 + 8mp_3p_4 + 4mp_4^2 + 8a(2p_1 - p_2)],$$

$$s_5^* = 2s_5 - 8a[2sp_1p_5 + 8mp_1p_6 + 4sp_2p_3 + sp_2p_4 + 3mp_2p_5 + 2mp_2p_6 + 2mp_3^2 + 7mp_3p_4 + 5mp_3p_5 + mp_4p_5 - 2a(p_1 - p_2 + p_3)],$$

$$s_6^* = 2s_6 - 2a[2sp_2p_5 + 8mp_2p_6 + 4sp_3^2 + 8mp_3p_5 + 16mp_3p_6 + mp_5^2 - 2a(p_2 - p_3)],$$

$$\bar{s}_1^* = s_2 - 2a[12sp_1p_2 + 8mp_1p_3 + 16mp_1p_4 + 3mp_2^2 + 2mp_2p_3 + 4mp_2p_4 - 2a(2p_1 - p_2)],$$

$$\bar{s}_2^* = 2s_4 - 4a[8sp_1p_4 + 8mp_1p_5 + 4sp_2^2 + 4mp_2p_3 + 10mp_2p_4 + 2mp_2p_5 + 4mp_3p_4 + 4mp_4^2 + a(4p_1 - 4p_2 + p_4)],$$

$$\bar{s}_3^* = s_5 - 4a[2sp_1p_5 + 8mp_1p_6 + 3sp_2p_3 + sp_2p_4 + 3mp_2p_5 + 2mp_2p_6 + 2mp_3^2 + 6mp_3p_4 + 3mp_3p_5 + 2mp_4p_5 - a(2p_1 - p_2 + p_3 + p_4)],$$

$$\bar{s}_4^* = 3s_7 - 8a[3sp_2p_4 + 3mp_2p_5 + 3mp_4^2 + 3mp_4p_5 + a(2p_2 - p_4)],$$

$$\bar{s}_5^* = 2s_8 - 4a[2sp_2p_5 + 8mp_2p_6 + 4sp_3p_4 + 4mp_3p_5 + 2sp_4^2 + 8mp_4p_5 + 8mp_4p_6 + 4mp_5^2 - a(2p_2 - 2p_3 - 3p_4)],$$

$$\bar{s}_6^* = s_9 - 2a[2sp_3p_5 + 8mp_3p_6 + 2sp_4p_5 + 8mp_4p_6 + 3mp_5^2 + 12mp_5p_6 - 2a(p_3 + p_4)]. \tag{24.13}$$

Here the sixth primary coefficient $p_6$ appears explicitly, as was to be expected, according to the discussion following equation (17.11). The starred coefficients are of course the *same* as those in (23.3). It should be mentioned that the equations (24.13) have not been reliably checked, and so may contain errors. Their complexity, as compared with (23.3), is striking. Moreover, the labour involved in their derivation is probably no less than that required to get the *seventh*-order equations (23.5). However, in practice one may be content with results which are approximate to the extent that we consider $\boldsymbol{\epsilon}'$ merely as the sum of the displacements of order $2n-1$ generated by the aberration function of that order alone; under which circumstances the discussion of Sections 19–21 would be *directly* relevant. In other words, one merely substitutes $\mathbf{y}_e'$ for $\boldsymbol{\sigma}$

and $\mathbf{y}_1$ for $\boldsymbol{\mu}$ in (17.10). In general terms this means that if any given ray be specified by the values of its coordinates $\mathbf{y}'_e$, $\mathbf{y}_1$, then the series (17.10) gives the displacement not for the ray in question but rather for that ray whose coordinates have the values $\mathbf{y}'_e+\delta\mathbf{y}'_e$, $\mathbf{y}_1+\delta\mathbf{y}_1$, where

$$\delta\mathbf{y}'_e = (-S\mathbf{y}'_e+Q\mathbf{y}_1)\,\Delta, \quad \delta\mathbf{y}_1 = (R\mathbf{y}'_e-P\mathbf{y}_1)\,\Delta, \tag{24.14}$$

with $1/\Delta = (1-P)(1-S)-QR$. The coordinate differences (24.14) are obviously at least of the third order; and we may expect the error committed by their omission to be small for reasonably well corrected systems. (See also the discussion following shortly after equation (69.5).)

Despite what has just been said the presence of the lower-order coefficients in the effective coefficients of a given order is of course important in principle. A strict discussion of the existence or otherwise of focal curves, for instance, plainly requires the use of the exact coefficients $u^{(n)}_{\mu\nu}, \bar{u}^{(n)}_{\mu\nu}$ of the various orders.

## 25. The total secondary displacement

Since the primary effective coefficients are those which enter into (19.2)—cf. (23.2)—the primary displacement continues to be given by (19.5). The secondary total displacement on the other hand is to be obtained from (23.1) and (23.3). One easily finds that

$$\left.\begin{aligned}
\epsilon'_{y5} = {} & 6(s_1+p_1)\rho^5\cos\theta+[(3s_2-8p_1+\tfrac{5}{2}p_2)+2(s_2-2p_1+p_2) \\
& \times\cos 2\theta]\rho^4h'+[(4s_3+2s_4+6p_1-5p_2+3p_3+p_4) \\
& +2(s_4-2p_2+p_4)\cos^2\theta]\cos\theta\,\rho^3h'^2+[(2s_5+\tfrac{3}{2}s_7 \\
& +3p_2-4p_3-3p_4+p_5)+(s_5+\tfrac{3}{2}s_7+\tfrac{3}{2}p_2-2p_3-3p_4 \\
& +\tfrac{1}{2}p_5)\cos 2\theta]\rho^2h'^3+(2s_6+2s_8+3p_3+3p_4-3p_5) \\
& \times\cos\theta\,\rho h'^4+(s_9+\tfrac{3}{2}p_5)\,h'^5, \\
\epsilon'_{z5} = {} & 6(s_1+p_1)\rho^5\sin\theta+2(s_2-2p_1+p_2)\rho^4h'\sin 2\theta \\
& +[(4s_3+2p_1-p_2+3p_3)+2(s_4-2p_2+p_4)\cos^2\theta] \\
& \times\sin\theta\,\rho^3h'^2+(s_5+\tfrac{1}{2}p_2-2p_3-p_4+\tfrac{1}{2}p_5) \\
& \times\sin 2\theta\,\rho^2h'^3+(2s_6+p_3-p_5)\sin\theta\,\rho h'^4.
\end{aligned}\right\} \tag{25.1}$$

The general character of the various partial displacements is essentially the same as that described in Section 20. This is not to say that all previous results remain unmodified. For example, one now has, in place of (20.6),

$$\kappa_{t5}/\kappa_{s5} = (5s_2 - 12p_1 + \tfrac{9}{2}p_2)/(s_2 - 4p_1 + \tfrac{1}{2}p_2). \qquad (25.2)$$

This kind of modification of previous results is evidently of a rather trivial nature. None the less, one should bear in mind that in the presence of (primary) spherical aberration one has, in principle at least, some *effective* secondary circular coma

$$\hat{y} = -4p_1(2+\cos 2\theta)\rho^4 h', \quad \hat{z} = -4p_1 \sin 2\theta\, \rho^4 h', \qquad (25.3)$$

which should be compared with (19.12).

In conclusion, it may be worth writing down the displacement for all $\rho$ and $\theta$, correct to the fifth order, when the displacement is known to vanish for all *meridional* rays. Writing down the various conditions which ensure this state of affairs, one finds that

$$\left.\begin{aligned} \epsilon'_y &= -2(s_4+p_4)\sin^2\theta\cos\theta\,\rho^3 h'^2 - (s_7-2p_4)\sin^2\theta\,\rho^2 h'^3, \\ \epsilon'_z &= -2p_4\sin\theta\,\rho h'^2 - [(4s_4+3p_4) - 2(s_4+p_4)\cos^2\theta]\sin\theta\,\rho^3 h'^2 \\ &\quad -(s_7-p_4)\sin 2\theta\,\rho^2 h'^3 + (2s_6-p_4)\sin\theta\,\rho h'^4. \end{aligned}\right\} \qquad (25.4)$$

These expressions are of interest because in the course of design a great deal of emphasis is often laid in practice on the performance of systems in the meridional section alone, supposedly as a measure of its over-all performance. It will be seen that even in the absence of curvature of field ($p_4 = s_6 = 0$) there is a residual secondary displacement, made up of certain amounts of oblique spherical aberration and elliptical coma.

## Problems

**P.4(i).** The reduced magnification associated with a pair of conjugate planes is $\hat{m}$. Determine their location, i.e. the quantities $q$ and $q'$, in terms of $s$, $m$ and $f$.

**P.4(ii).** Equations (14.5) hold for arbitrary rays if one understands $A, B, C, D$ to be functions of $\xi, \eta, \zeta$. Show that (14.7) is still valid.

**P.4(iii).** The ideal behaviour of a certain symmetric system $K$ is prescribed as follows. $K$ forms a sharp, but possibly distorted, plane image in $\mathscr{I}'$ of a plane object in $\mathscr{I}$. Show that $V_0$ has the generic form

$$V_0 = g(\zeta) - (d'^2 + \xi - 2D\eta + D^2\zeta)^{\frac{1}{2}},$$

where $D$ is a function of $\zeta$ only. What is the significance of $D(\zeta)$?

**P.4 (iv).** $K$ is a symmetric system of which it is desired that it should form a sharp image of a plane object in $\mathscr{I}$, this image lying in some surface of revolution $\mathscr{S}$. Show that $V_0$ has the generic form $V_0 = g(\zeta) - (d'^2 + \xi - 2D\eta + K)^{\frac{1}{2}}$, where $D$ and $K$ are certain functions of $\zeta$ alone. What is the significance of these functions? What is the curvature of $\mathscr{S}$ at its pole, if $k_1 = (dK/d\zeta)_{\zeta=0}$?

**P.4 (v).** Obtain the displacement $\boldsymbol{\epsilon}'$ in the ideal image plane $\mathscr{I}'$ for the characteristic function $V_0$ of the previous problem.

**P.4 (vi).** Justify equation (15.14).

**P.4 (vii).** Deduce the result (15.20).

**P.4 (viii).** Let the *ideal* imagery of a certain system $K$ be implicit in its ideal point characteristic (15.6). Suppose that its *actual* point characteristic $V$ is that given in problem P.4 (iii). Determine the third- and fifth-order aberration functions $v^{(2)}$ and $v^{(3)}$ of $K$. Also determine $v^{(2)}$ if its point characteristic $V$ is that given in problem P.4 (iv). (Note: set $d' = 1$ throughout.) Consider your results in the light of equations (23.2–3).

**P.4 (ix).** Justify equation (19.11).

**P.4 (x).** Obtain the result (21.7).

**P.4 (xi).** Determine the conditions which must be satisfied to ensure that, correctly to the fifth order, the image patch should be exactly symmetric about a line parallel to the $\hat{z}$-axis.

**P.4 (xii).** If $K$ is the system of example P.4 (iv), let it be further given that *effective* distortion of all orders is absent. Show that one must have

$$D = (1+\zeta)^{-\frac{1}{2}}(1+K)^{\frac{1}{2}}.$$

(Take $d' = 1$. Note that distortion refers to $\mathscr{I}'$.)

**P.4 (xiii).** Continuing the preceding problem, show that the equation of the image surface is

$$X' = d_1(Y'^2 + Z'^2) + (d_2 - d_1^2)(Y'^2 + Z'^2)^2 + O(6).$$

**P.4 (xiv).** The point characteristic of a certain system $K$ is

$$V = g(\zeta) + \tfrac{1}{2}f(u), \quad \text{where} \quad u = \xi - 2\eta + \zeta,$$

as usual. Show that all rays from a given object point $O$ which make a fixed angle with the axis of $K$ in the image space intersect $\mathscr{I}'$ in a circle. Is this circle concentric with $I'$?

CHAPTER 5

# THE SYMMETRIC SYSTEM (PART II)

## A. The sine-condition

### 26. Spherical aberration and circular coma

In certain situations which occur in practice the image height is so small that all aberrations varying non-linearly with $h'$ may be completely disregarded. This means that only spherical aberration and circular coma need to be considered. Bearing in mind that of these two types the second can be controlled by suitably adjusting the position of the stop, it is of importance to have at hand certain general results which can be used in a simple way to test for the presence of circular coma, without the need to examine anything but *axial* rays, i.e. rays through the axial point $O_0$ of $\mathscr{I}$.

To begin with, all terms of the aberration function varying non-linearly with $h'$ are now to be omitted, so that $v$ has the generic form

$$v = S(\xi) + \eta C(\xi), \tag{26.1}$$

where

$$S(\xi) = \sum_{n=1}^{\infty} s_n \xi^{n+1}, \quad C(\xi) = \sum_{n=1}^{\infty} c_n \xi^n. \tag{26.2}$$

Here $s_{n-1}$, $c_{n-1}$ are the (characteristic) coefficients of spherical aberration and circular coma, previously denoted by $v_{00}^{(n)}$ and $v_{10}^{(n)}$ respectively.

We shall frequently have to refer to the exact displacement induced by (26.1), and we proceed to write it down. To this end it is desirable to introduce a number of abbreviations. Thus let

$$\left.\begin{aligned} w = (1+\xi)^{\frac{1}{2}}, \quad P = 1 - 2w\dot{S}, \quad Q = 1 - 2w^3\dot{C}, \quad R = 1 + wC, \\ J = (w^2 - \xi P^2)^{-\frac{1}{2}} = (1 + 4w\xi\dot{S} - 4w^2\xi\dot{S}^2)^{-\frac{1}{2}}, \end{aligned}\right\} \tag{26.3}$$

where a dot denotes differentiation with respect to $\xi$. Note that the various functions just defined depend upon $\rho$ alone. Always rejecting terms depending non-linearly upon $h'$, we find that

$$\left.\begin{aligned} \boldsymbol{\beta}' &= \frac{1}{w}[-(P + Q\eta/w^2)\,\mathbf{y}' + R\mathbf{y}_1], \\ \boldsymbol{\beta} &= -2mg_\xi \mathbf{y}_1 - \frac{m}{w}(R\mathbf{y}' - \mathbf{y}_1). \end{aligned}\right\} \tag{26.4}$$

A somewhat tedious calculation, based directly on (17.1), then yields the required result

$$\boldsymbol{\epsilon}' = [(1-JP)+J^3(P^2R-Q)\eta]\,\mathbf{y}' - (1-JR)\,\mathbf{y}_1. \qquad (26.5)$$

If $\sigma$ stands for spherical aberration alone ($\theta = 0$), we thus have

$$\sigma = \rho(1-JP), \qquad (26.6)$$

whilst the tangential and sagittal circular coma are given by

$$\kappa_t = -h'[1-JR+\xi J^3(Q-P^2R)], \quad \kappa_s = -h'(1-JR). \quad (26.7)$$

We emphasize that equations (26.5)–(26.7) are exact, and all orders of aberrations are included.

## 27. The sine-relation

For the sake of orientation we first contemplate a pair of perfect conjugate planes. The point characteristic of $K$ associated with these is, as usual,

$$V = g(\zeta) - (1+u)^{\frac{1}{2}}, \qquad (27.1)$$

in the notation of Section 17. Then

$$\boldsymbol{\beta} - m\boldsymbol{\beta}' = -2m\mathbf{y}_1 g_\zeta(\zeta). \qquad (27.2)$$

For axial rays $\mathbf{y}_1 = 0$, and the right-hand member of (27.2) vanishes. If $\phi, \phi'$ are the angles which the initial and final rays make with the axis of $K$, one therefore has

$$\sin\phi = m\sin\phi'. \qquad (27.3)$$

This result constitutes the so-called *sine-relation*. It will evidently be valid under conditions much less stringent than those assumed above, for (27.3) will be unaffected by the addition to the right-hand member of (27.1) of any aberration function $v$ which depends non-linearly on $h'$, i.e. which is such that $\partial v/\partial \mathbf{y}'$ and $\partial v/\partial \mathbf{y}_1$ both vanish when $\mathbf{y}_1 = 0$. In other words, the sine-relation holds if $\mathscr{I}$ and $\mathscr{I}'$ are perfect in a region around $O_0'$, the linear dimensions of which are $O(1)$. In such a region only spherical aberration and circular coma (of all orders) play a part; and we conclude that the complete absence of these entails the validity of (27.3).

(27.3) requires a slight modification when the object is at infinity, which is most easily obtained by a limiting process. If a meridional axial ray intersects $\mathscr{E}$ in the point whose $\bar{y}$-coordinate is $y_e$, one has

$$\tan\phi = \frac{y_e}{d} = \frac{smy_e}{(m-s)f}, \tag{27.4}$$

in view of (14.28). Now divide throughout by $m$, and then let $m$ tend to zero. According to (27.4) $\tan\phi/m$ tends to the value $-y_e/f$, so that, when $m = 0$, (26.3) is to be replaced by

$$f\sin\phi' = -y_e. \tag{27.5}$$

## 28. The sine-condition

In view of the remarks following upon equation (27.3), we go on to investigate the consequences of the *sine-condition*, i.e. the condition that the sine-relation (27.3) should hold. We therefore consider the quantity

$$\Gamma(\rho) \equiv \sin\phi' - m^{-1}\sin\phi. \tag{28.1}$$

Since only axial rays are contemplated here, we may set $\theta = 0$, and of course $\mathbf{y}_1 = 0$. Then, at once, from (26.4),

$$\Gamma(\rho) = \rho(2\dot{S}+C) = \sum_{n=1}^{\infty}(2ns_n+c_n)\rho^{2n+1}. \tag{28.2}$$

If this is to vanish for all values of $\rho$ the sum $2\dot{S}+C$ must vanish; and this is the necessary and sufficient condition for the validity of (27.3).

In practice $\Gamma(\rho)$ will of course in general fail to vanish for all values of $\rho$, and then (28.2) is not very informative. Even when $\dot{S}$ vanishes for some value of $\rho$ it tells us only that $C(\rho^2) = 0$ for that value of $\rho$. This, however, does not lead to any exact statements about the true displacement. At best we can proceed approximately by considering the *pseudo-displacement* $^*\boldsymbol{\epsilon}'$ which is the displacement calculated by omitting from the effective coefficients of order $2n-1$ all coefficients of order less than $2n-1$. In the notation of Section 11,

$$^*\boldsymbol{\epsilon}' = \sum_{n=2}^{\infty}\boldsymbol{\epsilon}'_{2n-1}(v^{(n)}). \tag{28.3}$$

Continuing to reject all terms non-linear in $h'$,

$$\begin{aligned} ^*\epsilon'_y &= 2\cos\theta\rho\dot{S} + [(1+\cos 2\theta)\rho^2\dot{C}+C]h', \\ ^*\epsilon'_z &= 2\sin\theta\rho\dot{S} + \sin 2\theta\rho^2\dot{C}h'. \end{aligned} \tag{28.4}$$

The pseudo-displacement due to spherical aberration is $2\rho\dot{S}$, having set $\theta = 0$; whilst, on setting $\theta = \frac{1}{2}\pi$, the sagittal pseudo-displacement $^*\kappa_s$ due to circular coma turns out to be $Ch'$. We have thus inferred that if for some value of $\rho$ both $\Gamma$ and $\dot{S}$ vanish, then $^*\kappa_s$ also vanishes. This result is defective in two ways, first, because it relates only to pseudo-displacements, second, because two separate conditions must be satisfied before anything can be said about the vanishing of $C$. We shall therefore improve upon this situation in various ways in the following sections.

## 29. The modified sine-condition

We just saw that the vanishing of $\Gamma$ for a particular value of $\rho$ led merely to a conclusion which involved $\dot{S}$ and $C$ jointly, and was therefore of little practical value even if pseudo-displacements were regarded as a sufficient approximation to true displacements. A rather trivial modification of the sine-condition, however, allows us to consider $C$ separately from $\dot{S}$.

To this end, let an axial ray intersect $\mathscr{E}'$ in the point $D'$ and the axis of $K$ in the point $P'$, so that the previous angle $\phi'$ is the angle between the axis and the line $D'P'$. We may instead consider the angle $\bar{\phi}'$ between the axis and the line $D'O'_0$. By inspection,

$$\sin\bar{\phi}' = -\rho(1+\rho^2)^{-\frac{1}{2}},$$

so that
$$\sin\phi - m\sin\bar{\phi}' = -m\rho C(\rho^2). \tag{29.1}$$

In place of $\Gamma$, define the quantity

$$\Delta(\rho) = \sin\bar{\phi}' - m^{-1}\sin\phi = \rho C(\rho^2). \tag{29.2}$$

(The case of infinite object distance is easily accommodated as before.) Then the condition $\Delta(\rho) = 0$ will be called the *modified sine-condition*.

If now $\Delta(\rho) = 0$ for all values of $\rho$, then all (characteristic) coefficients of circular coma must be zero, irrespectively of the presence of spherical aberration; so that we have a much simpler theoretical situation than that which obtained with regard to the original sine-condition. Moreover, if $\Delta$ vanishes for some particular value of $\rho$, then the total sagittal circular comatic pseudo-displacement vanishes for that value of $\rho$. Of course, again nothing can be said about the value of the tangential coma.

## 30. Axial rays and approximate displacement

The displacement $\boldsymbol{\epsilon}'$ is implicit in the functions $S$ and $C$, the actual connection between them being provided by (26.5). In this section we investigate how one might obtain $\boldsymbol{\epsilon}'$ for all values of $\theta$, provided one is content with an approximation which amounts to the assumption that the effects of spherical aberration and circular coma of orders greater than $2k+1$ are relatively small, where $k$ is some integer which one will, in practice, take to be 2 or 3.

To begin with, observe that for axial rays equations (26.4) at once give

$$\dot{S} = \frac{\beta'}{2\rho} + \frac{1}{2w}, \quad C = -\frac{\beta}{m\rho} - \frac{1}{w}. \tag{30.1}$$

Now let $k$ rays, having $\rho = \rho_1, \rho_2, \ldots, \rho_k$ respectively, be accurately traced through $K$. For each such ray the corresponding values of $\dot{S}$ and $C$ ($\dot{S}_j$ and $C_j$, say, for the $j$th ray) may be read off from the traces. Next, truncate the series (26.2), each after its $k$th term. Then one has two separate sets of $k$ linear equations

$$\sum_{j=1}^{k} (n+1)\rho_j^{2n} s_n = \dot{S}_j, \quad \sum_{j=1}^{k} \rho_j^{2n} c_n = C_j, \tag{30.2}$$

whose solutions give the (approximate) values of the unknown characteristic coefficients $s_1, \ldots, s_k$ and $c_1, \ldots, c_k$ respectively. Once these are known, the effective coefficients of spherical aberration and circular coma may be obtained after the fashion of Section 23, the situation being relatively simple here since only the coefficients $u_{00}^{(n)}$, $u_{10}^{(n)}$ and $\bar{u}_{10}^{(n)}$ ($n = 2, \ldots, k+1$) are required here. (For $n = 2, 3, 4$ the results are already given by (23.2), (23.3) and (23.5).) In short, one now has an expression for the displacement which provides not only the sagittal circular coma, but the circular comatic displacement for rays having any value of $\theta$.

## 31. Offence against the sine-condition

We saw in Section 29 that the value of $\Delta(\rho)$ is an immediate measure of the extent to which a circular comatic term is present in the aberration function. It would therefore be appropriate to refer to $\Delta(\rho)$ as the 'offence against the modified sine-condition'. However, we also saw that a knowledge of the value of $C$ for some value of $\rho$ did not imply an accurate knowledge of comatic displacements, whether sagittal or otherwise; in the sense that we had to be content with pseudo-displacements. From a practical point of view this is not a happy state of affairs, and we must seek to improve it.

In view of the close relationship between $\Delta$ on the one hand and $^*\kappa_s$ (rather than $^*\kappa_t$) on the other, we first turn our attention to the *exact* total sagittal circular comatic displacement $\kappa_s$. We shall usually refer to this simply as 'sagittal coma', but the true meaning of this term, as here intended, must constantly be borne in mind. Accordingly, reference to (26.7) shows that we are concerned with the quantity

$$K = JR - 1. \tag{31.1}$$

Now, when $\mathbf{y}_1 = 0$ (and $\theta = 0$),

$$w\beta' = -\rho P, \quad w\beta/m = -\rho R, \tag{31.2}$$

according to (26.4). From these it follows at once that

$$K = \frac{JP\beta}{m\beta'} - 1. \tag{31.3}$$

However, let $\delta$ be the longitudinal spherical aberration, i.e. the distance from $O_0'$ to the point in which the axial ray given by $\rho$ intersects the axis. Then we have

$$\delta = \sigma/(\rho - \sigma) = 1/JP - 1, \tag{31.4}$$

because of (26.6). (31.3) therefore becomes finally

$$K = -1 + \frac{\sin\phi}{m(1+\delta)\sin\phi'}. \tag{31.5}$$

It is easily confirmed that this may be written in the equivalent, simple form

$$K = -1 - \frac{\beta}{m\rho\alpha'}. \tag{31.6}$$

It is remarkable that, except for an over-all change of sign, $K$ is exactly the quantity called '*the offence against the sine-condition*' by A. E. Conrady in Part I of his work *Applied Optics and Optical Design* (Dover Publications, 1957, chapter 7, p. 370), and denoted by him by the composite symbol OSC$'$. It is rather striking that so simple a relation as $K = -\text{OSC}'$ should obtain exactly, since the details of Conrady's work are quite difficult to compare with what has been done above. This is due in part to the fact that his 'coma' is not directly defined in terms of displacements in the ideal image plane. Again, we have taken sagittal rays to be those for which $\theta = \pm\frac{1}{2}\pi$, but Conrady defines them to be rays which have $\theta_e = \pm\frac{1}{2}\pi$, where $\theta_e$ is an angular coordinate in the *entrance* pupil; and these alternative definitions are not exactly equivalent.

At any rate, one must be careful not to interpret the phrase 'offence against the sine-condition' in the light of (28.1); for when $K = 0$ the quantity $\Gamma(\rho)$ of equation (28.1) will have the value $-\beta'\delta$, rather than zero. In other words, one is essentially concerned with another 'modified sine-condition' $\Gamma^*(\rho) = 0$, where $\Gamma^*$ is like $\Gamma$, but with the magnification $m$ replaced by the 'apparent magnification' $m(1+\delta)$. (With an arbitrary unit of length one must of course write $m(d'+\delta)$ here.) Thus

$$\Gamma^*(\rho) = \sin\phi' - [m(1+\delta)]^{-1}\sin\phi = -K\sin\phi'. \qquad (31.7)$$

In conclusion we restate the main result of this section: if $K$ has been calculated from (31.5) for an axial ray given by some value of $\rho$, then, irrespectively of the presence of spherical aberration, the circular comatic displacement $\kappa_s$ of sagittal rays, specified by the same value of $\rho$ and by the ideal image height $h'$, is *exactly* $Kh'$.

## 32. Offence against the sine-condition and tangential circular coma

We have just seen that the trace of a single axial ray of 'semi-aperture' $\rho$ allows one to calculate the offence against the sine-condition $K$, and that this is exactly equivalent to a knowledge of the total circular coma $\kappa_s$ for *sagittal* rays of the same semi-aperture. Unfortunately it does not tell us anything of substance about the *tangential* circular coma $\kappa_t$. It is easy to see why this should be so. According to (31.2) a single axial ray trace allows us to evaluate the functions $P$ and $R$, and hence $K$, for the value of $\rho$ in question. It cannot, however, yield the value of $Q$; yet this quantity enters into $\kappa_t$. Evidently, to calculate $\kappa_t$ we require, at least implicitly, the value of the *derivative* of $C$. In effect this means that the best we can do is to trace *several* axial rays, and obtain the value of $\dot{C}$, or of some quantity equivalent to it, by a process of interpolation. Although one is thus confronted with a problem somewhat more involved than that relating to the calculation of $\kappa_s$, it should be borne in mind that we continue to operate with *axial* ray traces alone. We proceed to develop the required details.

According to (26.7), if $\bar{K} = \kappa_t/h'$,

$$\bar{K} = \xi J^3(P^2R - Q) + K. \tag{32.1}$$

It is easily confirmed that

$$wJ = 1/\alpha' = \sec\phi', \tag{32.2}$$

so that, using the definition of $J$,

$$\rho JP = -\tan\phi'. \tag{32.3}$$

Again, $$Q = 1 - 2w^3\dot{C} = -2w^3 d(R/w)/d\xi,$$

whence $$\begin{aligned}\xi J^3 Q &= -2w^3J^3\xi d(R/w)/d\xi \\ &= -\sec^3\phi'\rho d[(K+1)\cos\phi']/d\rho,\end{aligned} \tag{32.4}$$

because of (31.1) and (32.2). (32.1) now becomes

$$\bar{K} = \tan^2\phi' + K\sec^2\phi' + \sec^3\phi'\rho d[(K+1)\cos\phi']/d\rho. \tag{32.5}$$

This equation can be rewritten in an illuminating form if one introduces the longitudinal spherical aberration $\delta$, though this quantity is strictly speaking redundant here. Thus one has

$$\delta = -1 - \rho\cot\phi', \tag{32.6}$$

which is consistent with (31.4) and (32.3). One then obtains in place of (32.5) the equivalent equation

$$\bar{K} = \sec^3\phi'\left(\frac{d}{d\rho}(\rho\cos\phi' K) - \sin^3\phi'\frac{d\delta}{d\rho}\right), \tag{32.7}$$

which constitutes the desired result; see also equation (33.5).

To calculate $\bar{K}$ one has to plot $\rho\cos\phi' K$ and $\delta$ against $\rho$ by reading these quantities off from a number of exact *axial* ray traces. (These will usually have been obtained in any event, since one is not likely to be content with the value of $K$ for one single value of $\rho$.) The slopes of the curves can be read off directly for any chosen value of $\rho$, and then the value of $\bar{K}$ follows immediately.

Having thus calculated the values of $K$ and $\bar{K}$ the total displacement for *all* values of $\rho$ and $\theta$ is given by

$$\left.\begin{aligned}\epsilon_y' &= \sigma\cos\theta + \tfrac{1}{2}[(\bar{K}+K) + (\bar{K}-K)\cos 2\theta]\,h', \\ \epsilon_z' &= \sigma\sin\theta + \tfrac{1}{2}(\bar{K}-K)\sin 2\theta\, h',\end{aligned}\right\} \tag{32.8}$$

terms not depending linearly on $h'$ having of course been rejected as before. In conclusion we note that when spherical aberration is absent (32.7) reduces to

$$\bar{K} = \frac{d(\beta' K)}{d\beta'}. \tag{32.9}$$

## 33. Two theorems relating to circular coma of all orders

This section concerns itself with two results, the first of which follows trivially from (32.7), but which is nevertheless of considerable interest.

### (i) *On isoplanatism and aplanatism*

A system is called *isoplanatic* if circular coma is completely absent (i.e. there is no asymmetry of the image for sufficiently small values of $h'$); and it is called *aplanatic* if, in addition, spherical aberration is completely absent. (An aplanatic system therefore forms a perfect image of a plane object in $\mathscr{I}$ lying in a sufficiently small neighbourhood of $O_0$.) (32.7) now shows that absence of circular coma, i.e. $K = \bar{K} = 0$ for all values of $\rho$, requires that $\delta = 0$ for all $\rho$. Therefore *an isoplanatic system is necessarily aplanatic.*

It appears that with our strict interpretation of the terms 'aplanatism' and 'isoplanatism' the latter must be regarded as redundant. The distinction arose in the first place by a restriction to third-order displacements alone, for under these circumstances isoplanatism only requires that $p_2$ be zero, whereas aplanatism requires that $p_1$ vanish as well. Likewise, proceeding to the fifth order, (25.1) shows that isoplanatism merely entails that $s_2 = p_1 = p_2 = 0$; but the vanishing of $s_1$ is not required. More generally, if one consistently goes to order $2n-1$ but no further, the presence of the factor $\sin^3\phi' (= O(3))$ in the second term on the right of (32.7) brings with it that isoplanatism imposes no condition on the coefficient of spherical aberration of order $2n-1$. However, the rather artificial nature of this result is evident.

An alternative possibility is to take isoplanatism to mean the absence of the circular comatic term from the aberration function, i.e. $C = 0$. In view of (28.4) there is then no circular comatic pseudo-displacement, and in the primary domain the classical situation obtains. There is, of course, a residual effective comatic displacement. In fact, one finds that

$$\bar{K} = J^3 - 1, \quad K = J - 1. \tag{33.1}$$

Provided $J \neq 1$ the comatic ratio for any value of $\rho$ is therefore

$$\bar{K}/K = J^2 + J + 1. \tag{33.2}$$

For sufficiently small values of $\rho$ the right-hand member of this has the value 3, consistently with (25.3); but the present result is more general since it remains valid even if spherical aberration is absent to some finite order (cf. equation (25.3)).

(ii) *Sine-condition and the comatic ratio of order $2n-1$*

In this section, let $\chi_{2n-1}$ denote the exact circular comatic ratio $\kappa_{t2n-1}/\kappa_{s2n-1}$. We have already seen—for instance in equation (25.2)—that this does not in general have the value $2n-1$ since the coefficients of spherical aberration and circular coma of the lower orders enter into it. The question naturally arises whether a system can be such that $\chi_{2n-1} = 2n-1$ exactly, for all $n$; and under what conditions this situation will obtain. To solve this problem we first rewrite equation (32.7), introducing for this purpose the 'sine-ratio' $\Phi$, defined as

$$\Phi = \sin\phi'/m\sin\phi; \tag{33.3}$$

so that

$$\rho(K+1) = -\Phi\tan\phi'. \tag{33.4}$$

The first term on the left of (32.7) may be transformed as follows:

$$\sec^3\phi'\frac{d}{d\rho}(\rho\cos\phi' K) = \frac{d(\rho K)}{d\rho} - \tan^2\phi'\frac{d(\Phi\tan\phi')}{d\rho}$$
$$-\tan^2\phi' + \rho K\sec^3\phi'\frac{d\cos\phi'}{d\rho}.$$

Then (32.7) becomes

$$\bar{K} = \frac{d(\rho K)}{d\rho} - \tan^3\phi'\frac{d\Phi}{d\rho} + \Big(\rho(K+1)\tan\phi'\frac{d\tan\phi'}{d\rho} - \tan^2\phi'$$
$$+\rho K\sec^3\phi'\frac{d\cos\phi'}{d\rho} + \tan^3\phi'\frac{d(\rho\cot\phi')}{d\rho}\Big).$$

It is readily confirmed that the expression in the large parentheses vanishes identically, so that

$$\bar{K} = \frac{d(\rho K)}{d\rho} - \tan^3\phi'\frac{d\Phi}{d\rho}; \tag{33.5}$$

and the relation

$$\Phi = (1+K)(1+\delta) \tag{33.6}$$

should be borne in mind.

Now the power series for $K$ has the generic form

$$K = \sum_{r=1}^{\infty} K_r \rho^{2r}, \tag{33·7}$$

where $K_r$ relates to order $2r+1$. If $\chi_{2n-1}$ is to have the value $2n-1$ for all $n$, $\bar{K}$, written as a power series, must be

$$\bar{K} = \sum_{r=1}^{\infty} (2r+1) K_r \rho^{2r}. \tag{33·8}$$

When this is the case, however,

$$\bar{K} = \frac{d(\rho K)}{d\rho}, \tag{33·9}$$

by inspection of (33·7) and (33·8). According to (33·5) it then follows that $\Phi$ = const. = 1; conversely, given $\Phi = 1$, (33·8) follows from (33·7). On the other hand the condition $\Phi = 1$ means precisely that the sine-condition is fulfilled. Accordingly we have proved that *the circular comatic ratio of order $2n-1$ has exactly the value $2n-1$ for all $n$ if and only if the sine-condition is satisfied for all values of $\rho$.*

We already know from (28.2) that if $\Phi = 1$ then $2\dot{S}+C = 0$, so that the comatic displacement is implicit in spherical aberration. In fact, recalling (33·6) and (33·9), we see that, exactly,

$$K = -\sigma/\rho, \quad \bar{K} = -d\sigma/d\rho. \tag{33·10}$$

## 34. Cosine-relations and cosine-conditions

The preceding discussion of the sine-condition was presented at some length largely because it has been a useful tool in design problems in the past; and partly to analyze clearly its function and meaning. In particular we wanted to bring out clearly its direct connection with total sagittal rather than tangential coma; the way the presence of spherical aberration is allowed for; and the fact that it concerns itself with circular coma alone, and so has nothing to do with non-linear comatic types.

It is possible to obtain generalizations of the sine-condition which are intended to say something about non-linear comatic types. On the other hand, they seem to lack any sort of usefulness in practice, and we therefore do not deal with them. However, for the sake of completeness we shall briefly discuss the so-called cosine-relations.

As in Section 27, consider first a pair of perfect conjugate planes, so that equation (27.2) obtains. The right-hand member of this depends on $\mathbf{y}_1$ only. It follows that for all rays through any point $O$ of the object

$$\beta - m\beta' = \text{const.}, \quad \gamma - m\gamma' = \text{const.}, \tag{34.1}$$

and these are the *cosine-relations*. The requirement that (34.1) should hold for all rays through any given point of $\mathscr{I}$ constitutes the *cosine-conditions*. They require that $(\partial V/\partial \mathbf{y}') + (\partial V/\partial \mathbf{y}_1)$ depend upon $\mathbf{y}_1$ only, i.e.

$$\mathbf{y}'(2V_\xi + V_\eta) + \mathbf{y}_1(V_\eta + 2V_\zeta) = \text{function of } \mathbf{y}_1, \tag{34.2}$$

for arbitrary values of $\mathbf{y}_1$. (34.2) splits up into the two separate conditions

$$2V_\xi + V_\eta = 0, \quad V_\eta + 2V_\zeta = \text{function of } \zeta, \tag{34.3}$$

and these are satisfied if, and only if, $V$ is the sum of a function of $\zeta$ alone and a function of $u\,(= \xi - 2\eta + \zeta)$ alone. Accordingly

$$V = g(\zeta) - (1+u)^{\frac{1}{2}} + v(u). \tag{34.4}$$

Evidently satisfaction of the cosine-conditions does not imply absence of aberrations. Given (34.4), the exact displacement is

$$\boldsymbol{\epsilon}' = (\mathbf{y}' - \mathbf{y}_1)\,[1 - (1-q)(1 + 2uq - uq^2)^{-\frac{1}{2}}], \tag{34.5}$$

where $q = 2(1+u)^{\frac{1}{2}}\,dv/du$. If $\sigma(\rho)$ is the usual spherical aberration $\epsilon'_y\,(\rho, h' = 0, \theta = 0)$, (34.5) may be written

$$\boldsymbol{\epsilon}' = (\mathbf{y}' - \mathbf{y}_1)\,u^{-\frac{1}{2}}\,\sigma(u^{\frac{1}{2}}), \tag{34.6}$$

though we might equally well have introduced distortion, for instance. At any rate, in the absence of spherical aberration satisfaction of the cosine-conditions ensures perfect imagery for the conjugate planes in question.

The last result is a generalization of the condition $2\dot{S} + C = 0$ which we met in Section 28. It is instructive to recover it here. We have

$$v(u) = v(\xi) - 2v_\xi(\xi)\,\eta + \ldots,$$

i.e. $S(\xi) = v(\xi)$, $C(\xi) = -2v_\xi(\xi)$, and so $2\dot{S} + C = 0$, as was to be shown.

## B. The spherical point characteristic $V^{\dagger}$, wavefronts and aberration functions

### 35. The characteristic function $V(x', y', z', y, z)$

Hitherto, from Section 6 onwards, the explicit appearance of the variable $x'$ in the point characteristic $V$ was prevented by the convention of choosing a fixed posterior base-plane, and so in effect simply setting $x' = 0$. On various occasions it is, however, important to know how $V$ in fact depends upon $x'$. For example, given this dependence, that of the aberration coefficients (on $x'$) is also known. By this is meant the following. If the origin of coordinates in the image space is taken at the previous base-point $B'$, we may contemplate a new base-point $\hat{B}'$ which is at a distance $\hat{x}'$ from $B'$. If the coordinates of points in the new base-plane are denoted by $\hat{\mathbf{y}}'$, and the corresponding rotational invariants by $\hat{\xi}, \hat{\eta}, \hat{\zeta}$ (where $\hat{\zeta} \equiv \zeta$, of course) then

$$\hat{V} = g(\hat{\zeta}) - [(1-\hat{x}')^2 + \hat{u}]^{\frac{1}{2}} + \hat{v}(\hat{x}'; \hat{\xi}, \hat{\eta}, \hat{\zeta}). \tag{35.1}$$

$\hat{V}$ is physically the optical distance between points in the object plane and the *new* base-plane. The function $g$ has been taken as fixed by prescription. This is permissible since the paraxial term of $g$ is in any event independent of the position of $B'$, whilst the part of $\hat{v}$ which depends upon $\hat{\zeta}$ alone, i.e. $\hat{v}(\hat{x}'; 0, 0, \hat{\zeta})$, will take care of all other terms. Except for the ubiquitous provision of circumflexes, the representation of $\hat{v}$ as a power series is formally the same as before. The aberration coefficients are thus $\hat{v}_{\mu\nu}^{(n)}(\hat{x}')$, their dependence upon $\hat{x}'$ now being indicated explicitly. In particular,

$$v_{\mu\nu}^{(n)} = \hat{v}_{\mu\nu}^{(n)}(0). \tag{35.2}$$

One way of proceeding from here would be to insert (35.1) in the first of the differential equations (3.6). The result is a sequence of comparatively simple *ordinary* differential equations for the $\hat{v}_{\mu\nu}^{(n)}$, obtained by equating the factor multiplying $\hat{\xi}^{n-\mu}\hat{\eta}^{\mu-\nu}\hat{\zeta}^{\nu}$ to zero. These equations must then be solved, subject to the initial conditions (35.2). It turns out, however, that this method is unnecessarily cumbersome, and we abandon it.

A more advantageous procedure is the following. Independently of the variables on which the functions $\hat{V}$ and $V$ depend, we may write

$$\hat{V} = (V+\hat{x}')/\alpha'. \tag{35·3}$$

On using (35.1) this gives

$$\hat{v} = v+[(1-\hat{x}')^2+\hat{u}]^{\frac{1}{2}}-(1+u)^{\frac{1}{2}}+\hat{x}'/\alpha'. \tag{35·4}$$

Though we are ultimately aiming at $\hat{v}$, regarded as a function of $\hat{\xi}$, $\hat{\eta}$, $\hat{\zeta}$, we evaluate the right-hand member in the first place as a function of $\xi$, $\eta$, $\zeta$. We temporarily introduce the abbreviations

$$A = 2(\mathbf{y}_1-\mathbf{y}').\boldsymbol{\epsilon}', \quad B = \boldsymbol{\epsilon}'.\boldsymbol{\epsilon}' \tag{35·5}$$

in terms of the notation suggested just after equation (14.4). Note that $A$ and $B$ are $O(4)$ and $O(6)$ respectively. In the first place we now have

$$1/\alpha' = (1+u+A+B)^{\frac{1}{2}}, \tag{35·6}$$

since

$$\boldsymbol{\beta}'/\alpha' = \mathbf{y}_1-\mathbf{y}'+\boldsymbol{\epsilon}', \tag{35·7}$$

which is just (17.1). Again

$$\hat{\mathbf{y}}'-\mathbf{y}_1 = (1-\hat{x}')(\mathbf{y}'-\mathbf{y}_1)+\hat{x}'\boldsymbol{\epsilon}', \tag{35·8}$$

so that

$$\hat{u} = (1-\hat{x}')^2(u+qA+q^2B), \tag{35·9}$$

where

$$q = -\hat{x}'/(1-\hat{x}'). \tag{35·10}$$

(35·4) now becomes

$$\hat{v} = v+(1+u)^{\frac{1}{2}}\{(1-\hat{x}')[1+(1+u)^{-1}(qA+q^2B)]^{\frac{1}{2}}$$
$$+\hat{x}'[1+(1+u)^{-1}(A+B)]^{\frac{1}{2}}-1\}.$$

Every square root of the generic form $(1+X)^{\frac{1}{2}}$ is now to be written as $1+\frac{1}{2}X-\frac{1}{8}X^2\ldots$; and it is after *this* step that the various orders are to be sorted out. The constant terms of $\hat{v}-v$ cancel out, as do the terms linear in $A$. One thus gets

$$\hat{v} = v-\tfrac{1}{2}q(\epsilon_{y3}'^2+\epsilon_{z3}'^2)+O(8), \tag{35·11}$$

which implies reasonable simplicity for orders three and five; and we confine our attention to these.

It remains to express $\xi$, $\eta$, $\zeta$ in terms of $\hat{\xi}$, $\hat{\eta}$, $\hat{\zeta}$. At this stage we recognize the virtues of having allowed $\boldsymbol{\epsilon}'$ to appear on the right of (35.11). At first sight its presence might seem to be a great complication here, bearing in mind that $\boldsymbol{\epsilon}'$ is related to the derivatives of $v$ in a complicated way. However, the situation is in fact quite simple. In the first place, the displacement is a quantity independent of the manner in which it is expressed; so that $\boldsymbol{\epsilon}'$ may simply be replaced by $\hat{\boldsymbol{\epsilon}}'$. This means, explicitly, that $\boldsymbol{\epsilon}' = 4d'p_1\mathbf{y}'\xi + \dots$ (cf. (19.2)) is to be replaced by $\hat{\boldsymbol{\epsilon}}' = 4\hat{d}'\hat{p}_1\hat{\mathbf{y}}'\hat{\xi} + \dots$. In other words, since here $d' = 1$ and $\hat{d}' = 1 - \hat{x}'$, $\hat{\boldsymbol{\epsilon}}'$ is formally obtained from $\boldsymbol{\epsilon}'$ by supplying *all* its variables and coefficients with a circumflex, and providing one common additional factor $1 - \hat{x}'$. How $\hat{\boldsymbol{\epsilon}}'$ depends on the $\hat{v}^{(n)}_{\mu\nu}$ is already known from the work of Section 23. On the other hand, the terms of lowest degree, i.e. those of degree 4, appear on the right of (35.11) only in its first term; so that this alone needs to be retained to determine the $\hat{v}^{(2)}_{\mu\nu}$. Going to terms of degree six, the coefficients which occur in the *second* term on the right of (35.11) are already known. Evidently one has a comparatively simple iterative process for getting the $\hat{v}^{(3)}_{\mu\nu}$, $\hat{v}^{(4)}_{\mu\nu}$, ... in turn, each step making the greatest possible use of the work done at preceding steps.

From equation (35.8) we easily find that

$$\left.\begin{aligned} \xi &= c^2\hat{\xi} + 2cq\hat{\eta} + q^2\hat{\zeta} + 2q(c\hat{\mathbf{y}}' + q\mathbf{y}_1)\,.\,\hat{\boldsymbol{\epsilon}}' + q^2\hat{\boldsymbol{\epsilon}}'\,.\,\hat{\boldsymbol{\epsilon}}', \\ \eta &= c\hat{\eta} + q\hat{\zeta} + q\hat{\boldsymbol{\epsilon}}'\,.\,\mathbf{y}_1, \\ \zeta &= \hat{\zeta}, \end{aligned}\right\} \tag{35.12}$$

where $c = (1 - \hat{x}')^{-1} = 1 - q$.

As far as the primary coefficients are concerned, the terms involving $\hat{\boldsymbol{\epsilon}}'$ in (35.11) and (35.12) are to be ignored, and one is left with a simple linear substitution. This at once leads to the relations

$$\left.\begin{aligned} \hat{p}_1 &= c^4 p_1, \\ \hat{p}_2 &= c^3(p_2 + 4qp_1), \\ \hat{p}_3 &= c^2(p_3 + qp_2 + 2q^2p_1), \\ \hat{p}_4 &= c^2(p_4 + 2qp_2 + 4q^2p_1), \\ \hat{p}_5 &= c(p_5 + 2qp_4 + 2qp_3 + 3q^2p_2 + 4q^3p_1), \\ \hat{p}_6 &= (p_6 + qp_5 + q^2p_4 + q^2p_3 + q^3p_2 + q^4p_1). \end{aligned}\right\} \tag{35.13}$$

Proceeding to the fifth-order coefficients, the terms of degree 4 in (35.12) must be retained. The equations corresponding to (35.13) become very lengthy and we therefore give the results for

$s_1$ and $s_2$ only. Thus

$$\hat{s}_1 = c^6(s_1 + 8qp_1^2),$$
$$\hat{s}_2 = c^5[(s_2 + 6qs_1) + 48q^2p_1^2 + 12qp_1p_2]. \tag{35.14}$$

We shall have occasion to contemplate these equations later in a different context. At any rate, we have now the means systematically to determine the functions $\hat{v}_{\mu\nu}^{(n)}(\hat{x}')$ to whatever order required.

Let this have been done. Then we change the notation as follows: we omit the circumflex everywhere, for it is now redundant. However, some distinguishing mark has to be added to the constants previously denoted by $v_{\mu\nu}^{(n)}$ and we write them here as $\mathring{v}_{\mu\nu}^{(n)}$; that is to say, the $\mathring{v}_{\mu\nu}^{(n)}$ are the aberration coefficients relating to the usual base-plane $\bar{x}' = 0$; and one should therefore now write $\mathring{v}$ in place of $v$. In short, what we have before us is exactly the point characteristic $V(x', y', z', y, z)$, with its dependence on $x'$ explicitly restored. Thus

$$V(x', y', z', y, z) = g(\zeta) - [(1-x')^2 + \xi - 2\eta + \zeta]^{\frac{1}{2}}$$
$$+ \sum_{n=2}^{\infty} \sum_{\mu=0}^{n} \sum_{\nu=0}^{\mu} v_{\mu\nu}^{(n)}(x')\, \xi^{n-\mu}\eta^{\mu-\nu}\zeta^{\nu}. \tag{35.15}$$

## 36. Arbitrary posterior base-surfaces

The conventional point characteristic $V(y', z', y, z)$ is recovered from (35.15) by simply setting $x' = 0$. There is, however, no reason why we should be restricted to taking some normal plane to serve as posterior reference surface—if, indeed, we want to choose such a surface at all. On the contrary, we are not prevented from setting

$$x' = \chi(y', z', y_1, z_1) \tag{36.1}$$

in (35.15), where $\chi$ can be chosen in any convenient way at all, subject, of course, to the usual condition of non-conjugacy. Any ray is then specified by the values of $\mathbf{y}_1$ and of the coordinates $\mathbf{y}'$ of its point of intersection $D'$ with the surface $\mathscr{B}^{*\prime}$ whose equation is $\bar{x}' = \chi(\bar{y}', \bar{z}', y_1, z_1)$. Then, using (36.1) and taking $\bar{x} = 0$ as usual, (4.1) becomes

$$dV_1 = \alpha'\left(\frac{\partial\chi}{\partial y'}dy' + \frac{\partial\chi}{\partial z'}dz' + \frac{\partial\chi}{\partial y}dy + \frac{\partial\chi}{\partial z}dz\right)$$
$$+ \beta' dy' + \gamma' dz' - \beta\, dy - \gamma\, dz,$$

the subscript on the left serving as a reminder that the base-surface is now $\mathscr{B}^{*\prime}$. In place of (6.1) we have here

$$\boldsymbol{\beta}' = \frac{\partial V_1}{\partial \mathbf{y}'} - \alpha' \frac{\partial \chi}{\partial \mathbf{y}'}, \quad \boldsymbol{\beta} = -\frac{\partial V_1}{\partial \mathbf{y}} + \alpha' \frac{\partial \chi}{\partial \mathbf{y}}. \tag{36.2}$$

Further, the aberration function $v_1$ is defined in the usual way by the equation

$$V_1 = g(\zeta) - \{[1 - \chi(y', z', y_1, z_1)]^2 + \xi - 2\eta + \zeta\}^{\frac{1}{2}} + v_1. \tag{36.3}$$

It should be noted that $\chi$ need not be expressible as a power series in $\xi$, $\eta$, $\zeta$, though, when it is not, much of the formal simplicity of the theory of symmetric systems will be lost. This would be the case, for instance, if $\mathscr{B}^{*\prime}$ were taken to be a *fixed* plane not normal to the axis of $K$ (cf. the discussion of Section 8, relating to the idea of symmetry). Simplicity can, however, be restored by taking such a plane to be 'movable'; i.e. one might require it to pass through $E'$, with its normal pointing in a direction parallel to the line $E'I'$, for then one has simply $\chi = -\eta$. At any rate, we see the reason for letting $\chi$ depend also on $\mathbf{y}_1$ in (36.1).

In (36.3) we have followed the usual prescription of defining the aberration function as the difference between the actual and the ideal characteristic functions. One might think of the ideal characteristic as 'modified' by the addition of the aberration function. It is obviously not mandatory that such a modification should be of just this kind; and in place of (36.3) one could, for instance, write

$$V_1 = g(\zeta) - \{[1 - \chi(y', z', y_1, z_1)]^2 + \xi - 2\eta + \zeta - 2\tilde{v}_1\}^{\frac{1}{2}}. \tag{36.4}$$

This, then, is one possible definition of a *modified aberration function* $\tilde{v}_1$; and we shall meet a special case of it at the end of the next section.

## 37. The spherical point characteristic $V^\dagger$

Inspection of (36.3) at once suggests an advantageous choice of base-surface, namely that which, in effect, removes the awkward square root altogether. This obviously requires that we take $\mathscr{B}^{*\prime}$ to be a spherical surface with centre at $I'$; so that it is movable in the sense of the discussion at the end of Section 36. When it is normalized by the requirement that it pass through $E'$, we denote it by $\mathscr{W}_0$ and the corresponding characteristic function $V_1$ by $V^\dagger$. It is natural to call $V^\dagger$ the *spherical point characteristic.* (36.1) now reads specifically

$$x' = 1-(1-\xi+2\eta)^{\frac{1}{2}}. \tag{37.1}$$

On inserting this in (36.3) there comes

$$V^\dagger = g^\dagger(\zeta)+v^\dagger, \tag{37.2}$$

where $g^\dagger(\zeta) = g(\zeta)-(1+\zeta)^{\frac{1}{2}}$. The relations (36.2) become

$$\boldsymbol{\beta}' = \partial v^\dagger/\partial \mathbf{y}' - \alpha'(1-\xi+2\eta)^{-\frac{1}{2}}(\mathbf{y}'-\mathbf{y}_1). \tag{37.3}$$

As usual $\boldsymbol{\epsilon}' = \mathbf{y}'-\mathbf{y}_1+(1-x')\boldsymbol{\beta}'/\alpha'$, and in view of (37.1) and (37.3) we get the result

$$\boldsymbol{\epsilon}' = X'\frac{\partial v^\dagger}{\partial \mathbf{y}'}, \tag{37.4}$$

where

$$X' = (1-x')/\alpha'. \tag{37.5}$$

The simplicity of (37.4) is striking. In the first place, the exact displacement is just the derivative of $v^\dagger$, all terms of the power series for which are merely to be multiplied by the *common* factor $X'$. Moreover, geometrically this is just the distance from $D'$ to $O'$ (see also Fig. 5.1), which must be nearly constant in practice if an image of reasonable quality is to be formed in $\mathscr{I}'$. Usually it will therefore be entirely adequate when calculating the total displacement to replace $X'$ by the *constant* factor $R'$, which is the distance from $E'$ to $I'$. (Of course $R'$ is a different constant for different values of $\zeta$.)

Even when the separate orders which go to make up $\boldsymbol{\epsilon}'$ are contemplated in the usual way, the situation is rather simple. Thus we have

$$X' = [(1-x')^2+(Y'-y')^2+(Z'-z')^2]^{\frac{1}{2}} = (1+\zeta+A+B)^{\frac{1}{2}}, \tag{37.6}$$

in the notation of equations (35.5). It follows that

$$X' = (1+\zeta)^{\frac{1}{2}}+\tfrac{1}{2}A+O(6). \tag{37.7}$$

The equations for the effective coefficients corresponding to (23.2) and (23.3) are obtained very easily here. The power series for $v^\dagger$ is written in the usual way, all coefficients having a dagger as additional superscript. The relations (23.2) can be taken over as they stand, provided one writes $p_\alpha^\dagger$ in place of $p_\alpha$ ($\alpha = 1, \ldots, 5$), at the same time replacing the asterisk by a 'double-dagger' ‡. This notation is required by the fact that the present effective coefficients relate to a series which is generically of the same form as (23.1), but in which the coordinates $\mathbf{y}'$ are not standard, i.e. do not relate to $\mathscr{E}'$. Hence one cannot write $u_{\mu\nu}^{(n)}, \bar{u}_{\mu\nu}^{(n)}$, or $p_\alpha^*, \bar{p}_\alpha^*, s_\alpha^*, \ldots$ for the effective coefficients here also; and strictly speaking $\mathbf{y}'$, $\xi$, $\eta$, $\zeta$ also should all be supplied with a dagger.

As regards the secondary relations, the primary coefficients which enter into them are induced solely by the simple term $\frac{1}{2}\zeta\boldsymbol{\epsilon}_3'$, so that, virtually by inspection,

$$\left.\begin{aligned} s_1^\ddagger &= 6s_1^\dagger, & \bar{s}_1^\ddagger &= s_2^\dagger,\\ s_2^\ddagger &= 4s_2^\dagger, & \bar{s}_2^\ddagger &= 2s_4^\dagger,\\ s_3^\ddagger &= 4s_3^\dagger + 2p_1^\dagger, & \bar{s}_3^\ddagger &= s_5^\dagger + \tfrac{1}{2}p_2^\dagger,\\ s_4^\ddagger &= 2s_4^\dagger, & \bar{s}_4^\ddagger &= 3s_7^\dagger,\\ s_5^\ddagger &= 2s_5^\dagger + p_2^\dagger, & \bar{s}_5^\ddagger &= 2s_8^\dagger + p_4^\dagger,\\ s_6^\ddagger &= 2s_6^\dagger + p_3^\dagger, & \bar{s}_6^\ddagger &= s_9^\dagger + \tfrac{1}{2}p_5^\dagger. \end{aligned}\right\} \tag{37.8}$$

Again, the contributions by the third- and fifth-order coefficients to the effective coefficients of order seven arise from the comparatively very simple expression

$$[(\mathbf{y}_1 - \mathbf{y}')\cdot\boldsymbol{\epsilon}_3' - \tfrac{3}{8}\zeta^2]\,\boldsymbol{\epsilon}_3' + \tfrac{1}{2}\zeta\boldsymbol{\epsilon}_5',$$

bearing in mind that $\quad \partial v^{(3)\dagger}/\partial \mathbf{y}' = \boldsymbol{\epsilon}_5' - \tfrac{1}{2}\zeta\boldsymbol{\epsilon}_3'.$

Here one writes everything first in terms of the $p_\alpha^\ddagger, \ldots$, and only reverts to the $p_\alpha^\dagger, \ldots$, in the end. In this way one finds

$$\begin{aligned} t_1^\ddagger &= 8t_1^\dagger - 16p_1^{\dagger 2},\\ t_2^\ddagger &= 6t_2^\dagger + 16p_1^{\dagger 2} - 20p_1^\dagger p_2^\dagger,\\ t_3^\ddagger &= 6t_3^\dagger + 3s_1^\dagger + 4p_1^\dagger p_2^\dagger - 16p_1^\dagger p_3^\dagger,\\ t_4^\ddagger &= 4t_4^\dagger + 16p_1^\dagger p_2^\dagger - 8p_1^\dagger p_4^\dagger - 6p_2^{\dagger 2},\\ t_5^\ddagger &= 4t_5^\dagger + 2s_2^\dagger + 16p_1^\dagger p_3^\dagger + 8p_1^\dagger p_4^\dagger - 4p_1^\dagger p_5^\dagger + 2p_2^{\dagger 2} - 10p_2^\dagger p_3^\dagger,\\ t_6^\ddagger &= 4t_6^\dagger + 2s_3^\dagger + 4p_1^\dagger p_5^\dagger + 2p_2^\dagger p_3^\dagger - 4p_3^{\dagger 2} - \tfrac{1}{2}p_1^\dagger,\\ t_7^\ddagger &= 2t_7^\dagger + 4p_2^{\dagger 2} - 4p_2^\dagger p_4^\dagger,\\ t_8^\ddagger &= 2t_8^\dagger + s_4^\dagger + 8p_2^\dagger p_3^\dagger + 4p_2^\dagger p_4^\dagger - 2p_2^\dagger p_5^\dagger - 4p_3^\dagger p_4^\dagger, \end{aligned}$$

$$t_9^{\ddagger} = 2t_9^{\dagger} + s_5^{\dagger} + 2p_2^{\dagger}p_5^{\dagger} + 4p_3^{\dagger 2} + 4p_3^{\dagger}p_4^{\dagger} - 2p_3^{\dagger}p_5^{\dagger} - \tfrac{1}{4}p_2^{\dagger},$$
$$t_{10}^{\ddagger} = 2t_{10}^{\dagger} + s_6^{\dagger} + 2p_3^{\dagger}p_5^{\dagger} - \tfrac{1}{4}p_3^{\dagger},$$

$$\bar{t}_1^{\ddagger} = t_2^{\dagger} - 4p_1^{\dagger}p_2^{\dagger},$$
$$\bar{t}_2^{\ddagger} = 2t_4^{\dagger} + 4p_1^{\dagger}p_2^{\dagger} - 8p_1^{\dagger}p_4^{\dagger} - 3p_2^{\dagger 2},$$
$$\bar{t}_3^{\ddagger} = t_5^{\dagger} + \tfrac{1}{2}s_2^{\dagger} - 4p_1^{\dagger}p_5^{\dagger} + p_2^{\dagger 2} - 2p_2^{\dagger}p_3^{\dagger},$$
$$\bar{t}_4^{\ddagger} = 3t_7^{\dagger} + 8p_1^{\dagger}p_4^{\dagger} + 2p_2^{\dagger 2} - 8p_2^{\dagger}p_4^{\dagger},$$
$$\bar{t}_5^{\ddagger} = 2t_8^{\dagger} + s_4^{\dagger} + 4p_1^{\dagger}p_5^{\dagger} + 2p_2^{\dagger}p_3^{\dagger} + 4p_2^{\dagger}p_4^{\dagger} - 4p_2^{\dagger}p_5^{\dagger} - 4p_3^{\dagger}p_4^{\dagger},$$
$$\bar{t}_6^{\ddagger} = t_9^{\dagger} + \tfrac{1}{2}s_5^{\dagger} + 2p_2^{\dagger}p_5^{\dagger} - 2p_3^{\dagger}p_5^{\dagger} - \tfrac{1}{8}p_2^{\dagger},$$
$$\bar{t}_7^{\ddagger} = 4t_{11}^{\dagger} + 4p_2^{\dagger}p_4^{\dagger} - 4p_4^{\dagger 2},$$
$$\bar{t}_8^{\ddagger} = 3t_{12}^{\dagger} + \tfrac{3}{2}s_7^{\dagger} + 2p_2^{\dagger}p_5^{\dagger} + 4p_3^{\dagger}p_4^{\dagger} + 4p_4^{\dagger 2} - 4p_4^{\dagger}p_5^{\dagger},$$
$$\bar{t}_9^{\ddagger} = 2t_{13}^{\dagger} + s_8^{\dagger} + 2p_3^{\dagger}p_5^{\dagger} + 4p_4^{\dagger}p_5^{\dagger} - p_5^{\dagger 2} - \tfrac{1}{4}p_4^{\dagger},$$
$$\bar{t}_{10}^{\ddagger} = t_{14}^{\dagger} + \tfrac{1}{2}s_9^{\dagger} + p_5^{\dagger 2} - \tfrac{1}{8}p_5^{\dagger}. \qquad (37{\cdot}9)$$

The relations (37.8) and (37.9) correspond exactly to (23.3) and (23.4), but are evidently much simpler than these. It may also be noted, with regard to the identities referred to at the end of Section 23, that equation (23.8) is valid here even in the secondary domain, on account of (37.7). The tertiary identities also are still quite simple, e.g. $t_2^{\ddagger} - 6\bar{t}_1^{\ddagger} = 16p_1^2 + 4p_1p_2$.

The examination and classification of aberrations presented earlier could of course equally well have been undertaken on the basis of $v^{\dagger}$ rather than $v$, without any effect whatever on the substance of our discussion. On the other hand the use of $V^{\dagger}$ rather than $V$ commends itself for a variety of reasons, not the least of which is that the pseudo-displacement, regarded as a function of coordinates in $\mathscr{W}_0$, is likely to be a much closer approximation to the true displacement than when it is considered as a function of coordinates in $\mathscr{E}'$. This point is vividly brought out by contemplating the modified aberration function $\tilde{v}^{\dagger}$ of equation (36.4) in place of $v^{\dagger}$. One confirms easily that

$$\tilde{v}^{\dagger} = (1+\zeta)^{\frac{1}{2}}v^{\dagger} - \tfrac{1}{2}v^{\dagger 2}, \qquad (37{\cdot}10)$$

and then (37.4) becomes

$$\boldsymbol{\epsilon}' = \frac{\partial \tilde{v}^{\dagger}}{\partial \mathbf{y}'} + O(7). \qquad (37{\cdot}11)$$

If the characteristic coefficients are now understood to refer to $\tilde{v}^{\dagger}$, there will be no primary coefficients in the equations (37.8), whilst the secondary coefficients and the terms linear in the primary coefficients will be absent from (37.9). In particular, up to the *fifth* order the true and pseudo-displacements are exactly equal. Other reasons for the preferential standing of $V^\dagger$ will appear later.

## 38. Equations of the wavefront, and the interpretation of the aberration functions

Recall from Section 1 that the rays from a given object point constitute a normal congruence throughout the optical system; that is to say, they are orthogonal to a set of surfaces, here called *wavefronts*. We focus our attention on the particular wavefront $\mathscr{W}$ which passes through $E'$. Evidently, when we have a pair of perfect conjugate planes, $\mathscr{W}$ will coincide with the surface $\mathscr{W}_0$ of Section 37; and $\mathscr{W}_0$ may therefore appropriately be referred to as the *ideal wavefront*. (The 'ideality of the wavefront', regarded as a function of $\mathbf{y}_1$, has the same inherent degree of arbitrariness as that in the aberration function, as discussed at the end of Section 7.) The non-vanishing of the aberration functions goes hand in hand with deviations of $\mathscr{W}$ from $\mathscr{W}_0$, and it is therefore of interest formally to examine the relationship between this deviation and the aberration functions. A number of alternatives arise, and we consider them in turn. Considerable care has to be exercised here, on account of the fact that at various times one is simultaneously contemplating the specification of rays by coordinates referring to either $\mathscr{E}'$, or $\mathscr{W}$, or $\mathscr{W}_0$, and this is a situation which can very easily give rise to endless confusion, since one is inclined to write $\mathbf{y}'$ every time for the coordinates in question. It would probably be better to write $\mathbf{y}'$, $\hat{\mathbf{y}}'$, and $\mathbf{y}^{\dagger\prime}$, respectively, but we shall use these symbols only for occasional reference.

### (i) *The equation* $V(x',y',z',y,z) = $ *constant*

Referring to equation (35.15), we see immediately that any wavefront has the equation $V(x',y',z',y,z) = \text{const.}$ The equation of $\mathscr{W}$ is therefore

$$V(x',y',z',y,z) = V(0,0,0,y,z). \tag{38.1}$$

This may be solved for $x'$ as a function of $\xi, \eta, \zeta$ (see the end of this section). We imagine the function $v(0;0,0,\zeta)$ to have been absorbed in $g(\zeta)$; and then (38.1) becomes

$$[(1-x')^2+u]^{\frac{1}{2}} = (1+\zeta)^{\frac{1}{2}}+v(x';\xi,\eta,\zeta). \tag{38.2}$$

Since the coordinates of $Q'$ are $x', y', z'$ (see Fig. 5.1 which, for the sake of clarity, has been drawn for $z' = z_1 = 0$) the left-hand member of (38.2) is just the distance $Q'I'$. The first term on the right is

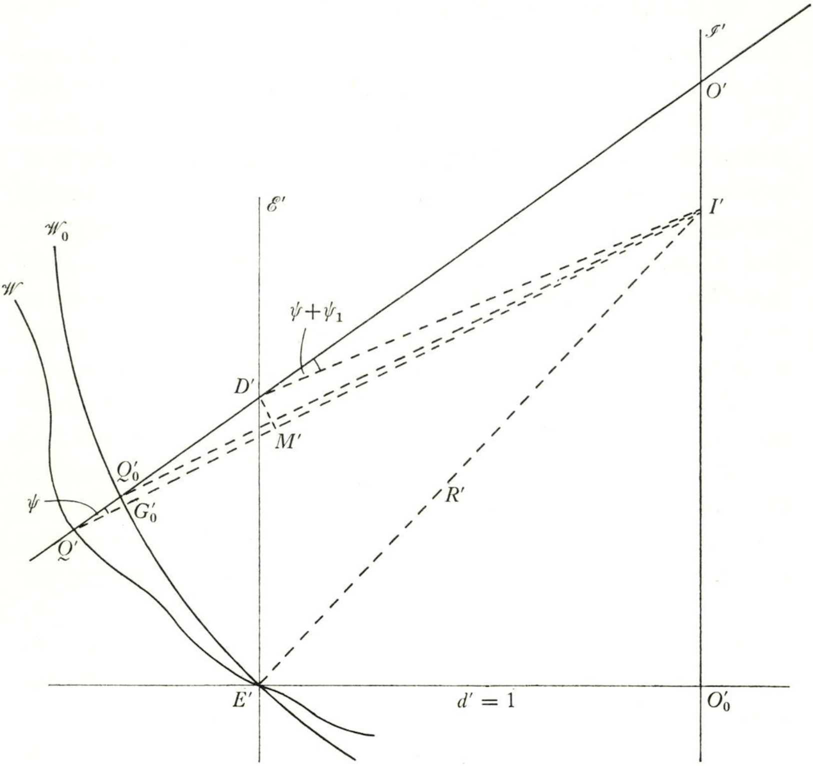


Fig. 5.1

$E'I' = G'_0 I'$, whence the geometrical interpretation of the function $v$ in (38.2) is given by

$$v(x';\xi,\eta,\zeta) = Q'G'_0, \tag{38.3}$$

the distance on the right being, of course, reduced, as always. It should be noted that in (38.3) $x'$ is itself to be written in terms of $\xi$, $\eta$, $\zeta$, i.e. by using the solution of (38.2) for $x'$.

(ii) *The equation* $V(\xi, \eta, \zeta) + s = constant$

Reverting to the ordinary point characteristic $V$ the optical distance from $O$ to any point on a ray $\mathscr{R}$ is $V+s$, where $s$ is measured along $\mathscr{R}$ from its point of intersection $D'$ with $\mathscr{E}'$. On the wavefront $V+s$ is therefore constant, so that the equation of $\mathscr{W}$ is

$$V(\xi, \eta, \zeta) + s = V(0, 0, \zeta). \tag{38.4}$$

Again absorbing $v(0, 0, \zeta)$ in $g(\zeta)$, (38.4) becomes

$$s = (1+u)^{\frac{1}{2}} - (1+\zeta)^{\frac{1}{2}} - \mathring{v}(\xi, \eta, \zeta). \tag{38.5}$$

This equation for $\mathscr{W}$ has a most inconvenient form from a practical point of view, the variables here referring to $\mathscr{E}'$. However, this need not concern us, since we are mainly interested now in obtaining a result corresponding to (38.3). To this end, note that $s = -Q'D'$ by definition, whilst the first and second terms on the right of (38.5) are $D'I'$ and $G'_0I'$ respectively. Hence

$$\mathring{v} = Q'D' + D'I' - G'_0 I'. \tag{38.6}$$

This curious result reflects all the complexities of $\mathring{v}$ as revealed in the cumbersome nature of (23.3) and (23.5). We notice in passing that $g(\zeta)$ is the sum of $E'I'$ and the (optical) distance from $O$ to $E'$.

(iii) *The equation* $V^{\dagger}(\xi, \eta, \zeta) + s = constant$

We proceed exactly as in the case of $V$, and the equation of $\mathscr{W}$ is now

$$s = v^{\dagger}(\xi, \eta, \zeta). \tag{38.7}$$

Bearing in mind the definition of $V^{\dagger}$, we see at once that

$$v^{\dagger} = Q'Q'_0. \tag{38.8}$$

The simplicity of this result as compared with (38.6) is striking. One must not forget, of course, that in (38.3) $v$ appears as a function of $\hat{\mathbf{y}}'$, $\mathring{v}$ in (38.6) as a function of $\mathbf{y}'$, whilst $v^{\dagger}$ in (38.8) is a function of $\mathbf{y}^{\dagger\prime}$. Of course, if desired, we may eventually express any one of them in terms of whatever independent variables we please.

(iv) *Deformation and retardation of the wavefront*

Without referring to any characteristic function in the first place, we can always imagine the equation of the wavefront to be written in the form

$$(1-x')^2 + \xi - 2\eta = 1 + 2D(\xi, \eta, \zeta). \tag{38.9}$$

The function $D$ thus defined will be called the *deformation of the wavefront*. It should be noted that $x'$ does not occur in it explicitly. Evidently $\mathbf{y}'$ really stands for $\hat{\mathbf{y}}'$ here. The direction ratios of the wave-normals, i.e. of the rays, are $-(1-x')$, $\mathbf{y}'-\mathbf{y}_1-\partial D/\partial \mathbf{y}'$, so that the equations of the ray are

$$\bar{x}' = x'-(1-x')\,t, \quad \bar{\mathbf{y}}' = \mathbf{y}'+(\mathbf{y}'-\mathbf{y}_1-\partial D/\partial \mathbf{y}')\,t, \quad (38.10)$$

where $t$ is a parameter. Setting $t = -1$, it follows directly that

$$\boldsymbol{\epsilon}' = \partial D/\partial \mathbf{y}'; \quad (38.11)$$

and this equation is exact. In view of (37.11) $D$ must evidently resemble $\tilde{v}^{\dagger}$ very closely (cf. equation (39.10)); but we must not forget that these functions depend, in the first place, upon $\hat{\mathbf{y}}'$ and $\mathbf{y}^{\dagger\prime}$ respectively. As a matter of fact, $D$ may be looked upon as a modified aberration function in the following sense. In (36.1) choose

$$\chi = 1-(1-\xi+2\eta+2D)^{\frac{1}{2}}, \quad (38.12)$$

i.e. $\mathscr{B}^{*\prime}$ is taken to be $\mathscr{W}$ itself, in view of (38.9). Then (36.4) becomes

$$V_1 = g(\zeta)-(1+\zeta+2D-2\tilde{v}_1)^{\frac{1}{2}}; \quad (38.13)$$

where we must clearly understand that the *actual* wavefront $\mathscr{W}$ is being maintained as base-surface. On the other hand $V_1$ is of course a function of $\zeta$ only, so that we must have $D = \tilde{v}_1$; and the remark above is thus justified. Geometrically

$$2D = (Q'I')^2-(G_0'I')^2. \quad (38.14)$$

Of the various small displacements we have encountered we take $Q'Q_0'$ to be the most important, partly because of the prominent position it occupies in diffraction theory. We call it the *retardation of the wavefront*, and denote it by the symbol $\Delta$.

Equations (38.2) and (38.9) are of course intimately related to one another. In fact, (38.9) constitutes just the solution of (38.2) for the 'unknown' $x'$. If in the latter one therefore replaces $x'$ everywhere by the right-hand member of (38.12), one has an identity from which the relations between the $\mathring{v}^{(n)}_{\mu\nu}$ and the coefficients of the power series of $D$ may be read off.

## 39. The relationship between $\Delta$ and the aberration functions

It is of some importance to have information at hand about the differences which exist between the retardation on the one hand, and $D$ and the various aberration functions on the other. More exactly, we wish to obtain expressions for the dominant terms of these differences, so that we shall know up to what order they may in fact be strictly ignored.

With regard to $Q'G'_0$ we may proceed very simply as follows. Since to a sufficient approximation $Q'_0G'_0$ is normal to $Q'G'_0$,

$$Q'G'_0 = \Delta - \tfrac{1}{2}\psi^2\Delta + \dots, \tag{39.1}$$

where $\psi$ is the angle between $Q'O'$ and $Q'I'$. Writing simply $\epsilon'$ for $(\boldsymbol{\epsilon}'.\boldsymbol{\epsilon}')^{\frac{1}{2}}$, we also have

$$\psi = \epsilon'/R' + \dots. \tag{39.2}$$

Now $\epsilon' = \epsilon'_3 + O(5)$ and $\Delta = O(4)$, so that the second term on the right of (39.1) is $O(10)$. We therefore have

$$v = \Delta - (\epsilon_3'^2/2R'^2)\,\Delta + O(12). \tag{39.3}$$

Thus no distinction need be drawn between $v$ and $\Delta$ as regards the displacement of orders 3, 5 and 7.

Next we come to $\mathring{v}$, where the state of affairs is not quite so straightforward. We first write (38.6) in the form

$$\mathring{v} = (Q'D' - Q'M') + (D'I' - M'I') + Q'G'_0, \tag{39.4}$$

where $M'$ is the foot of the normal from $D'$ onto $Q'I'$. Let $\psi_1$ be the angle between $D'I'$ and $Q'I'$, so that $\psi + \psi_1$ is the angle between $D'O'$ and $D'I'$, i.e.

$$\psi_1 + \psi = \epsilon'/S' + O(5), \tag{39.5}$$

where

$$S' = M'I'. \tag{39.6}$$

Then (39.4) becomes

$$\begin{aligned}\mathring{v} &= Q'M'(\sec\psi - 1) + S'(\sec\psi_1 - 1) + v \\ &= R'(\sec\psi - 1) + S'(\sec\psi_1 - \sec\psi) + v\sec\psi,\end{aligned} \tag{39.7}$$

since $Q'M' = R' + v - M'I'$. So far everything is exact. Now $\psi$ and $\psi_1$ are both $O(3)$, and $v = \Delta + O(10)$, so that (39.7) yields

$$\begin{aligned}\mathring{v} &= \Delta + \tfrac{1}{2}R'\psi^2 + \tfrac{1}{2}S'(\psi_1^2 - \psi^2) + O(10) \\ &= \Delta + \tfrac{1}{2}\epsilon'^2(S'^{-1} - R'^{-1}) + O(10),\end{aligned}$$

because of (39.2) and (39.5). However,

$$S' = D'I' + O(6) = 1 + \tfrac{1}{2}u + O(4),$$

whilst $R' = 1 + \tfrac{1}{2}\zeta + O(4)$. Hence finally

$$\mathring{v} = \Delta - \tfrac{1}{4}(\xi - 2\eta)\,\epsilon_3'^2 + O(10). \tag{39.8}$$

We see that $\mathring{v}$ does not approximate $\Delta$ quite as well as $v$ does.

$v^\dagger$ is of course exactly $\Delta$, by definition, whilst, according to (37.10),

$$\tilde{v}^\dagger = R'\Delta - \tfrac{1}{2}\Delta^2 = \Delta + \tfrac{1}{2}\zeta\Delta + O(8). \tag{39.9}$$

It remains to consider $D$. From (38.14)

$$2D = (R' + v)^2 - R'^2,$$

whence, using (39.3),

$$D = \Delta + \tfrac{1}{2}\zeta\Delta + O(8) = \tilde{v}^\dagger + O(10). \tag{39.10}$$

With regard to the aberration functions $t$ and $w_1$ it will be seen that one cannot deal with them in the same way as with $v$ or $v^\dagger$. The reason for this is that a wavefront corresponds to rays which originate from some object point $O$. However, the condition that they should do so is not easily accommodated when contemplating, say, the constancy of $T + U' - U$, this being $V$, but expressed as a function of the 'wrong' coordinates $\boldsymbol{\beta}'$, $\boldsymbol{\beta}$.

## 40. Remarks on an integral involving Δ. Circle polynomials

In the diffraction theory of aberrations one is naturally confronted with integrals of functions of $\Delta$, extended over a certain part of $\mathscr{W}_0$. Such integrals also occur in the purely geometric theory. A prominent example is the root-mean-square retardation of the wavefront, $\sqrt{Q}$, say. It is usual to use as coordinates $\mathbf{y}^{\dagger\prime}$, or rather a pair closely related to these. ($\mathbf{y}_1$ is left understood throughout.) They are then to be replaced by appropriate polar variables $\rho$ and $\theta$. At any rate, granted that the exit pupil is unvignetted and that the inclination of the line $E'I'$ to the axis (Fig. 5.1) is not too large, one is led to an integral of the form

$$Q = \pi^{-1}\int_0^1\int_0^{2\pi} t\Delta^2\,dt\,d\theta, \tag{40.1}$$

where $t = \rho/\rho_0$ is a normalized radial variable, and $\Delta$ is to be thought of as a function of this and of $\theta$. In place of $\sqrt{Q}$ one may take the root-mean-square deviation $\sqrt{Q_1}$ of the retardation from its mean $\bar{\Delta}$:

$$\bar{\Delta} = \pi^{-1}\int_0^1\int_0^{2\pi} t\Delta\, dt\, d\theta; \tag{40.2}$$

so that $Q_1$ is given by an integral like that in (40.1) but with $\Delta$ replaced by $\Delta - \bar{\Delta}$.

In the geometrical theory $Q$ or $Q_1$ is often referred to as a *merit function* or *performance number*, and the practical problem is to minimize it with respect to the parameters defining the constitution of the system $K$. There appears to be no compelling theoretical foundation for such a procedure. Rather, one is just formulating the intuitional notion that provided $Q$ can be made sufficiently small, $\epsilon'$ will be sufficiently small, and therefore the performance of $K$ adequate. This is certainly true, but the rub lies in the phrase 'sufficiently small'. However, $Q_1$ also appears, this time quite naturally, in the theory of diffraction-limited systems, i.e. systems such that

$$2\pi\Delta \gtrsim \lambda, \tag{40.3}$$

where $\lambda$ is the wavelength of the light forming the image. In fact, $1-(2\pi/\lambda)^2 Q_1$ is then the ratio of the intensity of light at the point where this is a maximum, i.e. at the *diffraction focus*, to what it would be at $I'$ if aberrations were absent. This ratio bears the name *Strehl intensity*.

At any rate, it is obvious that the usual form of $\Delta$ as a power series in $t$, with coefficients which are polynomials in $\cos\theta$, leads to a very unwieldy integration in the context of (40.1). To begin with, the integration over $\theta$ will be greatly simplified if the various powers of $\cos\theta$ just referred to be replaced by $\cos m\theta$ $(m = 1, 2, \ldots)$, bearing in mind that

$$\int_0^{2\pi} \cos m_1\theta \cos m_2\theta\, d\theta = \begin{cases} 0\,(m_1 \neq m_2) \\ \pi\,(m_1 = m_2 \neq 0) \\ 2\pi\,(m_1 = m_2 = 0). \end{cases} \tag{40.4}$$

The factor multiplying $\cos m\theta$ is a power series in $t$; and here again great simplicity will be attained with regard to the integration over $t$ if this factor be expressed in terms of orthogonal polynomials

$R_{nm}(t)$, i.e. polynomials which are such that

$$\int_0^1 tR_{n_1 m}(t)\, R_{nm}(t)\, dt = \begin{cases} 0 & (n_1 \neq n) \\ r_n & (n_1 = n), \end{cases} \tag{40.5}$$

where the $r_n$ are positive numbers still at our disposal. In the light of subsequent experience the choice

$$r_n = 1/[2(n+1)] \tag{40.6}$$

is a good one, in that it causes the coefficients of $R_{nm}$ to be integral. Evidently, granted that the highest power of $t$ in $R_{nm}$ is to be $n$, the sign of $t^n$ is not yet fixed, and we require it to be positive. We now write

$$\Delta = \sum_{n=0}^{\infty} \sum_{m=0}^{n}{}' A_{nm} R_{nm}(t) \cos m\theta, \tag{40.7}$$

where the prime is intended to indicate that when writing *this* series out in full every $A_{n0}$ is to be replaced by $2^{-\frac{1}{2}}A_{n0}$. (The reason for doing so is evident from (40.4).) The virtue of the form (40.7) of $\Delta$ is now obvious from the simplicity of the result

$$Q = \tfrac{1}{2} \sum_{n=0}^{\infty} \sum_{m=0}^{n} A_{nm}^2/(n+1). \tag{40.8}$$

(Here $A_{n0}$ of course really stands for $A_{n0}$.) $\bar{\Delta}$ is very easily evaluated if one remembers that $R_{00} = 1$, in view of $r_0$ having the value $\frac{1}{2}$. In fact

$$\bar{\Delta} = 2^{-\frac{1}{2}} A_{00}, \tag{40.9}$$

and then $Q_1$ is given by (40.8) with the term which has $n = m = 0$ omitted.

We desire to continue in an entirely elementary way, and show how the $R_{nm}$ may be found explicitly by means of a step-by-step method. To begin with, from a formal point of view any point given by $t$, $\theta$ is also given by $-t, \theta+\pi$. It follows that $R_{nm}$ must be even or odd according as $m$ is even or odd. Since the highest power of $t$ in $R_{nm}$ was to be $n$, it follows that

$$R_{nm}(t) = 0 \quad (n-m \text{ odd}), \tag{40.10}$$

and many terms of (40.8) are in fact absent as a consequence. Moreover, $\cos\theta$ appears everywhere only on account of its occurrence in $\eta$, where it has a factor $t$. Hence $\cos^m\theta$ and therefore $\cos m\theta$ must have $t^m$ as a factor, i.e. the power of the term of lowest degree in $R_{nm}$ is $m$. Consequently $R_{nn} = a_0 t^n$, and (40.5) shows that $a_0 = 1$:

$$R_{nn} = t^n. \tag{40.11}$$

Now, writing $R_{n,\,n-2} = a_1 t^n + b_1 t^{n-2}$, take $m = n-2$, and alternatively $n_1 = n$ and $n-2$ in (40.5). Using (40.11) one gets two conditions to determine $a_1$ and $b_1$, and in this way we get

$$R_{n,n-2} = nt^n - (n-1)\,t^{n-2}. \tag{40.12}$$

Again, given (40.11) and (40.12), the relation (40.5) with $m = n-4$ and $n_1 = n, n-2, n-4$ in turn allows one to calculate the three coefficients of $R_{n,n-4}$. We thus find that

$$R_{n,\,n-4} = \tfrac{1}{2}n\,(n-1)\,t^n - (n-1)\,(n-2)\,t^{n-2} + \tfrac{1}{2}(n-2)\,(n-3)\,t^{n-4}. \tag{40.13}$$

It should be clear by now how we can successively obtain

$$R_{n,n-6},\quad R_{n,n-8},\ \ldots$$

in this fashion. The explicit results above already give all the $R_{nm}$ for $n$ and $m$ not exceeding 6, with the sole exception of $R_{60}$ and this turns out to be $20t^6 - 30t^4 + 12t^2 - 1$. We therefore have the following table:

TABLE 5.1

| | $n$ | | | | | | |
|---|---|---|---|---|---|---|---|
| $m$ | 0 | 1 | 2 | 3 | 4 | 5 | 6 |
| 0 | 1 | . | $2t^2-1$ | . | $6t^4-6t^2+1$ | . | $20t^6-30t^4+12t^2-1$ |
| 1 | . | $t$ | . | $3t^3-2t$ | . | $10t^5-12t^3+3t$ | . |
| 2 | . | . | $t^2$ | . | $4t^4-3t^2$ | . | $15t^6-20t^4+6t^2$ |
| 3 | . | . | . | $t^3$ | . | $5t^5-4t^3$ | . |
| 4 | . | . | . | . | $t^4$ | . | $6t^6-5t^4$ |
| 5 | . | . | . | . | . | $t^5$ | . |
| 6 | . | . | . | . | . | . | $t^6$ |

The polynomials $R_{nm}(t)$ are the so-called *Zernike circle polynomials*. We have seen how they can be constructed in an elementary way; but they are in fact merely a special example from the theory of classical orthogonal polynomials. Explicitly, they are, in effect, special Jacobi polynomials; so that recursion relations, generating functions, and the like are easily written down. However, we have no need here to concern ourselves with these.

If we restore the explicit dependence of $\Delta$ upon $h'$, $A_{nm}$ becomes a power series in $h'$:

$$A_{nm} = \sum_{l=0}^{\infty} A_{nml} h'^{2l+m}. \quad (40.14)$$

Since $R_{nm}$ is of degree $n$ in $t$, any coefficient $A_{nml}$ may be thought of as relating to an aberration of 'order' $n+m+2l-1$; but, in fact, the polynomial character of $R_{nm}$ entails that such an aberration is a mixture of aberrations of order equal to *and less than* that just given, if the term 'order' is now understood in the sense ascribed to it hitherto. With regard to the relations between the old and the new aberration coefficients it will suffice to consider those of the third order. $\Delta$ is then given by the right-hand member of (19.1), together with (19.4) and $\rho = \rho_0 t$. Picking out the relevant terms from (40.7) and (40.14), direct comparison gives, with (19.6),

$$A_{400} = (\tfrac{1}{24}\sqrt{2})\rho_0^4\sigma_1, \quad A_{310} = \tfrac{1}{3}\rho_0^3\sigma_2, \quad A_{220} = \tfrac{1}{2}\rho_0^2\sigma_3,$$
$$A_{201} = (\tfrac{1}{4}\sqrt{2})\rho_0^2(2\sigma_3+\sigma_4), \quad A_{111} = \rho_0\sigma_5. \quad (40.15)$$

At any rate, the 'wavefront aberrations' are preferably to be taken as governed by the aberration coefficients $A_{nml}$, on account of the relatively transparent role these play in diffraction theory. The various 'types' of aberrations evidently differ to some extent from those studied earlier, at least with regard to the details of the displacements which they induce.

## C. The dependence of the aberrations on the positions of the object and of the stop

### 41. Stop shift and point characteristic

A shift of the stop causes the axial point $E'$ of the plane of the paraxial exit pupil to move through a certain distance, $\hat{x}'$ say. The customary interpretation of the aberrations on the basis of $V$ requires the posterior base-plane $\mathscr{B}'$ to coincide with $\mathscr{E}'$, so that when the latter is moved, $V$ has to be recalculated for the new plane of the exit pupil $\hat{\mathscr{E}}'$, say. This problem has, however, already been fully considered in Section 35. Accordingly, equations such as (35.13) and (35.14) give exactly the new (characteristic) coefficients in terms of the old.

With regard to the effective coefficients one must not forget that $\hat{d}' = 1 - \hat{x}' = c^{-1}$ is not equal to unity, see (23.4). Accordingly, one has, for instance

$$\hat{p}_1^* = 4c^{-1}\hat{p}_1 = 4c^3 p_1 = c^3 p_1^*, \tag{41.1}$$

and so on. Again, recalling (23.4),

$$s_1^* = 6(s_1 + p_1), \quad \hat{s}_1^* = 6(c^{-1}\hat{s}_1 + c\hat{p}_1),$$

from which one deduces, by means of (41.1) and the first member of (35.14), that

$$\hat{s}_1^* = c^5(s_1^* + 3qp_1^{*2}). \tag{41.2}$$

Of course it is not at all necessary to proceed via the characteristic coefficients to obtain (41.2).

The ubiquitous appearance of powers of $c$ is a nuisance. If desired it can be prevented in the following way. Let a ray intersect $\hat{\mathscr{E}}'$ in $\hat{D}'$, as usual. Lines drawn through $\hat{D}'$ and $I'$ on the one hand, and through $\hat{E}'$ and $I'$ on the other, intersect $\mathscr{E}'$ in points $D_1'$ and $D_p'$, respectively. Denote the differences between the coordinates of $D_1'$ and $D_p'$ by $\tilde{\mathbf{y}}'$. Then we may specify the ray in question by $\tilde{\mathbf{y}}'$ rather than $\hat{\mathbf{y}}'$. From elementary considerations it follows that

$$\tilde{\mathbf{y}}' = c\hat{\mathbf{y}}'. \tag{41.3}$$

Provided equations such as (35.13) be now interpreted as relating to the new coordinates—one should write $\tilde{p}_\alpha$ instead of $\hat{p}_\alpha$, and so on—all the powers of $c$ are to be omitted from them. This procedure ensures that, to a sufficient approximation, equal maximum values of the usual polar variables, $\rho$ and $\tilde{\rho}$ say, correspond to the same *angular* aperture of the respective pencils of rays admitted by $K$.

We note in passing that the quantity $c^{-2}(2\hat{p}_3 - \hat{p}_4)$ has a fixed value for all positions of the stop; a result which illuminates the definition of $\sigma_4$ in (19.6), and the generalization of which will occupy us in Sections 46ff.

One would surely like also to get rid of terms such as that in (41.2) which is quadratic in $p_1^*$. There is no very easy way of doing this here. It may, however, be mentioned that they can be eliminated, provided one goes over from $\hat{\mathbf{y}}'$ to another pair of variables in the *object* space, e.g. the coordinates of the point of intersection of the ray with the plane of the entrance pupil. There is no need to enter into the details of this process here, especially as the problem of stop shifts will be considered afresh in Section 44 in the context of the angle characteristic.

## 42. Object shift and point characteristic

Let the object plane $\mathscr{I}$ be shifted to a new position through a distance $\hat{x}$. The new object plane is naturally denoted by $\hat{\mathscr{I}}$, and the ideal image plane conjugate to it by $\hat{\mathscr{I}}'$. The distance between $E'$—which has remained fixed—and $\hat{O}_0'$ is $\hat{d}'$. The magnification associated with $\hat{\mathscr{I}}$ and $\hat{\mathscr{I}}'$ is $\hat{m}$, whereas that associated with $\mathscr{E}$ and $\mathscr{E}'$, i.e. $s$, is unaltered. Essentially we now need to repeat the work of Section 35, in the sense that we require the explicit dependence of the point characteristic upon $x$. Unfortunately this is a more involved problem than the earlier one, mainly on account of the non-constancy of $m$. We therefore consider it only in the barest outline, since the effects of object shift will be treated in some detail later on on the basis of the angle characteristic.

Independently of the variables on which the old and new characteristics depend, we may write

$$\hat{V} = V - \hat{x}/\alpha, \tag{42.1}$$

which may be compared with (35.3). This gives the equation corresponding to (35.4), i.e.

$$\hat{v}(\hat{x};\hat{\xi},\hat{\eta},\hat{\zeta}) = v(\xi,\eta,\zeta) + (\hat{d}'^2+\hat{u})^{\frac{1}{2}} - (d'^2+u)^{\frac{1}{2}} - \hat{x}/\alpha$$
$$+ (1/2\hat{m}f)\,\hat{\zeta} - (1/2mf)\,\zeta + \hat{x} - \hat{d}' + d'. \tag{42.2}$$

All non-linear terms of $g$ and $\hat{g}$ have been absorbed in $v$ and $\hat{v}$ respectively, the linear terms having coefficients quoted just after equation (15.4). Also, in view of (14.27), (14.28),

$$d' = (s-m)f, \quad \hat{d}' = (s-\hat{m})f, \quad \hat{x} = d - \hat{d} = [(1/\hat{m})-(1/m)]f. \tag{42.3}$$

It remains to express $\xi$, $\eta$, $\zeta$ on the right of (42.2) through $\hat{\xi}, \hat{\eta}, \hat{\zeta}$. To this end one uses the equation

$$\begin{aligned}\boldsymbol{\beta} &= -m\,\partial V/\partial \mathbf{y}_1\\ &= -m(d'^2+u)^{-\frac{1}{2}}(\mathbf{y}'-\mathbf{y}_1) + \mathbf{y}_1/f - m(\mathbf{y}'v_\xi + 2\mathbf{y}_1 v_\eta),\end{aligned} \tag{42.4}$$

and thence $1/\alpha$ and

$$\hat{\mathbf{y}}_1 = (\hat{m}/m)\,\mathbf{y}_1 + \hat{m}\hat{x}\boldsymbol{\beta}/\alpha \tag{42.5}$$

are obtained. The whole process is rather cumbersome, and we pursue it no further. In any event, one may deal directly with the effective coefficients; but then it must be borne in mind that $\boldsymbol{\epsilon}'$ and $\hat{\boldsymbol{\epsilon}}'$ now do *not* denote the same displacement merely regarded as functions of alternative sets of variables.

## 43. On a linear substitution involving direction cosines

This section deals with a simple algebraic problem which arises in various contexts. We first bring back to mind the rotational invariants $\xi, \eta, \zeta$ which occupy so prominent a place in the context of the angle characteristic. They are linear functions of the basic invariants $\bar{\xi}, \bar{\eta}, \bar{\zeta}$, according to equations (15.11) and (15.12). The following question now arises. Suppose we make a linear substitution with constant coefficients

$$\boldsymbol{\beta} \to \bar{a}\boldsymbol{\beta} + \bar{b}\boldsymbol{\beta}', \quad \boldsymbol{\beta}' \to \bar{c}\boldsymbol{\beta} + \bar{d}\boldsymbol{\beta}', \tag{43.1}$$

arbitrary except for the condition

$$g \equiv \bar{b}\bar{c} - \bar{a}\bar{d} \neq 0. \tag{43.2}$$

Then, what is the effect of (43.1) on $\xi, \eta, \zeta$?

If we momentarily write $a_1 = \bar{a} - s\bar{c}$, $b_1 = \bar{b} - s\bar{d}$, $c_1 = \bar{a} - m\bar{c}$, $d_1 = \bar{b} - m\bar{d}$ we find easily that, as a consequence of (43.1),

$$\left.\begin{aligned} \xi &\to b_1^2\bar{\xi} + 2a_1 b_1 \bar{\eta} + a_1^2 \bar{\zeta}, \\ \eta &\to b_1 d_1 \bar{\xi} + (a_1 d_1 + b_1 c_1)\bar{\eta} + a_1 c_1 \bar{\zeta}, \\ \zeta &\to d_1^2 \bar{\xi} + c_1 d_1 \bar{\eta} + c_1^2 \bar{\zeta}. \end{aligned}\right\} \tag{43.3}$$

On the right $\bar{\xi}, \bar{\eta}, \bar{\zeta}$ may be eliminated in favour of $\xi, \eta, \zeta$ by means of (15.12). Denoting the resulting expressions by $\hat{\xi}, \hat{\eta}, \hat{\zeta}$ we have the desired result

$$\left.\begin{aligned} \xi \to \hat{\xi} &\equiv \hat{a}^2\xi - 2\hat{a}\hat{b}\eta + \hat{b}^2\zeta, \\ \eta \to \hat{\eta} &\equiv \hat{a}\hat{c}\xi - (\hat{a}\hat{d} + \hat{b}\hat{c})\eta + \hat{b}\hat{d}\zeta, \\ \zeta \to \hat{\zeta} &\equiv \hat{c}^2\xi - 2\hat{c}\hat{d}\eta + \hat{d}^2\zeta, \end{aligned}\right\} \tag{43.4}$$

where

$$\left.\begin{aligned} \hat{a} &= (s-m)^{-1}[(m\bar{a} + \bar{b}) - s(m\bar{c} + \bar{d})], \\ \hat{b} &= (s-m)^{-1}[(s\bar{a} + \bar{b}) - s(s\bar{c} + \bar{d})], \\ \hat{c} &= (s-m)^{-1}[(m\bar{a} + \bar{b}) - m(m\bar{c} + \bar{d})], \\ \hat{d} &= (s-m)^{-1}[(s\bar{a} + \bar{b}) - m(s\bar{c} + \bar{d})]; \end{aligned}\right\} \tag{43.5}$$

and $$\hat{b}\hat{c}-\hat{a}\hat{d} = -g, \quad A = g^3, \tag{43.6}$$

where $A$ is the discriminant of (43.4). The inverse of (43.4) is

$$\left.\begin{aligned} \xi &= g^{-2}(\hat{d}^2\hat{\xi}-2\hat{b}\hat{d}\hat{\eta}+\hat{b}^2\hat{\zeta}), \\ \eta &= g^{-2}[\hat{c}\hat{d}\hat{\xi}-(\hat{b}\hat{c}+\hat{a}\hat{d})\,\hat{\eta}+\hat{a}\hat{b}\hat{\zeta}], \\ \zeta &= g^{-2}\,(\hat{c}^2\hat{\xi}-2\hat{a}\hat{c}\hat{\eta}+\hat{a}^2\hat{\zeta}). \end{aligned}\right\} \tag{43.7}$$

## 44. Stop shift and angle characteristic

When the stop is shifted, the pupil planes will move, and the magnification associated with them will change from $s$ to $\hat{s}$, say. The base-points being taken at $O_0$ and $O_0'$ as usual, the position of the stop is therefore contained in $T$ only through the constant $s$. We again absorb all non-linear terms of $g(\zeta)$ in the aberration function, so that according to (15.13) and (14.35)

$$T = g_0+\zeta/2m+t, \tag{44.1}$$

the focal length $f$ having been taken as unity. (Note that $f$ is a constant of the system, unlike $d'$.) Since $T$, regarded as an optical distance, is independent of the position of the stop, we thus have the identity

$$\hat{\zeta}/2m+\hat{t}(\hat{\xi},\hat{\eta},\hat{\zeta}) = \zeta/2m+t(\xi,\eta,\zeta), \tag{44.2}$$

where $\xi$, $\eta$, $\zeta$ refer to $s$, and $\hat{\xi}$, $\hat{\eta}$, $\hat{\zeta}$ to $\hat{s}$. We can use the results of the preceding section, for the change from the first to the second set of invariants corresponds to choosing the constants in (43.1) in such a way that $\boldsymbol{\beta}-s\boldsymbol{\beta}' \to \boldsymbol{\beta}-\hat{s}\boldsymbol{\beta}'$ whilst $\boldsymbol{\beta}-m\boldsymbol{\beta}'$ remains fixed. This requires that

$$\bar{a} = 1, \quad \bar{b} = mq/(1-q), \quad \bar{c} = 0, \quad \bar{d} = 1/(1-q), \tag{44.3}$$

where $$q = (\hat{s}-s)/(\hat{s}-m). \tag{44.4}$$

Then, from (43.5),

$$\hat{a} = -1/(1-q), \quad \hat{b} = -q/(1-q), \quad \hat{c} = 0, \quad \hat{d} = 1. \tag{44.5}$$

In (35.10) $q$ denoted the quantity $-\hat{x}'/(d'-\hat{x}')$. Now $\hat{x}' = d'-\hat{d}'$, and so, from (14.27), $q = -(s-\hat{s})/(\hat{s}-m)$, which is identical with the right-hand member of (44.4). It will be convenient to retain also the previous abbreviation $c = 1-q$.

The relations (43.7) here become, since $g^2 = (1-q)^{-2}$,

$$\xi = c^2\hat{\xi} + 2cq\hat{\eta} + q^2\hat{\zeta}, \quad \eta = c\hat{\eta} + q\hat{\zeta}, \quad \zeta = \hat{\zeta}. \tag{44.6}$$

This is formally the same as (35.12), provided all terms not quadratic in the coordinates are omitted from the latter. To this extent one therefore has a much simpler situation here.

Since $\zeta = \hat{\zeta}$, (44.2) reduces to

$$\sum\sum \hat{t}^{(n)}_{\mu\nu}\hat{\xi}^{n-\mu}\hat{\eta}^{\mu-\nu}\hat{\zeta}^{\nu} \equiv \sum\sum t^{(n)}_{\mu\nu}\xi^{n-\mu}\eta^{\mu-\nu}\zeta^{\nu}, \tag{44.7}$$

and it only remains to insert (44.6) on the right. We use the usual notation for the coefficients of low orders, e.g. $t^{(2)} = p_1\xi^2 + p_2\xi\eta + \ldots$, so that incidentally the initial factor $(s-m)^{-1}$ in (19.24) is to be omitted here. For the primary coefficients we then get exactly the equations (35.13) again. Then, for instance,

$$\hat{p}_1^* = (\hat{s}-m)\hat{p}_1 = c^4(\hat{s}-m)p_1 = c^3p_1,$$

consistently with (41.4). The equations for the secondary coefficients are

$$\begin{aligned}
\hat{s}_1 &= c^6 s_1,\\
\hat{s}_2 &= c^5(s_2 + 6qs_1),\\
\hat{s}_3 &= c^4(s_3 + qs_2 + 3q^2s_1),\\
\hat{s}_4 &= c^4(s_4 + 4qs_2 + 12q^2s_1),\\
\hat{s}_5 &= c^3(s_5 + 2qs_4 + 4qs_3 + 6q^2s_2 + 12q^3s_1),\\
\hat{s}_6 &= c^2(s_6 + qs_5 + q^2s_4 + 2q^2s_3 + 2q^3s_2 + 3q^4s_1),\\
\hat{s}_7 &= c^3(s_7 + 2qs_4 + 4q^2s_2 + 8q^3s_1),\\
\hat{s}_8 &= c^2(s_8 + 3qs_7 + 2qs_5 + 5q^2s_4 + 4q^2s_3 + 8q^3s_2 + 12q^4s_1),\\
\hat{s}_9 &= c(s_9 + 2qs_8 + 3q^2s_7 + 2qs_6 + 3q^2s_5 + 4q^3s_4 + 4q^3s_3 + 5q^4s_2 + 6q^5s_1),\\
\hat{s}_{10} &= s_{10} + qs_9 + q^2s_8 + q^3s_7 + q^2s_6 + q^3s_5 + q^4s_4 + q^4s_3 + q^5s_2 + q^6s_1.
\end{aligned} \tag{44.8}$$

These are exact; and to derive those relating to seventh and higher orders is evidently a matter of little difficulty. We give the results for the first eleven coefficients of order $2n-1$, writing these simply as $k_1, \ldots, k_{11}$, taken in the usual order. This means that

$$t^{(n)}_{\mu\nu} \to k_\alpha, \quad \alpha = \tfrac{1}{2}\mu(\mu+1) + \nu + 1, \tag{44.9}$$

for all $n, \mu$ and $\nu$ such that $\nu \leqslant \mu \leqslant n$. Then, absorbing a factor $c^{-2n+\mu+\nu}$ in $\hat{k}_\alpha$ (cf. the discussion following equation (41.2)) one finds that

$$
\begin{aligned}
\hat{k}_1 &= k_1, \\
\hat{k}_2 &= k_2 + 2nqk_1, \\
\hat{k}_3 &= k_3 + qk_2 + nq^2k_1, \\
\hat{k}_4 &= k_4 + 2(n-1)\,qk_2 + 2n(n-1)\,qk_1, \\
\hat{k}_5 &= k_5 + 2qk_4 + (n-1)\,(2qk_3 + 3q^2k_2) + 2n(n-1)\,q^3k_1, \\
\hat{k}_6 &= k_6 + qk_5 + q^2k_4 + (n-1)\,(q^2k_3 + q^3k_2) + \tfrac{1}{2}n(n-1)\,q^4k_1, \\
\hat{k}_7 &= k_7 + 2(n-2)\,qk_4 + 2(n-1)\,(n-2)\,q^2k_2 + \tfrac{4}{3}n(n-1)\,(n-2)\,q^3k_1, \\
\hat{k}_8 &= k_8 + 3qk_7 + (n-2)\,(2qk_5 + 5q^2k_4) + 2(n-1)\,(n-2)\,(q^2k_3 + 2q^3k_2) \\
&\qquad + 2n(n-1)\,(n-2)\,q^4k_1, \\
\hat{k}_9 &= k_9 + 2qk_8 + 3q^2k_7 + (n-2)\,(2qk_6 + 3q^2k_5 + 4q^3k_4) \\
&\qquad + (n-1)\,(n-2)\,(2q^3k_3 + \tfrac{5}{2}q^4k_2) + n(n-1)\,(n-2)\,q^5k_1, \\
\hat{k}_{10} &= k_{10} + qk_9 + q^2k_8 + q^3k_7 + (n-2)\,(q^2k_6 + q^3k_5 + q^4k_4) \\
&\qquad + \tfrac{1}{2}(n-1)\,(n-2)\,(q^4k_3 + q^5k_2) + \tfrac{1}{6}n(n-1)\,(n-2)\,q^6k_1, \\
\hat{k}_{11} &= k_{11} + 2(n-3)\,qk_7 + 2(n-2)\,(n-3)\,q^2k_4 + \tfrac{4}{3}(n-1)\,(n-2) \\
&\qquad \times (n-3)\,q^3k_2 + \tfrac{2}{3}n(n-1)\,(n-2)\,(n-3)\,q^4k_1.
\end{aligned} \tag{44.10}
$$

The relations (44.9) are of course contained in the first ten of these.

In Section 41 it was noted that $c^{-2}(2\hat{p}_3 - \hat{p}_4)$ has a fixed value for all positions of the stop. From (44.10) we now see by inspection that $2(n-1)\,\hat{k}_3 - \hat{k}_4 = 2(n-1)\,k_3 - k_4$. Restoring the appropriate power of $c$ this means that the quantity

$$\sigma_{2n-1,1} = 2(n-2)!\,(s-m)^{2(n-1)}[2(n-1)\,k_3 - k_4] \tag{44.11}$$

has a value independent of $s$; so that this is a generalization to all orders of the result previously obtained (the factor $2(n-1)!$ has been inserted for later convenience). The coefficients which occur in (44.11) govern oblique spherical aberration of order $2n-1$. We shall have occasion to reconsider such 'invariant' expressions later on in full generality.

It certainly appears at first sight as if the theory of the effects of stop shifts is much simpler here than the corresponding developments of Section 41. This simplicity is, alas, illusory, to the extent that all the complexities of detail reappear as soon as one enquires into the effects of stop shifts on the *effective* coefficients; for these are, after all, what one is ultimately interested in, at any rate in the geometrical theory. As a matter of fact, quite a lengthy calculation is required merely to confirm that (41.2) can be reproduced on the basis of (35.13), (44.8) and the first member of (24.13). In short, by using the angle characteristic one sweeps all the nasty details under the carpet; and one does well to keep this constantly in mind when going through the otherwise elegant results which we are about to derive.

## 45. Object shift and angle characteristic

We now go on to investigate the effect of shifting the object plane through a distance $\hat{x}$, the conjugate image plane shifting through a corresponding distance $\hat{x}'$. It will emerge that the theory is here genuinely simpler than that of Section 42. We have, to begin with,

$$\hat{T} = T - \hat{x}\alpha + \hat{x}'\alpha', \tag{45.1}$$

with

$$\hat{x} = -(\hat{m}-m)/m\hat{m}, \quad \hat{x}' = -(\hat{m}-m). \tag{45.2}$$

As usual all non-linear terms of $g$ are to be absorbed in the aberration function, and then (45.1) becomes

$$\hat{t}(\hat{\xi},\hat{\eta},\hat{\zeta}) = t(\xi,\eta,\zeta) - (\hat{\zeta}/2\hat{m}) + (\zeta/2m) + [(\hat{m}-m)/m\hat{m}]\,[(1-\zeta)^{\frac{1}{2}}-1]$$
$$-(\hat{m}-m)\,[(1-\bar{\xi})^{\frac{1}{2}}-1]. \tag{45.3}$$

We proceed exactly as in the context of stop shifts and determine the constants in (43.1) in such a way that $\boldsymbol{\beta}-m\boldsymbol{\beta}' \to \boldsymbol{\beta}-\hat{m}\boldsymbol{\beta}'$ whilst $\boldsymbol{\beta}-s\boldsymbol{\beta}'$ remains fixed. This requires that

$$\bar{a} = 1, \quad \bar{b} = sp/(1-p), \quad \bar{c} = 0, \quad \bar{d} = 1/(1-p), \tag{45.4}$$

with

$$p = -(\hat{m}-m)/(s-\hat{m}). \tag{45.5}$$

Then from (43.5)

$$\hat{a} = -1, \quad \hat{b} = 0, \quad \hat{c} = p/(1-p), \quad \hat{d} = 1/(1-p), \tag{45.6}$$

whence, writing $b = 1-p$,

$$\xi = \hat{\xi}, \quad \eta = p\hat{\xi} + b\hat{\eta}, \quad \zeta = p^2\hat{\xi} + 2bp\hat{\eta} + b^2\hat{\zeta}. \tag{45.7}$$

Equations (45.4)–(45.7) evidently run entirely parallel to (44.3)–(44.6). On the right of (45.3) we also have $\bar{\xi}$ and $\bar{\zeta}$ explicitly, but these may be expressed in terms of $\hat{\xi}, \hat{\eta}, \hat{\zeta}$ by means of (15.3), i.e.

$$\bar{\xi} = (s-\hat{m})^{-2}(\hat{\xi}-2\hat{\eta}+\hat{\zeta}), \quad \bar{\zeta} = (s-\hat{m})^{-2}(\hat{m}^2\hat{\xi}-2s\hat{m}\hat{\eta}+s^2\hat{\zeta}). \tag{45.8}$$

The constant terms, and the terms linear in the rotational invariants correctly cancel out in (45·3). The quadratic terms then give the following relations.

$$\left.\begin{aligned}\hat{p}_1 &= p_1+pp_2+p^2p_3+p^2p_4+p^3p_5+p^4p_6+j_1(1-\hat{m}^3/m),\\ \hat{p}_2 &= b(p_2+2pp_3+2pp_4+3p^2p_5+4p^3p_6)-4j_1(1-s\hat{m}^2/m),\\ \hat{p}_3 &= b^2(p_3+pp_5+2p^2p_6)+2j_1(1-s^2\hat{m}/m),\\ \hat{p}_4 &= b^2(p_4+2pp_5+4p^2p_6)+4j_1(1-s^2\hat{m}/m),\\ \hat{p}_5 &= b^3(p_5+4pp_6)-4j_1(1-s^3/m),\\ \hat{p}_6 &= b^4p_6+j_1(1-s^4/m\hat{m}),\end{aligned}\right\} \tag{45.9}$$

where
$$j_1 = \tfrac{1}{8}(\hat{m}-m)(s-\hat{m})^{-4}, \tag{45.10}$$

and $\hat{m} = (m-sp)/b$, in view of (45·5).

The equations (45.9) already suffice to show that one cannot have more than one distinct pair of perfect planes. To see this most simply, take $s = 0$, since the actual value of $s$ is clearly irrelevant. Then the equations $p_\alpha = \hat{p}_\alpha = 0\,(\alpha \neq 6)$ are obviously mutually inconsistent.

The $\hat{p}_\alpha$ now all involve $p_6$, whereas in the context of stop shifts the $\hat{p}_\alpha\,(\alpha < 6)$ did not. Quite generally, in order $2n-1$,

$$\hat{k}_\alpha \quad [\alpha \neq \tfrac{1}{2}(n+1)(n+2)]$$

requires that we know the value of $t_{nn}^{(n)}$ when the object is shifted, but not when the stop is shifted. We also see that primary spherical aberration can vanish for at most four positions of the object. If it vanishes for some given position, then it will vanish for neighbouring positions obtained from the first member of (45·9) by taking $p$ to be infinitesimal. This gives

$$p_2 = \tfrac{1}{8}(s-m)^{-3}(1-m^2), \tag{45.11}$$

that is to say, $\hat{p}_1$ can vanish together with $p_1$ only if the coefficient of primary coma has a definite value depending upon the values of the magnifications $s$ and $m$ alone.

(45·11) constitutes *Hockin's condition* (also known as Herschel's condition) in the primary domain. It is easily extended to all orders.

If we reject all terms not linear in $p$, and at the same time set $\hat{\eta} = \hat{\zeta} = 0$ (since at present we are only interested in those terms of $\hat{T}$ which depend on $\hat{\xi}$ alone) we have to put

$$\xi \to \hat{\xi}, \quad \eta \to p\hat{\xi}, \quad \zeta \to 0, \quad \bar{\xi} \to (s-m)^{-2}\hat{\xi}, \quad \bar{\zeta} \to m^2(s-m)^{-2}\hat{\xi}$$

on the right of (45.3). Writing

$$t(\xi, \eta, \zeta) = S(\xi) + \eta C(\xi) + \ldots \tag{45.12}$$

in analogy with (26.1), we find

$$\xi C(\xi) = m^{-2}[(s-m)^2 - m^2\xi]^{\frac{1}{2}} - [(s-m)^2 - \xi]^{\frac{1}{2}} + (s-m)(1 - 1/m^2). \tag{45.13}$$

If we recall that $C(\xi) = p_2\xi + s_2\xi^2 + \ldots$ we see at once that (45.11) is correctly contained in (45.13). Evidently spherical aberration *and* circular coma can vanish together for neighbouring positions of $\mathcal{I}$ only if $m^2 = 1$.

From (45.9) one deduces very easily that $2\hat{p}_3 - \hat{p}_4 = b^2(2p_3 - p_4)$ so that the quantity $(s-m)^2(2p_3 - p_4)$ has a fixed value for all positions of the object. Recalling a previous result, it thus has the same value for all stop *and* object positions; a state of affairs which we shall shortly consider in greater generality.

The secondary equations corresponding to (45.9) are readily written down. We do not do so here since they will be contained in the more general results of the next section.

## 46. Joint shifts of stop and object

It is of considerable interest to examine the consequences of changing the positions of the object and of the stop simultaneously by arbitrary amounts. Our previous results will then be merely special cases of those about to be derived; but they are valuable for general orientation if nothing else.

The constants in (43.1) now have to have values such that

$$\boldsymbol{\beta} - s\boldsymbol{\beta}' \to \boldsymbol{\beta} - \hat{s}\boldsymbol{\beta}' \quad \text{and} \quad \boldsymbol{\beta} - m\boldsymbol{\beta}' \to \boldsymbol{\beta} - \hat{m}\boldsymbol{\beta}'.$$

This means taking

$$\bar{a} = 1, \quad \bar{b} = \frac{m\hat{s} - \hat{m}s}{s-m}, \quad \bar{c} = 0, \quad \bar{d} = \frac{\hat{s} - \hat{m}}{s-m}.$$

The definitions (44.4) and (45.5) of $q$ and $p$ may appropriately be retained here, and then

$$\bar{a} = 1, \quad \bar{b} = \frac{sp+mq-(s+m)pq}{(1-p)(1-q)}, \quad \bar{c} = 0, \quad \bar{d} = \frac{1-pq}{(1-p)(1-q)}. \tag{46.1}$$

From these we infer in turn that

$$\hat{a} = -1/(1-q), \quad \hat{b} = -q/(1-q), \quad \hat{c} = p/(1-p), \quad \hat{d} = 1/(1-p). \tag{46.2}$$

The definitions of the constants $b$ and $c$ may be somewhat generalized to

$$b = \frac{1-p}{1-pq}, \quad c = \frac{1-q}{1-pq}, \tag{46.3}$$

and then

$$\left.\begin{aligned} \xi &= c^2\hat{\xi} + 2bcq\hat{\eta} + b^2q^2\hat{\zeta}, \\ \eta &= c^2p\hat{\xi} + bc(1+pq)\hat{\eta} + b^2q\hat{\zeta}, \\ \zeta &= c^2p^2\hat{\xi} + 2bcp\hat{\eta} + b^2\hat{\zeta}. \end{aligned}\right\} \tag{46.4}$$

Even the appearance of the constants $b$ and $c$ could be prevented here by 'rescaling' $\rho$ and $h'$ each time $s$ or $m$ or both are changed; that is to say, by understanding $\rho$ to stand for $c\rho$ and $h'$ for $bh'$ (cf. the discussion following upon equation (41.2)). However, one is then apt to forget that $\rho$, for instance, is not the actual radial coordinate in whatever exit pupil one has in hand, but only proportional to it. We therefore desist from employing this device in the present context.

The identity (45.3) remains as it stands, but we now have to use (46.4) in place of (45.7), whilst in (45.8) $s$ has to be replaced by $\hat{s}$. The equations (45.9) then have the following generalized counterparts:

$$\hat{p}_1 = c^4(p_1 + pp_2 + p^2p_3 + p^2p_4 + p^3p_5 + p^4p_6) + j_1(1 - \hat{m}^3/m),$$

$$\begin{aligned} \hat{p}_2 = bc^3[&4qp_1 + (1+3pq)p_2 + 2p(1+pq)p_3 + 2p(1+pq)p_4 \\ &+ p^2(3+pq)p_5 + 4p^3p_6] - 4j_1(1 - \hat{s}\hat{m}^2/m), \end{aligned}$$

$$\begin{aligned} \hat{p}_3 = b^2c^2[&2q^2p_1 + q(1+pq)p_2 + (1+p^2q^2)p_3 + 2pqp_4 + p(1+pq)p_5 \\ &+ 2p^2p_6] + 2j_1(1 - \hat{s}^2\hat{m}/m), \end{aligned}$$

$$\hat{p}_4 = b^2c^2[4q^2p_1+2q(1+pq)p_2+4pqp_3+(1+pq)^2p_4 + 2p(1+pq)p_5+4p^2p_6]+4j_1(1-\hat{s}^2\hat{m}/m),$$

$$\hat{p}_5 = b^3c[4q^3p_1+q^2(3+pq)p_2+2q(1+pq)p_3+2q(1+pq)p_4 + (1+3pq)p_5+4pp_6]-4j_1(1-\hat{s}^3/m),$$

$$\hat{p}_6 = b^4(q^4p_1+q^3p_2+q^2p_3+q^2p_4+qp_5+p_6)+j_1(1-\hat{s}^4/m\hat{m}), \quad (46.5)$$

where now $$j_1 = \tfrac{1}{8}(\hat{m}-m)(\hat{s}-\hat{m})^{-4}. \quad (46.6)$$

The combination $2\hat{p}_3-\hat{p}_4$ is again proportional to $2p_3-p_4$, the factor of proportionality being

$$[(1-p)(1-q)/(1-pq)]^2 = (s-m)^2(\hat{s}-\hat{m})^{-2}.$$

It follows that $$(s-m)^2(2p_3-p_4) = \text{const.}, \quad (46.7)$$

i.e. the expression on the left has a value entirely independent of $s$ *and* $m$; as was shown previously.

Though rather lengthy, the secondary relations will now be exhibited in full. With

$$j_2 = \tfrac{1}{16}(\hat{m}-m)(\hat{s}-\hat{m})^{-6}, \quad (46.8)$$

they are

$$\hat{s}_1 = c^6(s_1+ps_2+p^2s_3+p^2s_4+p^3s_5+p^4s_6+p^3s_7+p^4s_8+p^5s_9+p^6s_{10}) + j_2(1-\hat{m}^5/m),$$

$$\hat{s}_2 = bc^5[6qs_1+(1+5pq)s_2+2p(1+2pq)s_3+2p(1+2pq)s_4 + 3p^2(1+pq)s_5+2p^3(2+pq)s_6+3p^2(1+pq)s_7 + 2p^3(2+pq)s_8+p^4(5+pq)s_9+6p^5s_{10}]-6j_2(1-\hat{s}\hat{m}^4/m),$$

$$\hat{s}_3 = b^2c^4[3q^2s_1+q(1+2pq)s_2+(1+2p^2q^2)s_3+pq(2+pq)s_4 + p(1+pq+p^2q^2)s_5+p^2(2+p^2q^2)s_6+3p^2qs_7+p^2(1+2pq)s_8 + p^3(2+pq)s_9+3p^4s_{10}]+3j_2(1-\hat{s}^2\hat{m}^3/m),$$

$$\hat{s}_4 = b^2c^4[12q^2s_1+4q(1+2pq)s_2+4pq(2+pq)s_3+(1+pq)(1+5pq)s_4 + 2p(1+4pq+p^2q^2)s_5+4p^2(1+2pq)s_6+3p(1+pq)^2s_7 + p^2(1+pq)(5+pq)s_8+4p^3(2+pq)s_9+12p^4s_{10}] + 12j_2(1-\hat{s}^2\hat{m}^3/m),$$

$$\hat{s}_5 = b^3c^3[12q^3s_1+6q^2(1+pq)s_2+4q(1+pq+p^2q^2)s_3+2q(1+4pq+p^2q^2)s_4+(1+pq)(1+4pq+p^2q^2)s_5+4p(1+pq+p^2q^2)s_6+6pq(1+pq)s_7+2p(1+4pq+p^2q^2)s_8+6p^2(1+pq)s_9+12p^3s_{10}]-12j_2(1-\hat{s}^3\hat{m}^2/m),$$

$$\hat{s}_6 = b^4c^2[3q^4s_1+q^3(2+pq)s_2+q^2(2+p^2q^2)s_3+q^2(1+2pq)s_4+q(1+pq+p^2q^2)s_5+(1+2p^2q^2)s_6+3pq^2s_7+pq(2+pq)s_8+p(1+2pq)s_9+3p^2s_{10}]+3j_2(1-\hat{s}^4\hat{m}/m),$$

$$\hat{s}_7 = b^3c^3[8q^3s_1+4q^2(1+pq)s_2+8pq^2s_3+2q(1+pq)^2s_4+4pq(1+pq)s_5+8p^2qs_6+(1+pq)^3s_7+2p(1+pq)^2s_8+4p^2(1+pq)s_9+8p^3s_{10}]-8j_2(1-\hat{s}^3\hat{m}^2/m),$$

$$\hat{s}_8 = b^4c^2[12q^4s_1+4q^3(2+pq)s_2+4q^2(1+2pq)s_3+q^2(1+pq)\times(5+pq)s_4+2q(1+4pq+p^2q^2)s_5+4pq(2+pq)s_6+3q(1+pq)^2s_7+(1+pq)(1+5pq)s_8+4p(1+2pq)s_9+12p^2s_{10}]+12j_2(1-\hat{s}^4\hat{m}/m),$$

$$\hat{s}_9 = b^5c[6q^5s_1+q^4(5+pq)s_2+2q^3(2+pq)s_3+2q^3(2+pq)s_4+3q^2(1+pq)s_5+2q(1+2pq)s_6+3q^2(1+pq)s_7+2q(1+2pq)s_8+(1+5pq)s_9+6ps_{10}]-6j_2(1-\hat{s}^5/m),$$

$$\hat{s}_{10} = b^6(q^6s_1+q^5s_2+q^4s_3+q^4s_4+q^3s_5+q^2s_6+q^3s_7+q^2s_8+qs_9+s_{10})+j_2(1-\hat{s}^6/m\hat{m}). \quad (46.9)$$

The ratio of the last terms in the third and fourth of these equations is a number, independent of $s$ and $m$, and the same is true of the fifth and seventh, and the sixth and eighth equations. It is therefore natural to consider the combinations $4\hat{s}_3-\hat{s}_4$, $2\hat{s}_5-3\hat{s}_7$, $4\hat{s}_6-\hat{s}_8$. One finds that

$$4\hat{s}_3-\hat{s}_4 = (1-q)^4(1-p)^2(1-pq)^{-4}[(4s_3-s_4)+p(2s_5-3s_7)+p^2(4s_6-s_8)],$$

$$2\hat{s}_5-3\hat{s}_7 = (1-q)^3(1-p)^3(1-pq)^{-4}[2q(4s_3-s_4)+(1+pq)(2s_5-3s_7)+2p(4s_6-s_8)],$$

$$4\hat{s}_6-\hat{s}_8 = (1-q)^2(1-p)^4(1-pq)^{-4}[q^2(4s_3-s_4)+q(2s_5-3s_7)+(4s_6-s_8)]. \quad (46.10)$$

This result means that under any change of $s$ and $m$ $4s_3-s_4$, $2s_5-3s_7$, $4s_6-s_8$ transform amongst themselves, independently of all other coefficients; so that if all three of these combinations vanish for some particular value of $s$ and of $m$ they will do so for all values of $s$ and $m$. In particular, $(s-m)^4(4s_3-s_4)$ is independent of $s$ (cf. (44.11)), whilst $(s-m)^4(4s_6-s_8)$ is independent of $m$.

Conclusions of the kind we have just drawn are of sufficient interest to make us seek some simple general method which will allow us to arrive at similar results governing the coefficients of any order; and this we now proceed to develop.

## D. Invariant and semi-invariant aberrations

### 47. The focal angle characteristic

The angle characteristic $T$ which entered into the considerations of the three preceding sections was taken with respect to the base-points $O_0$ and $O'_0$. Whereas from a formal viewpoint the dependence of $T$ on $s$ was inessential, in the sense that $s$ enters into $T$ only through the particular rotational invariants which happen to be chosen, its dependence on $m$ is essential, for a change of $m$ entails a change of $T$, no matter how one chooses the rotational invariants. This means that the $t^{(n)}_{\mu\nu}$ are necessarily dependent upon $m$, that is to say, they are not 'absolute' constants of the system. For our present purposes it is therefore convenient to write $T$ as the sum of two parts, of which the first is the angle characteristic referred to base-points which are, as it were, inherent in the structure of $K$. The principal foci $F$ and $F'$ suggest themselves, and we adopt them. The angle characteristic $\bar{T}$ so defined is called the *focal angle characteristic*. Paraxially $\bar{T}$ is remarkably simple:

$$\bar{T} = -\bar{\eta}, \tag{47.1}$$

leaving aside a trivial additive constant. The value unity has been attached to $f$ as usual. The coefficients $\bar{t}^{(n)}_{\mu\nu}$ are now absolute constants of $K$, i.e. they involve neither $s$ nor $m$. If $O'_0$ and $O_0$ lie at distances $x'$ and $x$ to the right of $F'$ and $F$ respectively, $x'$ is given by (14.27) with $s = 0$, and likewise $-x$ is given by (14.28) with $1/s = 0$. It follows that

$$T = \bar{T} - m\alpha' - m^{-1}\alpha. \tag{47.2}$$

Here the various functions may still be taken to depend upon variables of our choice, and we adopt (47.2) in the form

$$T(\xi,\eta,\zeta) = \bar{T}(\bar{\xi},\bar{\eta},\bar{\zeta}) - m(1-\bar{\xi})^{\frac{1}{2}} - m^{-1}(1-\bar{\zeta})^{\frac{1}{2}}. \tag{47.3}$$

We also write this occasionally as

$$T = \tilde{T} + D. \tag{47.4}$$

Here, then, the right-hand member does not involve $s$ at all, whilst $m$ enters into it only through $D$. In principle we could now simply express $\bar{\xi}, \bar{\eta}, \bar{\zeta}$ in terms of $\xi, \eta, \zeta$ on the right of (47.4), and so, in effect, recover our previous equations such as (44.8) or (45.9), for example, though differently expressed. Our present objective, however, is to attain, as quickly and systematically as we can, results of the kind represented by (44.11) or (46.7); and we shall see that this can be done without having first to write down the equations giving the explicit dependence of the aberration coefficients upon $s$ and $m$.

## 48. The idea of invariant and semi-invariant aberrations

Before attempting to derive specific results, let us become quite clear about what we are trying to do. To this end, consider the dependence of the primary aberration coefficients upon $s$, as given by (35.13). The latter represents a linear transformation of these coefficients, from the $p_\alpha$ to the $\hat{p}_\alpha$; and one can write down six combinations of coefficients (i.e. as many as there are coefficients), the values of which are independent of $s$. A simple example is the expression $(s-m)^3(4p_1+p_2)$; for, since $c+q = 1$,

$$4\hat{p}_1 + \hat{p}_2 = c^3(4p_1+p_2). \tag{48.1}$$

Bearing in mind that $c = (s-m)/(\hat{s}-m)$ it follows that

$$(\hat{s}-m)^3(4\hat{p}_1+\hat{p}_2) = (s-m)^3(4p_1+p_2), \tag{48.2}$$

or, in other words, $(s-m)^3(4p_1+p_2)$ has the same value for every position of the stop. One might therefore say that this expression is invariant under stop-shifts, or '$s$-invariant'; and this has indeed sometimes been done. However, there is not much point in such a terminology here, because it reflects *nothing* but the content of the first two members of (35.13). In any event, this kind of '$s$-invariance' has no useful significance in the context of the displacement. Thus, with regard to our example, the meridional partial displacement is $(s-m)(4p_1\rho^3+3p_2\rho^2h')$; and (48.2) has evidently no direct useful implications. The situation would be otherwise if $p_1$ and $p_2$ multiplied the same powers of $\rho$ and $h'$, i.e. jointly governed a particular type of aberration, as do $p_3$ and $p_4$. In that case the partial displacement could be subdivided into two parts, one of which would have the property that if it vanished for one value of $s$, then it would do so for all values of $s$, independently of $\rho$ and $h'$. It is also characteristic of the state of affairs just described that results of the kind (48.2) are coordinate-dependent, in the sense that, upon a mere change of scale of one coordinate, the individual coefficients will have to be supplied with further numerical factors.

The upshot of the preceding remarks is that we should principally concern ourselves only with combinations of coefficients governing aberrations of a given order and type. Any such combination will have the generic form

$$\omega_\lambda^{(n)} = \sum a_{\mu\nu}^{(n)} t_{\mu\nu}^{(n)}, \tag{48.3}$$

where the $a_{\mu\nu}^{(n)}$ are purely numerical coefficients, and the sum goes over values of $\mu$ and $\nu$ such that $\mu+\nu = \text{const.} = \lambda$, say. If there exists an integer $r$ such that $(s-m)^r\,\omega_\lambda^{(n)}$ is independent of $s$ *and* $m$, then we call this expression an (absolute) *invariant*, or also an invariant aberration (the factor $\xi^{n-\mu}\eta^{\mu-\nu}\zeta^{\nu}$ then being left understood); when it depends only upon $s$ *or* $m$, but not upon both together, we call it a *semi-invariant* (aberration). More precisely, when it is independent of $s$ only it is an *s-invariant*, and when it is independent of $m$ only it is an *m-invariant*. Thus, recalling (46.7) and (44.11), $(s-m)^2(2p_3-p_4)$ is a third-order absolute invariant, and $(s-m)^{2(n-1)}[2(n-1)k_3-k_4]$ is an $s$-invariant of order $2n-1$, where $n > 2$. Our next task is to enumerate, and as far as possible exhibit explicitly, all the invariants and semi-invariants of the various orders.

## 49. Generators of invariants and semi-invariants

In consequence of the linear relations (15.11) and (15.12) any differentiation with respect to the unbarred variables can at once be replaced by differentiation with respect to the barred variables and vice versa. Let us put

$$S = (s-m)^2\frac{\partial}{\partial\xi}, \qquad \bar{S} = \frac{\partial}{\partial\bar{\xi}}+m\frac{\partial}{\partial\bar{\eta}}+m^2\frac{\partial}{\partial\bar{\zeta}}, \tag{49.1}$$

$$M = (s-m)^2\frac{\partial}{\partial\zeta}, \qquad \bar{M} = \frac{\partial}{\partial\bar{\xi}}+s\frac{\partial}{\partial\bar{\eta}}+s^2\frac{\partial}{\partial\bar{\zeta}}, \tag{49.2}$$

$$A = (s-m)^2\left(4\frac{\partial^2}{\partial\xi\partial\zeta}-\frac{\partial^2}{\partial\eta^2}\right), \quad \bar{A} = 4\frac{\partial^2}{\partial\bar{\xi}\,\partial\bar{\zeta}}-\frac{\partial^2}{\partial\bar{\eta}^2}. \tag{49.3}$$

Then it is an elementary exercise to confirm that

$$S = \bar{S}, \quad M = \bar{M}, \quad A = \bar{A}. \tag{49.4}$$

For reasons which will become clear shortly, we call the differential operators $S$, $M$ and $A$ *generators* of invariants (semi-invariants here being included in this term). $\bar{S}, \bar{M}, \bar{A}$ are effectively the same generators, though differently expressed, as we see from (49.4). We shall say that $\bar{S}$ and $S$, $\bar{M}$ and $M$, $\bar{A}$ and $A$ are *adjoint* to each other; and this terminology will be extended to products of powers of these operators.

We now take (47.4) order by order, and write

$$t^{(n)} = \bar{t}^{(n)} + d^{(n)}, \tag{49.5}$$

quantities with a superscript $(n)$ relating to order $2n-1$, as usual. However, exceptionally we shall speak of them as being of degree $n$, rather than $2n$, thus thinking in terms of the rotational invariants only. This is appropriate in the present context, since differentiation with respect to the ray-coordinates nowhere occurs.

Let us then consider first the action of some power of the generator $A$ on $t^{(n)}$. Clearly $A^\alpha t^{(n)}$ is a polynomial of degree $n-2\alpha$, so that for no choice of $\alpha$ will $A^\alpha t^{(n)}$ be a constant unless $n$ is even. When $n$ is even $(= 2N$, say $(N = 1, 2, \ldots))$, take $\alpha = N$. Then, on the left of (49.5), $A^N t^{(2N)}$ is $(s-m)^{2N}$ times a linear combination of aberration coefficients of order $4N-1$ which govern a particular type of aberration. On the right of (49.5) we apply the adjoint operator $\bar{A}^N$. $\bar{A}$ by itself completely annihilates $D$ in any event, whereas $\bar{A}^N \bar{t}^{(n)}$ reduces to a constant which, as we know, does not depend upon $s$ or $m$. Since all relations (49.1)–(49.5) are identities by nature, it follows that *the constant*

$$\alpha_{4N-1} = A^N t^{(2N)} \tag{49.6}$$

*is an absolute invariant of order* $4N-1$ $(N = 1, 2, \ldots)$.

Next, we come to the semi-invariants, and we take the $s$-invariants first. Inspection shows that $A^\alpha S^{n-2\alpha} t^{(n)}$ is a constant, where $\alpha$ is a positive integer or zero, such that $n - 2\alpha \geqslant 0$. Moreover, apart from a common factor $(s-m)^{2(n-\alpha)}$, this constant is a linear combination (with purely numerical coefficients) of aberration coefficients governing a particular type. On the other hand, since neither $\bar{A}$ nor $\bar{S}$ nor $d^{(n)}$ involves $s$, the application of the adjoint operator $\bar{A}^\alpha \bar{S}^{n-2\alpha}$

to the right-hand member of (49.5) yields a constant independent of $s$. Thus *the constants*

$$\sigma_{2n-1,\alpha} = A^{\alpha} S^{n-2\alpha} t^{(n)} \tag{49.7}$$

*are s-invariants*, where $\alpha = 0, 1, \ldots, \frac{1}{2}n$ when $n$ is even or $\frac{1}{2}(n-1)$ when $n$ is odd.

It remains to consider the $m$-invariants. These are evidently given by an expression like that on the right of (49.7), with $S$ replaced by $M$. However, in this case one must exclude the value 0 of $\alpha$, since otherwise on applying $\bar{S}^n$ to $\bar{t}^{(n)} + d^{(n)}$, the term $d^{(n)}$ will fail to be annihilated, and the constant which has been generated will then depend upon $m$. Accordingly, *the constants*

$$\mu_{2n-1,\alpha} = A^{\alpha} M^{n-2\alpha} t^{(n)} \tag{49.8}$$

*are m-invariants*, where $\alpha = 1, 2, \ldots, \frac{1}{2}n$ when $n$ is even or $\frac{1}{2}(n-1)$ when $n$ is odd.

The number of invariant aberrations of order $2n-1$ is easily counted. In doing so we shall adopt the convention to exclude the value 0 of $\alpha$ also in the case of the $s$-invariants, on the grounds that this relates to the somewhat trivial $s$-invariance of spherical aberration. We thus get Table 5.2 for the number of invariants of orders $4N \pm 1$, where $N = 1, 2, 3, \ldots$.

TABLE 5.2

| Order | Absolute invariants | $s$-invariants | $m$-invariants | Total number |
|---|---|---|---|---|
| $4N-1$ | 1 | $N-1$ | $N-1$ | $2N-1$ |
| $4N+1$ | 0 | $N$ | $N$ | $2N$ |

There are no invariants or semi-invariants beyond those just enumerated. Bearing in mind that the phrase 'aberration coefficients governing a certain type' is to be understood as relating to the pseudo-displacement, any generator must be homogeneous, i.e. a linear combination, with fixed coefficients, of operators $S^{n-\mu} J^{\mu-\nu} M^{\nu}$ ($\mu + \nu = \text{const.} = \lambda$, say), where

$$J = (s-m)^2 (\partial/\partial\eta). \tag{49.9}$$

This ensures that in $t^{(n)}$ all terms not annihilated by the generators have $\mu+\nu = \lambda$ (cf. Section 21). If this generator be now expressed in terms of the adjoints $\bar{S}$, $\bar{M}$ and $\bar{J}$, where

$$\bar{J} = -2(\partial/\partial\bar{\xi})-(s+m)(\partial/\partial\bar{\eta})-2sm(\partial/\partial\bar{\zeta}),$$

either $s$ or $m$ is required to occur in the resulting expression at most through a common factor $(s-m)^r$. It is not difficult to convince oneself that this leaves just the generators considered above as the only possibilities.

## 50. On the absolute invariants

Using the simple machinery of Section 49 we can now write down the absolute invariants $\alpha_{4N-1}$ explicitly in terms of the $t_{\mu\nu}^{(2N)}$. From (49.6)

$$\begin{aligned}\alpha_{4N-1} &= A^N t^{(2N)} = (s-m)^{-2N}(4SM-J^2)^N t^{(2N)} \\ &= (s-m)^{-2N}\sum_{\lambda=0}^{N}(-1)^{\lambda}\,2^{2(N-\lambda)}\binom{N}{\lambda}S^{N-\lambda}J^{2\lambda}M^{N-\lambda}t^{(2N)}. \quad (50.1)\end{aligned}$$

For any particular value of $\lambda$ the operators annihilate all terms of $t^{(2N)}$ other than that which has $\mu = N+\lambda$ and $\nu = N-\lambda$. Carrying out the differentiations one then gets the desired result

$$\alpha_{4N-1} = N!(s-m)^{2N}\sum_{\lambda=0}^{N}(-1)^{\lambda}\,2^{2(N-\lambda)}(N-\lambda)!\,(2\lambda)!\,(\lambda!)^{-1}t_{N+\lambda,\,N-\lambda}^{(2N)}. \quad (50.2)$$

In particular one has, for $N = 1$, the well-known invariant

$$\alpha_3 = 2(s-m)^2(2t_{11}^{(2)}-t_{20}^{(2)}) \equiv 2(s-m)^2(2p_3-p_4). \quad (50.3)$$

Similarly, when $N = 2$,

$$\alpha_7 = 8(s-m)^4(8t_{22}^{(4)}-2t_{31}^{(4)}+3t_{40}^{(4)}) \equiv 8(s-m)^4(8t_6-2t_8+3t_{11}). \quad (50.4)$$

With increasing $N$ these invariants become more and more of merely academic interest, if for no other reason than that $\alpha_{4N-1}$ involves $N+1$ distinct coefficients. Effectively the best one can do is to write the partial (pseudo-) displacement as the sum of two terms, of which the first is

$$\epsilon'_{4N-1} = [2^{2N-1}(s-m)^{2N-1}N!\,(N-1)!]^{-1}\,\alpha_{4N-1}\rho^{2N-1}h'^{2N}\begin{cases}\cos\theta\\ \sin\theta,\end{cases} \quad (50.5)$$

whilst the second part is still governed jointly by $N$ coefficients, i.e. all those other than $t_{NN}^{(2N)}$. This decomposition of the *astigmatism of order* $4N-1$ *and degree* $2N-1$ is a generalization of that of primary curvature of field into Petzval and astigmatic curvature: the former vanishing for all values of $s$ and $m$ if it vanishes for any one pair of values.

## 51. On the semi-invariants

The semi-invariants, as explicit combinations of the $t_{\mu\nu}^{(n)}$, are obtained quite easily by the procedure of the preceding section. Thus, as regards the $s$-invariants,

$$\sigma_{2n-1,\alpha} = A^{\alpha}S^{n-2\alpha}t^{(n)}$$
$$= (s-m)^{-2\alpha}\sum_{\lambda=0}^{\alpha}(-1)^{\lambda}2^{2(\alpha-\lambda)}\binom{\alpha}{\lambda}S^{n-\alpha-\lambda}J^{2\lambda}M^{\alpha-\lambda}t^{(n)}.$$

In the $\lambda$th term on the right only the term of $t^{(n)}$ with $\mu = \alpha+\lambda$ and $\nu = \alpha-\lambda$ need be retained. Then, with $0 < 2\alpha < n$,

$$\sigma_{2n-1,\alpha} = \alpha!(s-m)^{2(n-\alpha)}\sum_{\lambda=0}^{\alpha}(-1)^{\lambda}2^{2(\alpha-\lambda)}$$
$$\times(n-\alpha-\lambda)!\,(2\lambda)!\,(\lambda!)^{-1}t_{\alpha+\lambda,\,\alpha-\lambda}^{(n)}. \quad (51.1)$$

Taking $\alpha = 1$, this gives exactly the expression (44.11). When $\alpha = 2$,

$$\sigma_{2n-1,2} = 8(n-4)!\,(s-m)^{2(n-2)}$$
$$\times[4(n-2)(n-3)k_6-2(n-3)k_8+3k_{11}], \quad (51.2)$$

in the notation of (44.9). This result will be found to be in agreement with (44.10). The coefficients which occur in (51.1) govern terms in the $(2n-1)$th-order displacement varying with $\rho$ and $h'$ according to $\rho^{2n-2\alpha-1}h'^{2\alpha}$, i.e. those relating to astigmatism of degree $2n-2\alpha-1 > n-1$.

The $m$-invariants, finally, are

$$\mu_{2n-1,\alpha} = A^{\alpha}M^{n-2\alpha}t^{(n)},$$

where now $0 < 2\alpha \leqslant n$. It turns out, not surprisingly, that $\mu_{2n-1,\alpha}$ is given by an expression exactly like (51.1) except that in the coefficient $t_{\alpha+\lambda,\,\alpha-\lambda}^{(n)}$ one has to replace $\alpha$ by $n-\alpha$.

The coefficients which appear in $\mu_{2n-1,\alpha}$ are those which govern astigmatism of degree $2\alpha-1 \leqslant n-1$. All the invariants and semi-invariants, taken together, thus relate just to all the various astigmatic types of aberrations. $\mu_{2n-1,1}$ alone is concerned with linear astigmatism. Indeed, temporarily writing

$$t^{(n)}_{n-1,\,n-1} = c, \quad t^{(n)}_{n,\,n-2} = \bar{c},$$
$$(s-m)^{2(n-1)}\bar{c} = c^*, \quad \mu_{2n-1,1}/2(n-2)! = \mu,$$

we have, on the one hand,

$$\mu = (s-m)^{2(n-1)}(2c-\bar{c}), \tag{51.3}$$

whilst the linear astigmatic partial displacement is

$$\epsilon'_y = 2(s-m)(c+\bar{c})\rho h'^{2n-2}\cos\theta, \; \epsilon'_z = 2(s-m)c\rho h'^{2n-2}\sin\theta. \tag{51.4}$$

Now absorb a factor $(s-m)$ in $\rho$, and a factor $(s-m)^{-1}$ in $h'$. Then (51.4) becomes

$$\epsilon'_y = (\mu+3c^*)\rho h'^{2n-2}\cos\theta, \; \epsilon'_z = (\mu+c^*)\rho h'^{2n-2}\sin\theta; \tag{51.5}$$

and the discussion of Section 19, based upon (19.14), may now be recalled. In a sense, therefore, $\mu$ is the $(2n-1)$th-order analogue of the Petzval curvature; nevertheless a closer examination of the absolute invariants $\alpha_{4N-1}$ shows them to be also to some extent analogous to the Petzval curvature, though in an entirely different sense, which derives from the way in which they depend on parameters which define the detailed structure of the system.

## 52. Conditional invariants and semi-invariants

In Section 48 we discussed in detail why relations such as (48.2) were to be rejected in the context of $s$-invariance. Indeed, this relation essentially merely restates the $s$-invariance of $p_1$, taken together with the form of the second member of (35.13). On the other hand, if $p_1 = 0$ for some value of $s$ (and therefore for every value of $s$), then $\hat{p}_2 = c^3 p_2$, so that *then* $(s-m)^3 p_2$ is an $s$-invariant. Similarly if $p_1$ *and* $p_2$ vanish simultaneously for some value of $s$ then $(s-m)^2 p_3$ will be an $s$-invariant; and so on, step-by-step down the set of equations (35.13), (44.8), .... Such conclusions are somewhat trivial in character. Yet they, when taken in conjunction with (46.10), provide a clue towards certain less trivial results.

Contemplate changes of $s$ alone, so that $p = 0$ in (46.10). $(s-m)^4 \times (4s_3 - s_4)$ is $s$-invariant, as we already know, so that, if $4s_3 - s_4$

vanishes, the second equation of (46.10) shows that $(s-m)^3 \times(2s_5-3s_7)$ is then also $s$-invariant; and here we have now only coefficients governing one type of aberration. We shall say that $(s-m)^3(2s_5-3s_7)$ is a *conditional s-invariant*. It will be noted that the coefficient of spherical aberration is not involved here, nor is that of circular coma, so that to this extent one has a state of affairs somewhat less trivial than that discussed above. Furthermore, we have now before us a result relating to a *comatic* aberration, namely elliptical coma. When, in particular, $2s_5-3s_7=0$ for one value of $s$, then it will vanish for all values of $s$, i.e. in (20.9) $k=2$ for every value of $s$, and the comatic flare is generated by ellipses whose eccentricity is independent of $s$.

Inspection of (46.10) shows that $(s-m)^3(2s_5-3s_7)$ is also a *conditional m-invariant*, but the condition implicit in this statement is a different one, namely that $4s_6-s_8$ should vanish for some value of $m$. If $4s_3-s_4$ *and* $4s_6-s_8$ vanish for some $s$ and $m$, then $(s-m)^3(2s_5-3s_7)$ will be independent of $s$ *and* $m$, and in this sense the quantity $(s-m)^3(2s_5-3s_7)$, which relates to one type of aberration alone, is a *conditional absolute invariant*.

The nature of these results, especially with regard to orders exceeding the fifth, is such that it is scarcely appropriate to pursue them here in any kind of detail. At any rate, it will suffice to remark that, as in Sections 50 and 51, it is not necessary to go through explicit equations such as (46.10). Instead one can use the method of generators after the fashion of Section 49. Here, however, one conveniently adds the following operators:

$$\left.\begin{aligned} S^* &= (s-m)^{-1}(2S+J), & \bar{S}^* &= -\left(\frac{\partial}{\partial\bar{\eta}}+2m\frac{\partial}{\partial\bar{\zeta}}\right),\\ M^* &= (s-m)^{-1}(2M+J), & \bar{M}^* &= \left(\frac{\partial}{\partial\bar{\eta}}+2s\frac{\partial}{\partial\bar{\zeta}}\right),\\ K &= (s-m)^{-2}(S+J+M), & \bar{K} &= \frac{\partial}{\partial\bar{\zeta}}. \end{aligned}\right\} \qquad (52.1)$$

Of course $\bar{S}^*=(s-m)^{-1}(2\bar{S}+\bar{J})$, and so on; but we have written the adjoints down explicitly so as to show that $\bar{S}^*$ is independent of $m$, $\bar{M}^*$ independent of $s$, and $\bar{K}$ independent of both $s$ and $m$. Also, it will be noted that $S^*+M^*=2(s-m)K$.

Now let $\Omega$ be an operator which is a product of the various generators, appropriately chosen; that is to say, in such a way that (i) the adjoint operator $\bar{\Omega}$ does not depend upon $s$ and $m$ simultaneously, (ii) $\Omega$ has $A$ and at least one of $S^*$, $M^*$, $K$ as a factor, and (iii) $\Omega t^{(n)}$ no longer depends on $\xi$, $\eta$, $\zeta$, nor does it vanish identically. Under these circumstances the application of $\bar{\Omega}$ to the right-hand member (49.5) yields a constant independent of $s$ or $m$ or both, as the case may be. It follows that $\Omega t^{(n)}$ is a linear combination of aberration coefficients which has the same properties; a combination, moreover, into which the coefficient of spherical aberration, at any rate, does not enter. The independence of $s$, or $m$, or both entails that the coefficients which enter into $\Omega t^{(n)}$ must transform linearly amongst themselves under changes, respectively, of $s$, or $m$, or both. The condition that certain of the coefficients in question vanish leaves the rest as conditional invariants or semi-invariants.

These remarks may be illustrated by some simple examples. In the primary domain $\Omega$ cannot contain any operators in addition to $A$, and so there are no primary conditional invariants of any kind. In the secondary domain we only have the possibilities $AS^*$, $AM^*$ and $AK$. One finds very easily that

$$\left.\begin{aligned} AS^*t^{(3)} &= (s-m)^3\,[4(4s_3-s_4)+2(2s_5-3s_7)], \\ AM^*t^{(3)} &= (s-m)^3\,[2(2s_5-3s_7)+4(4s_6-s_8)], \\ AKt^{(3)} &= 2(s-m)^2\,[(4s_3-s_4)+(2s_5-3s_7)+(4s_6-s_8)]. \end{aligned}\right\} \quad (52.2)$$

The three expressions on the right are independent of $s$, of $m$, and of both, respectively. On the other hand $(s-m)^3(2s_5-3s_7)$ is not independent of $s$, though $(s-m)^4(4s_3-s_4)$ is. Thus the vanishing of $4s_3-s_4$ for any value of $s$ entails the $s$-invariance of $(s-m)^3 \times(2s_5-3s_7)$. We see that we have recovered precisely the results discussed earlier in this section.

One final example, relating to tertiary coefficients, will suffice. Inspection of $ASS^*t^{(4)}$ shows that $(s-m)^5(4t_5-3t_7)$ is a conditional $s$-invariant, namely when $6t_3-t_4=0$. $AS^{*2}t^{(4)}$ and $ASKt^{(4)}$ yield two further conditional $s$-invariants. These are $(s-m)^4(8t_6-t_8)$ and $(s-m)^4(t_8-3t_{11})$, where now $6t_3-t_4$ and $4t_5-3t_7$ must both vanish (for some value of $s$). Again, three conditions must be satisfied for $(s-m)^3(4t_9-3t_{12})$ to be $s$-invariant; and this follows from considering $AS^*Kt^{(4)}$. Finally, upon evaluating $AK^2t^{(4)}$ it emerges that if, in addition, $4t_9 - 3t_{12} = 0$ then $6t_{10} - t_{13}$ is $s$-invariant. In short, under changes of $s$ the six groups $6t_3 - t_4$, $4t_5 - 3t_7$, $8t_6 - t_8$, $t_8 - 3t_{11}$, $4t_9 - 3t_{12}$ and $6t_{10} - t_{13}$ transform linearly amongst themselves; and this is in fact the case for simultaneous changes of $s$ and $m$, since $\bar{K}$ is independent of these. The six groups just enumerated, multiplied by appropriate powers of $s - m$ are evidently also conditional $m$-invariants. The absolute invariant $\alpha_7$ is implicit in these results, since one has, identically, $A = 4SK - S^{*2} = 4MK - M^{*2}$.

## Problems

**P.5 (i).** A system is free from spherical aberration of all orders, but circular coma of *all* orders is present. The circular comatic ratios have a fixed value for all orders. Show that

$$K = \frac{c\rho^2}{1+\rho^2},$$

where $c$ is a constant.

**P.5 (ii).** Obtain an analogue of the quantity $\Delta(\rho)$, intended to relate to that part of the aberration function which is responsible for *distortion.*

**P.5 (iii).** Show that unless $m = n$

$$\int_0^1 t^{m+1} R_{nm}(t)\, dt = 0.$$

**P.5 (iv).** Using (44.10) express the sum

$$4(n-2)(n-3)\hat{k}_6 - 2(n-3)\hat{k}_8 + 3\hat{k}_{11}$$

in terms of the $k_\alpha$.

**P.5 (v).** The primary aberrations associated with a given pair of conjugate planes, magnification $m$, are given to be zero. Show that primary spherical aberration and coma can vanish together for magnification $\hat{m}$ only if $\hat{m}^2 = 1$.

**P.5 (vi).** Using equation (4.5) show that the focal angle characteristic $\tilde{T}$ of a single spherical refracting surface is

$$\tilde{T} = \kappa^{-1}(1 + 2\kappa\chi)^{-\frac{1}{2}} + \alpha + \alpha',$$

where $\chi = 1 - (\alpha\alpha' + \beta\beta' + \gamma\gamma')$, and $f$ has been set equal to unity.

**P.5 (vii).** Find the spherical point characteristic $V^\dagger$ of a system which produces no aberrations other than distortion.

CHAPTER 6

# SYMMETRIC SYSTEMS WITH ADDITIONAL SYMMETRIES

## A. Reversible systems

### 53. Definition of the reversible system

A symmetric system $K$ will be called *reversible* if there exists a *normal* plane of symmetry $\mathscr{C}$; or, in other words, which is such that the part of $K$ which lies to the right of $\mathscr{C}$ is the mirror image of that part which lies to the left of it. For the purposes of this definition neither the object and image planes, nor the pupil planes are to be taken as being 'parts of $K$'. One might equivalently call a symmetric system reversible if there exists a line through the axis of $K$ and normal to it, such that $K$ is invariant under rotations through 180° about this line. This remark will appear less trivial if one notes that the alternative definitions are equivalent only on account of the symmetries which define the symmetric system. The situation is otherwise, for example, in the case of a system consisting of a single triangular cylinder; granted that the 'axis' is a line parallel to one side of the triangular cross-section. At any rate, it may be remarked that any symmetric system containing one reflecting surface is automatically reversible, granted that its refracting part is traversed twice by every ray. In this context the necessary imperfections of the imagery which will shortly be described are particularly interesting; and the reasons for using curved receiving surfaces are apparent.

The reversible system is a generalization of the traditional holosymmetric system, the latter being reversible in our sense with the additional stipulation that the stop must be central and the conjugate object and image planes symmetrically disposed; in other words, that $s = 1$ and $m = -1$. When these conditions are satisfied we shall call the system *completely reversible*. However, the absence of these restrictions makes the whole problem a good deal more interesting. Consequently, since general values of $s$ and $m$ are to be contemplated it is natural to carry out the present investigation on the basis of the angle characteristic. Except in the last part of Section 56 we therefore suppose $K$ throughout not to be telescopic: $f \neq \infty$; and we shall generally set $f = 1$, as usual.

We shall call $\mathscr{C}$ the *central plane*, and denote its axial point by $C$. Although we are ultimately interested in the angle characteristic $T(\xi, \eta, \zeta)$ referred to the base-points $O_0$ and $O_0'$, the present context dictates a temporary choice of base-points which are symmetrically disposed about $C$. Accordingly we choose $O_0'$ as posterior base-point, whilst as anterior base-point we take the axial point $O_0^*$

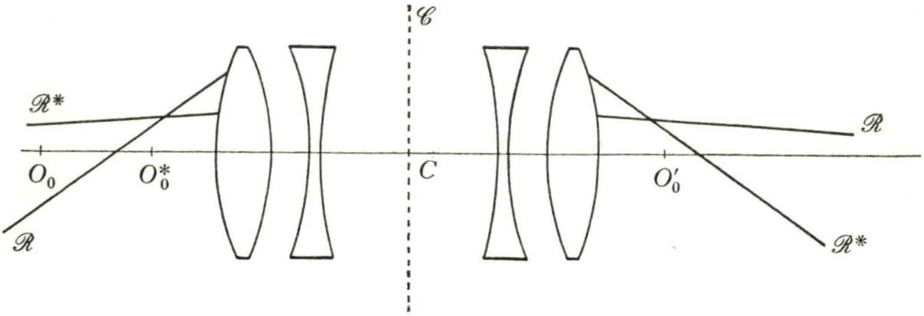


Fig. 6.1

which lies as far to the left of $C$ as $O_0'$ lies to the right of it. The corresponding angle characteristic, regarded as a function of $\bar{\xi}, \bar{\eta}, \bar{\zeta}$, will be denoted by $T^*$. Now let $\boldsymbol{\beta}, \boldsymbol{\beta}'$ be the values of the ray-coordinates of some ray $\mathscr{R}$ through $K$. With $\mathscr{R}$ we can associate the ray $\mathscr{R}^*$ which is the 'reflection' of $\mathscr{R}$ in $\mathscr{C}$. The values of the ray-coordinates of $\mathscr{R}^*$ are obviously $-\boldsymbol{\beta}'$, $-\boldsymbol{\beta}$. The optical distances between the feet of the normals drawn from the base-points on to the rays $\mathscr{R}$ and $\mathscr{R}^*$ are clearly the same for both, since the 'original' and the 'reflected' systems are indistinguishable; see Fig. 6.1. (We have chosen spherical refracting surfaces merely for the sake of illustration.) It follows that the basic identity

$$T^*(\bar{\xi}, \bar{\eta}, \bar{\zeta}) = T^*(\bar{\zeta}, \bar{\eta}, \bar{\xi}) \tag{53.1}$$

obtains; and this must ultimately yield all the special properties which reversibility entails. It should be added that (53.1) is a necessary but not a sufficient condition for reversibility. We shall see later, in Section 64, that concentric systems satisfy (53.1), but they need not be reversible.

The distance $d^* = O_0 O_0^*$ is easily calculated. If, for the moment, we set $s = 1$, $E$ and $E'$ will be symmetrically situated with respect to $C$, and $O_0^* E = E' O_0' = d'$. Then $d^* = d - d'$, or

$$d^* = -(1 - m^2)/m, \tag{53.2}$$

from (14.27) and (14.28), with $s = 1$.

We shall develop the theory of reversible systems *ab initio*, that is to say, for general values of $s$ and $m$. To a certain extent this entails some redundancy, for one could confine oneself in the first instance to the particularly simple case of complete reversibility (Section 59). Subsequently general values of $s$ and $m$ are then accounted for by using the results of Section 46, relating to joint changes of $s$ and $m$.

## 54. The basic identity ($m \neq 0$)

The stage has been reached for reverting to the usual base-points $O_0$, $O_0'$. It is convenient for the time being to require that $m$ be not zero, reserving the special case $m = 0$ for separate treatment in Section 60. Then

$$T = T^* + d^*\alpha. \tag{54.1}$$

Incidentally, for once we need not contemplate the use of reduced distances, on account of the fact that, necessarily, $N = N'$; so that we can set $N = N' = 1$, without loss of generality.

Now let $\xi, \eta, \zeta$ go over into $\hat{\xi}, \hat{\eta}, \hat{\zeta}$ when $\bar{\xi}$ and $\bar{\zeta}$ are interchanged. Then (53.1) and (54.1) jointly imply the basic identity

$$T(\xi, \eta, \zeta) - d^*(1 - \bar{\zeta})^{\frac{1}{2}} = T(\hat{\xi}, \hat{\eta}, \hat{\zeta}) - d^*(1 - \bar{\xi})^{\frac{1}{2}}. \tag{54.2}$$

By virtue of (53.2) the linear terms of this mutually cancel. Then, once again absorbing all non-linear terms of $g(\zeta)$ in the aberration function, (54.2) becomes

$$t^{(n)}(\xi, \eta, \zeta) - \kappa_n \bar{\zeta}^n \equiv t^{(n)}(\hat{\xi}, \hat{\eta}, \hat{\zeta}) - \kappa_n \bar{\xi}^n, \tag{54.3}$$

where
$$\kappa_n = (-1)^n \binom{\frac{1}{2}}{n} d^*. \tag{54.4}$$

The expressions for $\hat{\xi}, \hat{\eta}, \hat{\zeta}$ in terms of $\xi, \eta, \zeta$ may be obtained directly from the results of Section 43. Here $\bar{a} = 0, \bar{b} = -1, \bar{c} = -1, \bar{d} = 0$. Then, for instance,

$$\hat{\xi} = (s-m)^{-2}[(1-sm)^2\xi - 2(1-s)^2(1-sm)\eta + (1-s^2)^2\zeta], \tag{54.5}$$

with two similar equations for $\hat{\eta}$ and $\hat{\zeta}$. These, and (15.12), now need to be inserted in (54.3); and then the required relations between the aberration coefficients may be read off. However, it is immediately obvious that such relations will generally involve the coefficient $t^{(n)}_{nn}$. This, on the other hand, does not enter into the displacement (regarded as a function of the variables $\boldsymbol{\mu}$ and $\boldsymbol{\sigma}$ of Section 24). In other words this coefficient is to be eliminated, and this gives rise to tedious and irrelevant manipulations. We shall, indeed, see in the next section how they can be avoided.

## 55. Introduction of auxiliary variables

To circumvent the need to remove the unwanted coefficients $t^{(n)}_{nn}$ from the relations between the aberration coefficients which are implied by the reversibility of $K$, we introduce the auxiliary variables $L$, $M$, $N$, defined as

$$\left.\begin{aligned}(1-m^2)^2 L &= m^2\bar{\xi}-2m\bar{\eta}+\bar{\zeta},\\ (1-m^2)^2 M &= m\bar{\xi}-(1+m^2)\bar{\eta}+m\bar{\zeta},\\ (1-m^2)^2 N &= \bar{\xi}-2m\bar{\eta}+m^2\bar{\zeta},\end{aligned}\right\} \tag{55.1}$$

the inverse relations being

$$\bar{\xi} = m^2L-2mM+N, \quad \bar{\eta} = mL-(1+m^2)M+mN,$$
$$\bar{\zeta} = L-2mM+m^2N. \tag{55.2}$$

Here we have supposed that $m^2 \neq 1$, reserving the very special case $m^2 = 1$ for separate treatment in Section 59. Then, temporarily writing

$$u = 1-sm, \quad v = s-m, \quad w = 1-m^2, \tag{55.3}$$

we find that

$$\xi = u^2L+2uvM+v^2N, \quad \eta = w(uL+vM), \quad \zeta = w^2L. \tag{55.4}$$

The left-hand member of (54.3) now becomes

$$t^{(n)}\big(u^2L+2uvM+v^2N,\, w(uL+vM),\, w^2L\big)-\kappa_n(L-2mM+m^2N)^n. \tag{55.5}$$

It will be seen that through the introduction of the auxiliary variables $L$, $M$, $N$ a situation of considerable simplicity has been attained. Thus, the basic mutual interchange of $\bar{\xi}$ and $\bar{\zeta}$ merely implies the interchange of $L$ and $N$. This entails that $\hat{\xi}, \hat{\eta}, \hat{\zeta}$ arise from $\xi, \eta, \zeta$ merely by interchanging $L$ and $N$. Consequently the required relations, i.e. those which do not involve $t^{(n)}_{nn}$, are just those which express the equality of the factors of $L^{n-\mu}M^{\mu-\nu}N^{\nu}$ on the one hand and of $L^{\nu}M^{\mu-\nu}N^{n-\mu}$ on the other in (55.5), where $\mu \neq 0$ or $n$.

## 56. The third-order relation ($m^2 \neq 1$)

As usual, in particular in Section 44, we write

$$t^{(2)} = p_1\xi^2 + p_2\xi\eta + p_3\xi\zeta + p_4\eta^2 + p_5\eta\zeta + p_6\zeta^2, \qquad (56.1)$$

i.e. the initial factor $(s-m)^{-1}$ on the right of (19.24) is again omitted. (The Seidel coefficients are then given by

$$\sigma_1 = 4(s-m)p_1, \quad \dots, \quad \sigma_5 = (s-m)p_5.)$$

The expression (55.5) reads, for $n = 2$,

$$\begin{aligned} &p_1(u^2L + 2uvM + v^2N)^2 + p_2 w(u^2L + 2uvM + v^2N)(uL + vM) \\ &\quad + p_3 w^2(u^2L + 2uvM + v^2N)L + p_4 w^2(uL + vM)^2 \\ &\quad + p_5 w^3(uL + vM)L + p_6 w^4 L^2 + \tfrac{1}{8}d^*(L - 2mM + m^2N)^2. \qquad (56.2) \end{aligned}$$

Evidently only *one* relation is obtained from this, namely that which arises from the equality of the factors multiplying $LM$ and $MN$ respectively; but let it be recalled that $m^2 \neq 1$. Thus

$$\begin{aligned} &4uv(u^2 - v^2)p_1 + vw(3u^2 - v^2)p_2 + 2uvw^2(p_3 + p_4) \\ &\qquad + vw^3 p_5 = \tfrac{1}{2}d^* m(1 - m^2). \qquad (56.3) \end{aligned}$$

Using (53.2) and (55.3) this becomes

$$\begin{aligned} &4(1 - s^2)(1 - sm)p_1 + [3(1 - sm)^2 - (s - m)^2]p_2 \\ &\quad + 2(1 - m^2)(1 - sm)(p_3 + p_4) + (1 - m^2)^2 p_5 = -\tfrac{1}{2}(1 - m^2)/(s - m). \\ &\qquad (56.4) \end{aligned}$$

Finally we go over to the Seidel coefficients, and then

$$\begin{aligned} &(1 - s^2)(1 - sm)\sigma_1 + [3(1 - sm)^2 - (s - m)^2]\sigma_2 + (1 - m^2)(1 - sm) \\ &\qquad \times (3\sigma_3 + \sigma_4) + (1 - m^2)^2 \sigma_5 = -\tfrac{1}{2}(1 - m^2). \qquad (56.5) \end{aligned}$$

As a consequence of reversibility one therefore has *one* relation between the five Seidel coefficients. It is remarkable for the fact that, provided $m^2 \neq 1$, it is inhomogeneous. In other words, *when $m^2 \neq 1$ the primary aberrations of the symmetric reversible system* *cannot be simultaneously reduced to zero.* We shall see later that an analogous conclusion holds for the aberrations of any order.

When the stop is central (i.e. lies in $\mathscr{C}$) $s = 1$, and (56.5) reduces to

$$2\sigma_2+(1+m)(3\sigma_3+\sigma_4)+(1+m)^2\sigma_5 = -\tfrac{1}{2}(1+m)/(1-m). \quad (56.6)$$

Although we have supposed that $m \neq 0$ we may let $m \to 0$ here, with the intention of confirming later (in Section 60) that the result so obtained is correct. One thus gets

$$2\sigma_2+3\sigma_3+\sigma_4+\sigma_5 = -\tfrac{1}{2}. \quad (56.7)$$

This equation relates just to the conditions for which photographic objectives have sometimes been designed in the past. It is of course valid irrespectively of the particular structure which $K$ may possess. There is no indication that 'coma should be small'. Indeed, if tangential curvature of field and distortion have been removed, one is *necessarily* left with coma of amount

$$\sigma_2 = -\tfrac{1}{4}, \quad (56.8)$$

where of course a factor $f^{-2}$ has to be inserted on the right if $f \neq 1$. Similarly, if the system has been designed so as to produce a *sharp* image, then one will be left with distortion of amount

$$\sigma_5 = -\tfrac{1}{2}; \quad (56.9)$$

see also Section 62. At any rate one should bear in mind that by adhering strictly to reversibility one severely limits oneself from the outset with regard to the extent to which satisfactory imagery may be attained.

Although $p_6$ has so far been left out of account it is worth remarking that its value is determined by that of the other aberration coefficients. This is fortunate, bearing in mind that $p_6$ enters into the effective higher-order displacements. From the equality of the factors multiplying $L^2$ and $N^2$ respectively in (56.2), we find that

$$(1-s^2)(u^2+v^2)p_1+u^3p_2+u^2w(p_3+p_4)+uw^2p_5+w^3p_6 = \tfrac{1}{8}(1-m^4)/m. \quad (56.10)$$

It may be noted that for any given value of $m$ a particular value of $s$ gives results of great simplicity; see also Section 58. It is such that $E'$ lies as far to the right of $C$ as $O_0$ lies to the left of it. This condition requires that $d'+d^* = 0$, i.e. $s = 1/m$, or

$$u = 0. \quad (56.11)$$

In that case (56.5) reduces to a condition involving $\sigma_2$ and $\sigma_5$, i.e. the comatic coefficients, only:

$$\sigma_2/m^2 - \sigma_5 = \tfrac{1}{2}(1-m^2)^{-1}. \tag{56.12}$$

Finally, we must again draw attention to the fact that $K$ has throughout been supposed to be non-telescopic. In the contrary case, i.e. when $f=\infty$, we must abandon the angle characteristic. Consider therefore $V$, taken with respect to base-points $O_0$ and $B'$, where $B'$ shall lie as far to the right of $C$ as $O_0$ lies to the left of it. It is assumed that the point so defined does not happen to be conjugate to $O_0$. Under these conditions the linear function on the right of (14.37) must be invariant under the mutual interchange of $\xi$ and $\zeta$, which entails that

$$m^2 = 1. \tag{56.13}$$

It looks at first sight as if we were confronted with the state of affairs to be studied separately in Section 59. This is, however, not so: in the non-telescopic situation the case $m^2 = 1$ is 'special', because $O_0$ and $O_0'$ are then equidistant from $C$, whereas this need not be so here; we have, in fact, just excluded this special case. Since the ideal point characteristic is unaffected when $\xi$ and $\zeta$ are interchanged, reversibility simply requires

$$v^{(n)}(\xi,\eta,\zeta) \equiv v^{(n)}(\zeta,\eta,\xi). \tag{56.14}$$

In particular, in the primary domain, $\sigma_2 = \sigma_5$.

We are naturally inclined to wonder whether this result could not have been arrived at from those already obtained by means of some limiting process. It is instructive to see how this may, in fact, be done. Let $f$ therefore be assumed to be large but finite. Then the present choice of $B'$ implies that $d' = (1-m^2)f/m$. Accordingly, let $\epsilon$ be a parameter which is ultimately to tend to zero, and set

$$m^2 = 1+\epsilon, \quad s = 1/m, \quad f^2 = [(1+\epsilon)/\epsilon^2]\,d'^2; \tag{56.15}$$

in harmony with the relations between $s, m, f$ and $d'$. Restoring the factor $f^{-2}$ explicitly in the right-hand member of (56.12), the latter equation becomes

$$\sigma_2 - \sigma_5 = (\sigma_5 - \tfrac{1}{2}d'^{-2})\epsilon. \tag{56.16}$$

On allowing $\epsilon$ to tend to zero, one recovers the stated result

$\sigma_2 = \sigma_5$. We are thus in the fortunate, but hardly surprising, position of being able to deal with telescopic systems by means of this kind of limiting device.

## 57. The fifth-order relations

The secondary expression analogous to (56.2) is obtained by substituting from (55.2) and (55.4) in the polynomial

$$s_1\xi^3 + s_2\xi^2\eta + \ldots + s_{10}\zeta^3 + \tfrac{1}{16}d^*\bar{\zeta}^3.$$

The fifth-order relations are those which express the equality of the factors multiplying (i) $L^2M$ and $MN^2$, (ii) $L^2N$ and $LN^2$, (iii) $LM^2$ and $M^2N$, respectively. They are

$$\left.\begin{aligned}
&\text{(i)}\quad 6uv(u^2+v^2)(1-s^2)s_1 + v(5u^4-v^4)s_2 + 4u^3vw(s_3+s_4)\\
&\qquad + 3u^2vw^2(s_5+s_7) + 2uvw^3(s_6+s_8)\\
&\qquad + vw^4s_9 = -\tfrac{3}{8}(1-m^4),\\
&\text{(ii)}\quad 3u^2v^2(1-s^2)s_1 + uv^2(2u^2-v^2)s_2 + v^2w(2u^2-v^2)s_3\\
&\qquad + u^2v^2ws_4 + uv^2w^2s_5 + v^2w^3s_6 = \tfrac{3}{16}m(1-m^2),\\
&\text{(iii)}\quad 12u^2v^2(1-s^2)s_1 + 4uv^2(2u^2-v^2)s_2 + 4u^2v^2ws_3\\
&\qquad + v^2w(5u^2-v^2)s_4 + uv^2w^2(2s_5+3s_7)\\
&\qquad + v^2w^3s_8 = \tfrac{3}{4}m(1-m^2).
\end{aligned}\right\} \qquad (57.1)$$

By subtraction one finds from the last two equations that

$$(1-s^2)(4s_3-s_4) + (1-sm)(2s_5-3s_7) + (1-m^2)(4s_6-s_8) = 0; \qquad (57.2)$$

but no further substantial simplification appears possible. It is interesting, though not very surprising, that the (conditional) invariants which were studied in Section 52 reappear here. At any rate, at least one of the relations must be inhomogeneous, so that not all fifth-order coefficients can be simultaneously reduced to zero.

When the stop is central and $m$ is allowed to tend to zero, one gets the secondary counterparts to (56.7). After some simplification

$$\left.\begin{aligned}
3s_7 + 3s_8 + 2s_9 &= -\tfrac{3}{4},\\
2s_5 + 4s_6 - 3s_7 - s_8 &= 0,\\
s_2 + s_3 + s_4 + s_5 + s_6 &= 0.
\end{aligned}\right\} \qquad (57.3)$$

When the characteristic comatic coefficients have been reduced to zero, i.e. $s_2 = s_5 = s_7 = s_9$ one is then left with

$$s_3+s_4 = \tfrac{1}{16}, \quad s_6 = -\tfrac{1}{16}, \quad s_8 = -\tfrac{1}{4}, \tag{57.4}$$

with factors $f^{-4}$ on the right when $f \neq 1$. Under these conditions the secondary meridional pseudo-displacement is known exactly. In fact

$$\epsilon' = \tfrac{1}{8}f^{-4}(2\rho^3h'^2 - 5\rho h'^4). \tag{57.5}$$

When amongst the characteristic secondary coefficients that of distortion alone is non-zero it must have the value

$$s_9 = -\tfrac{3}{8}f^{-4}. \tag{57.6}$$

The following point is of interest. Suppose that all the primary and secondary aberration coefficients other than those of distortion vanish. Then, according to (24.13) the effective coefficients $\bar{p}_3^*, \bar{s}_5^*, s_6^*, \bar{s}_6^*$ will not be zero for general values of $s$ and $m$, bearing in mind that $p_6 \neq 0$, as we saw previously. The system can therefore not form a sharp image in $\mathscr{I}'$, except possibly when $m = 0$, for then $\bar{s}_5^* = s_6^* = 0$. We therefore have the interesting questions whether (when $m^2 \neq 1$) the system can at least form a sharp image in some surface other than $\mathscr{I}'$, and whether this surface, if it exists, can ever be plane. The answers to these will be provided in Sections 62 and 63.

## 58. The relations of all orders when $s = 1/m$

One can proceed to the seventh and higher orders after the fashion of Section 57, but it is hardly feasible to write out in full the relations of order $2n-1$ for general values of $s$ and $m$. However, it may be reflected that the effects of stop shifts on the coefficients of the angle characteristic are rather easily accounted for, as we saw in Section 44. Consequently, we may choose some convenient value of $s$ and find the general relations for this; and, if required, the stop may finally be shifted so as to correspond to the actual value of $s$ which may happen to be of interest. This procedure would thus be a special case of that outlined at the end of Section 53.

In view of the simplicity of (56.12) we choose $s = 1/m$, i.e. $u = 0$. Then, defining a set of purely numerical coefficients $b_{\mu\nu}^{(n)} (\equiv b_{n-\nu,n-\mu}^{(n)})$

by the equation

$$(L-2mM+m^2N)^n \equiv \sum_{\mu,\nu} b^{(n)}_{\mu\nu} m^{2n-\mu-\nu} N^{n-\mu} M^{\mu-\nu} L^\nu, \tag{58.1}$$

the left-hand member of (54.3) becomes

$$\sum_{\mu,\nu} [t^{(n)}_{\mu\nu} w^{2n} m^{-2n+\lambda} - \kappa_n b^{(n)}_{\mu\nu}] N^{n-\mu} M^{\mu-\nu} L^\nu, \tag{58.2}$$

where $\lambda = \mu+\nu$; bearing in mind that now $v = w/m$. Excluding again the case $m^2 = 1$, we immediately read off the whole set of required relations,

$$t^{(n)}_{\mu\nu} - m^{2n-2\lambda} t^{(n)}_{n-\nu,n-\mu} = (-1)^n \binom{\frac{1}{2}}{n} b^{(n)}_{\mu\nu} m^{2n-1}(1-m^2)^{-2n+1}(1-m^{2n-2\lambda}). \tag{58.3}$$

Here $\lambda = 1, 2, \ldots, n-1$, the value $\lambda = 0$ corresponding to a relation which involves the unwanted coefficient $t^{(n)}_{nn}$. Evidently, as $\lambda$ increases, successive relations involve just two coefficients at a time, *both* of which are alternately of comatic and astigmatic type; specifically, circular coma and distortion when $\lambda = 1$, oblique spherical aberration and linear astigmatism when $\lambda = 2$, cubic coma and elliptical coma when $\lambda = 3$, and so on. Since the number of relations of any order is not affected by the value of $s$, we can obtain it directly from (58.3) by counting the number of allowed sets of values of $\mu$ and $\nu$. $\lambda$, we know, ranges from 1 to $n-1$. For any value of $\lambda$, $\mu$ can then range from 0 to $\frac{1}{2}\lambda$ or $\frac{1}{2}(\lambda-1)$, according as $\lambda$ is even or odd. Thus, when $n$ is even, there are $2+3+\ldots+\frac{1}{2}n$ possibilities corresponding to even values of $\lambda$, and $1+2+\ldots+\frac{1}{2}n$ corresponding to odd values. Proceeding similarly when $n$ is odd we find in this way that the number of relations of order $2n-1$ is

$$\left.\begin{array}{ll} \frac{1}{4}(n^2+2n-4) & \textit{when } n \textit{ is even,} \\ \text{and} \quad \frac{1}{4}(n-1)(n+3) & \textit{when } n \textit{ is odd.} \end{array}\right\} \tag{58.4}$$

Since the number of coefficients is $\frac{1}{2}n(n+3)$, the number of independent coefficients of order $2n-1$ is

$$\left.\begin{array}{ll} \frac{1}{4}(n+2)^2 & \textit{when } n \textit{ is even,} \\ \text{and} \quad \frac{1}{4}(n+1)(n+3) & \textit{when } n \textit{ is odd.} \end{array}\right\} \tag{58.5}$$

Thus there are only 4, 6, 9, 12, ... independent coefficients of orders 3, 5, 7, 9, ....

## 59. The relations of all orders when $m^2 = 1$

The time has come to investigate the consequences of reversibility when $m^2 = 1$, for this case has hitherto been strictly excluded. $d^*$ now vanishes, and (54.3) reduces to

$$t^{(n)}(\xi, \eta, \zeta) \equiv t^{(n)}(\hat{\xi}, \hat{\eta}, \hat{\zeta}). \tag{59.1}$$

Using (43.4) and (43.5) one gets the simple result

$$\hat{\xi} = \xi - 2j\eta + j^2\zeta, \quad \hat{\eta} = -\eta + j\zeta, \quad \hat{\zeta} = \zeta, \tag{59.2}$$

where
$$j = 1 \pm s \quad \textit{when} \quad m = \pm 1. \tag{59.3}$$

A new important feature arises here on account of the fact that $t^{(n)}_{nn}$ disappears from (59.1), so that the need to eliminate it does not arise now. Accordingly one has in every order one additional relation, as compared with the number of relations which obtains when $m^2 \neq 1$, i.e. that given by (58.4).

The primary coefficients turn out to be subject to the relations

$$\left.\begin{aligned} 2jp_1 + p_2 &= 0, \\ j^2p_2 + 2j(p_3 + p_4) + 2p_5 &= 0. \end{aligned}\right\} \tag{59.4}$$

The first of these corresponds to (56.4), but the second is the additional primary relation just referred to. There are no inhomogeneous relations now, so that there is no *a priori* reason against the achievement of perfect imagery.

The most interesting situation is perhaps that of complete reversibility, $s = 1$ and $m = -1$. Then (59.2) shows that $t^{(n)}(\xi, \eta, \zeta)$ must be invariant under the reversal of the sign of $\eta$. Therefore the reversibility conditions are here simply

$$t^{(n)}_{\mu\nu} = 0 \quad \textit{when} \quad \mu - \nu \quad \textit{is odd}. \tag{59.5}$$

However, when $\mu - \nu$ is odd so is $\mu + \nu$, and reference to Section 22 tells us that this means exactly the vanishing of all characteristic coefficients governing aberrations of comatic type. In particular,

$$\left.\begin{aligned} &\text{(i) } \textit{in third order}: && p_2 = p_5 = 0, \\ &\text{(ii) } \textit{in fifth order}: && s_2 = s_5 = s_7 = s_9 = 0, \end{aligned}\right\} \tag{59.6}$$

and so on.

The following remarks concerning the limit $m \to -1$ (when $s = 1$) may not be out of place. It is clear that, on account of the factors $(1-m^2)^2$ which are present on the left of (55.1), this limit is not as tractable as the limit $m \to 0$. Suppose one simply sets $m = -1$ in (56.6) and (56.10)—with $s = 1$ in the latter—then the first of these gives $p_2 = 0$, but the second gives $p_2 = 0$ again, and not $p_5 = 0$ as one might possibly have expected. There is, however, no inconsistency, for, by simple subtraction, $p_2$, $p_3$, $p_4$ may be eliminated between the two equations in question. If a factor $(1+m)^2$ be removed from the resulting equation, a factor which becomes zero as $m \to -1$, one is left with

$$p_5 + 2(1+m)p_6 = \tfrac{1}{4}(1+m)/[m(1-m)^2], \tag{59.7}$$

and now, as $m \to -1$, one gets $p_5 = 0$.

A few words may also be added concerning the procedure outlined at the end of Section 53. It will be amply sufficient to do so in the context of the primary coefficients, since the generalization to all orders is obvious. In the equations of Section 46 we set $\hat{s} = 1$, $\hat{m} = -1$, so that the $\hat{p}_\alpha$ are the 'known' coefficients; and

$$\hat{p}_2 = \hat{p}_5 = 0. \tag{59.8}$$

Then
$$\left.\begin{aligned} p &= (1+m)/(1+s), & q &= (1-s)/(1-m), \\ b &= \tfrac{1}{2}(1-m), & c &= \tfrac{1}{2}(1+s), \end{aligned}\right\} \tag{59.9}$$

and
$$j_1 = -\tfrac{1}{128}(1+m). \tag{59.10}$$

(59.8) then imply two relations between the $p_\alpha$, and when $p_6$ is eliminated between them, one gets just (56.4); though the work involved in this process is quite tedious. If we take $s = 1$, however, the situation is somewhat simpler, for then $q = 0$. In fact one finds at once that $\hat{p}_5 = 0$ is exactly equivalent to equation (59.7).

## 60. The relations when the object is at infinity

Infinite object distance corresponds to $m$ being zero. In Section 56 we already dealt with this case by letting $m$ tend to zero from finite values. We want to confirm that the results thus obtained are in fact correct, limiting processes being now excluded. For this purpose we have to choose a new anterior base-plane, and we take this to be $\mathscr{E}$. According to equation (15.17) we then have

$$T = g(\zeta)+s^{-1}\eta(1-\zeta)^{-\frac{1}{2}}+t(\xi,\eta,\zeta). \tag{60.1}$$

The distance $O_0^*E$ may be found in the same way as $d^*$ was determined at the end of Section 53, and it turns out to be $1/s$. Hence

$$T^* = T+s^{-1}\alpha. \tag{60.2}$$

If all non-linear terms which depend upon $\zeta$ alone be now absorbed in $t$ as usual,

$$T^* = g_0-\bar{\eta}(1-\bar{\zeta})^{-\frac{1}{2}}+t(\xi,\eta,\zeta),$$

with $\quad \xi = s^2\bar{\xi}-2s\bar{\eta}+\bar{\zeta}, \quad \eta = -s\bar{\eta}+\bar{\zeta}, \quad \zeta = \bar{\zeta}.$

Accordingly

$$t^{(n)}(s^2\bar{\xi}+2s\bar{\eta}+\bar{\zeta},\bar{\eta}+s\bar{\zeta},\bar{\zeta})+(-1)^{n-1}\binom{-\frac{1}{2}}{n-1}\bar{\eta}\bar{\zeta}^{n-1} \tag{60.3}$$

must be invariant under the mutual interchange of $\bar{\xi}$ and $\bar{\zeta}$. Here the sign of $\bar{\eta}$ has been reversed as a matter of convenience. This is certainly permissible since the invariance of the expression (60.3) is indifferent to the sign of $\bar{\eta}$.

(60.3) may now be compared directly with (55.5). To the latter we arbitrarily add a term $\kappa_n L^n$ on the formal grounds that, as long as we disregard the factors multiplying $L^n$ and $N^n$, all conclusions relating to factors which do not involve $t_{nn}^{(n)}$ will be unaffected. Proceeding to the limit $m = 0$, one has, first, that $u = 1$, $v = s$, $w = 1$. Again

$$\begin{aligned}\lim_{m\to 0}\kappa_n\,&[(L-2mM+m^2N)^n-L^n]\\ &= \lim_{m\to 0}(-1)^{n-1}\binom{\frac{1}{2}}{n}(1-m^2)\,m^{-1}(-2nmL^{n-1}M+\dots)\\ &= -2(-1)^{n-1}\binom{\frac{1}{2}}{n}nL^{n-1}M = -(-1)^{n-1}\binom{-\frac{1}{2}}{n-1}L^{n-1}M.\end{aligned}$$

If one now formally identifies $L$ with $\bar{\zeta}$, $M$ with $\bar{\eta}$, and $N$ with $\bar{\xi}$, the limit of (55.5), modified by the addition of $\kappa_n L^n$, is identical with (60.3). It follows that all the results previously derived from (55.5) with $m \neq 0$ will give valid results in the limit $m \to 0$; excepting of course any relations containing the coefficients $t_{nn}^{(n)}$. From (60.3), with the new anterior base-point, the relation giving $t_{nn}^{(n)}$ is

$$\sum_{\mu,\nu} t_{\mu\nu}^{(n)} = s^{2n}t_{00}^{(n)}. \tag{60.4}$$

When $n = 2$ the same result follows from (56.10) if one simply omits the right-hand member from it. This formal device again reflects the difference between the present angle characteristic and that contemplated in Section 54.

## 61. Remark on the effective aberration coefficients

From very elementary geometrical considerations one infers easily that when the system is completely reversible the ray from $O$ which passes through $C$, i.e. the central point of the stop here, also passes through $I'$. For this reason one traditionally states that a completely reversible system produces no distortion at all. Let us therefore investigate this result in the light of the results obtained above.

In the primary domain, since $p_5 = 0$, the effective coefficient of distortion $\bar{p}_3^*$ also vanishes. Next, going on to the last member of (24.13), we find in the first place, since $s_9 = p_5 = 0$, that

$$\bar{s}_6^* = 2a[8(p_3+p_4)p_6+2a(p_3+p_4)].$$

However, since we are no longer including the factor $(s-m)^{-1}$ which appears on the right-hand side of (24.1), we have to supply every primary coefficient in the equation just quoted with an additional factor $s-m$ ($=2$, here). Also, $a = \frac{1}{4}$, so that we finally get

$$\bar{s}_6^* = \tfrac{1}{2}(p_3+p_4)(32p_6+1) \qquad (61.1)$$

for the effective fifth-order coefficient of distortion; and this in general fails to vanish.

The contradiction is only apparent, for it is brought about by diverse definitions of distortion. The traditional definition relates to the displacement of principal rays, i.e. rays through the axial point of the stop. In the work above we have throughout taken distortion to refer to the displacement of rays through the axial point of the paraxial exit pupil, and occasionally called *these* principal rays. The two alternative definitions are clearly in conflict with one another unless spherical aberration has been removed for the pupil planes.

To illustrate these remarks let us suppose that, in fact, not only spherical aberration, but all aberrations associated with the pupil planes have been removed. The angle characteristic referred to $E$ and $E'$ as base-points must therefore be a function $\phi$ of $\bar{\xi}-2\bar{\eta}+\bar{\zeta}$

alone, since $s = 1$. Hence the angle characteristic referred to $O_0$ and $O_0'$ as base-points is

$$T = \phi(\xi) + 2(1-\bar{\xi})^{\frac{1}{2}} + 2(1-\bar{\zeta})^{\frac{1}{2}},$$

since $d = d' = 2$; that is to say,

$$T = \phi(\xi) + 2[1-\tfrac{1}{4}(\xi-2\eta+\zeta)]^{\frac{1}{2}} + 2[1-\tfrac{1}{4}(\xi+2\eta+\zeta)]^{\frac{1}{2}}. \quad (61.2)$$

Under these conditions we therefore know the values of all aberration coefficients except those of spherical aberration. In particular, if $\phi = \phi_0 + \phi_1\xi + \phi_2\xi^2 + \ldots$, we have

$$t^{(2)} = (\phi_2 - \tfrac{1}{32})\,\xi^2 - \tfrac{1}{32}(2\xi\zeta + 4\eta^2 + \zeta^2), \quad (61.3)$$

i.e. $\quad p_2 = 0, \quad p_3 = -\tfrac{1}{16}, \quad p_4 = -\tfrac{1}{8}, \quad p_5 = 0, \quad p_6 = -\tfrac{1}{32}.$ (61.4)

This value of $p_6$ correctly reduces $\bar{s}_6^*$ to zero.

There will, in general, be residual comatic displacements of the fifth and higher orders (cf. Section 25). In other words, complete reversibility of $K$ does *not* guarantee the absence of coma of all kinds. Elementary demonstrations of the 'absence of coma' do not conflict with this conclusion, since they rest largely upon the consideration of certain meridional rays which intersect in the object space in points not in $\mathscr{I}$. A precise interpretation of such results is therefore difficult to attain, and it is hardly worth pursuing this topic any further.

## 62. On the attainability of a sharp image when $m^2 \neq 1$

The time has come to answer the question raised at the end of Section 57. It was this: granted that $m^2 \neq 1$, can a reversible system form a sharp image at all, and if so what can be said about the nature of the image surface? It is advisable to treat the two cases $m \neq 0$ and $m = 0$ separately, and we proceed to deal with the first of these. For this purpose it appears best to use the point characteristic, associated, as usual, with the base-points $O_0$ and $E'$. The value of $s$ is clearly irrelevant here, and we choose it in such a way that $E'$ lies as far to the right of $C$ as $O_0$ lies to the left of it, i.e. $s = 1/m$, according to (56.11). $d' \neq 0$ since $m^2 \neq 1$, and we therefore set it equal to unity as usual.

Take Cartesian axes with origin at $O_0'$, and let the coordinates of points on the image surface $\mathscr{I}^{*\prime}$ be $X'$, $Y'$, $Z'$; so that $\mathbf{Y}'$ does not, in general, refer to $\mathscr{I}'$ now. Formation of a sharp image lying in $\mathscr{I}^{*\prime}$ of a plane object in $\mathscr{I}$ requires that all rays from any point $O(\mathbf{y})$ of $\mathscr{I}$ shall intersect in just the point $O'(X', \mathbf{Y}')$ of $\mathscr{I}^{*\prime}$. If this is, indeed, to be possible, there must therefore exist functions $C(\zeta)$ and $D(\zeta)$ such that

$$X' = C(\zeta), \quad \mathbf{Y}' = \mathbf{y}D(\zeta), \tag{62.1}$$

$\xi$, $\eta$, $\zeta$ being the rotational invariants (13.1). It is obvious that one must have

$$C(0) = 0, \quad D(0) = m. \tag{62.2}$$

We are evidently looking for an ideal point characteristic which is, for once, defined with respect to a curved image surface. We can simply transcribe (7.4) into the present context. Because of symmetry, $g(y, z)$ must be a function of $\zeta$ alone, say $g(\zeta)$, as usual. $d'$ becomes $1 + C(\zeta)$ and in place of $m\mathbf{y}$ we must take $\mathbf{y}D(\zeta)$. Hence, at once,

$$V = g(\zeta) - [\phi(\xi, \zeta) - 2\eta D(\zeta)]^{\frac{1}{2}}, \tag{62.3}$$

where

$$\phi(\xi, \zeta) = 1 + \xi + 2C(\zeta) + C^2(\zeta) + \zeta D^2(\zeta). \tag{62.4}$$

It remains to impose the condition of reversibility. Since $\mathscr{I}$ and $\mathscr{E}'$ are, by choice, symmetrically situated with respect to $\mathscr{C}$, (62.3) must be invariant under the mutual interchange of $\mathbf{y}$ and $\mathbf{y}'$, or, what comes to the same thing, of $\xi$ and $\zeta$. In other words the equation

$$g(\zeta) - [\phi(\xi, \zeta) - 2\eta D(\zeta)]^{\frac{1}{2}} = g(\xi) - [\phi(\zeta, \xi) - 2\eta D(\xi)]^{\frac{1}{2}} \tag{62.5}$$

must be *identically* satisfied. Write both members of (62.5) as series in ascending powers of $\eta$. Then the factors multiplying $\eta^n$ show that one must have

$$\phi^{2n-1}(\zeta, \xi)\, D^{2n}(\zeta) = \phi^{2n-1}(\xi, \zeta)\, D^{2n}(\xi).$$

It suffices to take $n = 1$ and 2 in turn; and we conclude at once that

$$D(\xi) \equiv D(\zeta) = \text{const.} = m, \tag{62.6}$$

and

$$\phi(\xi, \zeta) \equiv \phi(\zeta, \xi). \tag{62.7}$$

It then follows from (62.5) that

$$g(\xi) \equiv g(\zeta) = \text{const.} = g_0, \tag{62.8}$$

say; whilst, in view of (62.4), the identity (62.7) reads

$$(1 - m^2)\,\xi - 2C(\xi) - C^2(\xi) \equiv (1 - m^2)\,\zeta - 2C(\zeta) - C^2(\zeta).$$

This is possible only if both members of this equation are equal to the same constant. The latter must be zero on account of (62.2), and so

$$C(\zeta) = [1+(1-m^2)\zeta]^{\frac{1}{2}}-1. \tag{62.9}$$

The point characteristic $V$ is thus fully determined:

$$V = g_0-(1+\xi-2m\eta+\zeta)^{\frac{1}{2}}. \tag{62.10}$$

Finally, from (62.1), (62.6) and (62.9), there follows *the equation for* $\mathscr{I}^{*\prime}$:

$$(X'+1)^2+(1-1/m^2)(Y'^2+Z'^2) = 1. \tag{62.11}$$

$\mathscr{I}^{*\prime}$ is therefore an ellipsoid or hyperboloid of revolution, according as $m^2 > 1$ or $m^2 < 1$. Since $m^2 \neq 1$, it is never a plane. It should not be forgotten that the unit of length was so chosen that

$$(1-m^2)f/m = 1.$$

It remains to investigate the case $m = 0$ which has hitherto been excluded. To this end we revert to the angle characteristic $T^*$ of Section 53; and this coincides here with the focal angle characteristic $\tilde{T}$ of Section 47. The equation of $\mathscr{I}^{*\prime}$ shall be

$$X' = C(\bar{\zeta}), \quad \mathbf{Y}' = \boldsymbol{\beta}D(\bar{\zeta}), \tag{62.12}$$

where the functions $C$ and $D$ are of course not the same as those in (62.1). Taking again $f = 1$, the counterparts to (62.2) are

$$C(0) = 0, \quad D(0) = 1, \tag{62.13}$$

bearing (15.14) in mind. Now

$$\mathbf{Y}' = \mathbf{y}'+X'\boldsymbol{\beta}'/\alpha',$$

so that one must have

$$\frac{\partial T^*}{\partial \boldsymbol{\beta}'} = C(\bar{\zeta})\,\boldsymbol{\beta}'/\alpha'-D(\bar{\zeta})\,\boldsymbol{\beta}$$

which on integration gives

$$T^* = -(1-\bar{\xi})^{\frac{1}{2}}\,C(\bar{\zeta})-\bar{\eta}D(\bar{\zeta})+g(\bar{\zeta}), \tag{62.14}$$

where $g$ is some function of $\bar{\zeta}$ alone. Reversibility of $K$ now requires that the identity

$$(1-\bar{\xi})^{\frac{1}{2}}\,C(\bar{\zeta})+\bar{\eta}D(\bar{\zeta})-g(\bar{\zeta}) \equiv (1-\bar{\zeta})^{\frac{1}{2}}\,C(\bar{\xi})+\bar{\eta}D(\bar{\xi})-g(\bar{\xi}) \tag{62.15}$$

be satisfied. In the first place one concludes at once that

$$D(\bar{\zeta}) = \text{const.} = 1. \tag{62.16}$$

Next, expand (62.15) in ascending powers of $\bar{\xi}$, writing

$$C(\bar{\xi}) = \tfrac{1}{2}k\bar{\xi}+\dots, \quad \text{and} \quad g(\bar{\xi}) = g_0+g_1\bar{\xi}+\dots.$$

One needs to consider only the terms independent of, and linear in, $\bar{\xi}$, and these give rise to the two equations

$$C(\bar{\zeta}) = k(1-\bar{\zeta})^{\frac{1}{2}}+2g_1, \quad g(\bar{\zeta}) = C(\bar{\zeta})+g_0. \tag{62.17}$$

Setting $\bar{\zeta} = 0$ in the first of these, it follows that $g_1 = -\frac{1}{2}k$; so that

$$C(\bar{\zeta}) = k[(1-\bar{\zeta})^{\frac{1}{2}}-1]. \tag{62.18}$$

Thus, now, $\quad T^* = g_0-\bar{\eta}-k[(1-\bar{\xi})^{\frac{1}{2}}-1]\,[(1-\bar{\zeta})^{\frac{1}{2}}-1]. \quad (62.19)$

*The equation of* $\mathscr{I}^{*\prime}$ follows from (62.16) and (62.18). It is

$$(X'/k+1)^2+Y'^2+Z'^2 = 1. \tag{62.20}$$

The image surface is therefore now in general an ellipsoid of revolution. Only in the exceptional case when $k = 0$ does this degenerate into a plane. However, one is of course still left with severe barrel distortion, for the displacement is then

$$\boldsymbol{\epsilon}' = \boldsymbol{\beta}(1-1/\alpha) = \mathbf{y}_1[(1+y_1^2+z_1^2)^{-\frac{1}{2}}-1]. \tag{62.21}$$

The actual image height $H'$ thus stands to the ideal image height $h'$ in the ratio $(1+h'^2)^{-\frac{1}{2}}$; and as $h'\to\infty$, $H'\to 1$; and generally the system transforms a line in the $y$, $z$-plane at infinity, whose ideal image is at a perpendicular distance $y_1$ from the axis of $K$, into the arc of an ellipse of eccentricity $(1+y_1^2)^{-\frac{1}{2}}$.

## 63. Generalization to curved object surfaces

The results of the preceding section surely suggest that one should also consider curved object surfaces; and to inquire in that case again into the attainability of a sharp image. Take $m \neq 0$, as before, and let the *object* surface $\mathscr{I}^*$ have the equation

$$x = K(\zeta), \tag{63.1}$$

where $\zeta = y^2+z^2$, and $\mathbf{y}$ denotes coordinates in $\mathscr{I}^*$. The equations of $\mathscr{I}^{*\prime}$ are (62.1) again. Also, we take $s = 1/m$ as before. However, instead of $\mathscr{E}'$ as posterior base-plane, we take a curved base-surface $\mathscr{B}^{*\prime}$ which is the 'reflection' of $\mathscr{I}^*$ in $\mathscr{C}$. This surface evidently passes through $E'$. It has the advantage that the point characteristic

$V_1$ must be invariant under the mutual interchange of $\xi$ and $\zeta$, $\mathbf{y}'$ being now coordinates in $\mathscr{B}^{*\prime}$, of course; but $\mathbf{y} = \text{const.}$ still characterizes a given point of the object. In terms of the coordinate system to which $X'$, $\mathbf{Y}'$ refer, the equation of $\mathscr{B}^{*\prime}$ is

$$x' = -1 - K(\xi). \tag{63.2}$$

Hence by the usual argument

$$V_1 = g(\zeta) - \{[1 + K(\xi) + C(\zeta)]^2 + \xi - 2\eta D(\zeta) + \zeta D^2(\zeta)\}^{\frac{1}{2}}. \tag{63.3}$$

Expanding this in ascending powers of $\eta$, invariance of $V_1$ under the mutual interchange of $\xi$ and $\zeta$ requires that $g(\zeta)$, $D(\zeta)$ and $V_1(\xi, 0, \zeta)$ be separately invariant. Thus

$$g(\zeta) = \text{const.} = g_0, \quad D(\zeta) = \text{const.} = m, \tag{63.4}$$

and

$$[1 + K(\xi) + C(\zeta)]^2 + \xi + m^2\zeta \equiv [1 + K(\zeta) + C(\xi)]^2 + \zeta + m^2\xi. \tag{63.5}$$

Now expand this in powers of $\xi$, writing $K(\xi) = K_1\xi + \ldots$, and $C(\xi) = C_1\xi + \ldots$. The terms independent of, and linear in, $\xi$ give

$$[1 + C(\zeta)]^2 + (m^2 - 1)\,\zeta = [1 + K(\zeta)]^2, \tag{63.6}$$

$$2K_1[1 + C(\zeta)] + 1 = 2C_1[1 + K(\zeta)] + m^2. \tag{63.7}$$

On setting $\zeta = 0$ in (63.7) it emerges that $K$ and $C$ are constant multiples of each other; say

$$K(\zeta) = kC(\zeta). \tag{63.8}$$

Then (63.6) yields

$$C(\zeta) = \frac{1}{1+k}\left[\left(1 - \frac{(1+k)(m^2-1)}{1-k}\,\zeta\right)^{\frac{1}{2}} - 1\right]. \tag{63.9}$$

The explicit form of $V_1$ is now at hand. To write it down compactly, put temporarily

$$a = \frac{1 - k^2m^2}{1 - k^2}, \quad b = \frac{2k}{(1+k)^2}, \quad c = \frac{(1+k)(m^2-1)}{1-k}, \tag{63.10}$$

and then

$$V_1 = g_0 - [1 + a(\xi + \zeta) - 2m\eta + b[(1 - c\xi)^{\frac{1}{2}}(1 - c\zeta)^{\frac{1}{2}} - 1]]^{\frac{1}{2}}. \tag{63.11}$$

This is of course invariant under the mutual interchange of $\xi$ and $\zeta$. We are now in a position to write down *the equation of* $\mathscr{I}^*$

$$\left(\frac{1+k}{k}\,x + 1\right)^2 + \frac{(1+k)(m^2-1)}{1-k}(y^2 + z^2) = 1, \tag{63.12}$$

and *the equation of* $\mathscr{I}^{*\prime}$ conjugate to it,

$$[(1+k)X'+1]^2+\frac{(1+k)(1-1/m^2)}{1-k}(Y'^2+Z'^2) = 1. \quad (63.13)$$

In general, therefore, either both surfaces are ellipsoids or they are both hyperboloids. A given ellipsoid can evidently have a sharp image only for one value of $m^2$. When $\mathscr{I}^*$ is spherical, then so is $\mathscr{I}^{*\prime}$, and this situation obtains when $k^2m^2 = 1$. At this point we restore, for the moment, an arbitrary unit of length, so that

$$d' = m^{-1}(1-m^2)f.$$

Then the centres of these anterior and posterior spheres are at $x = -(m^{-1}\mp 1)f$ and $X' = -(1\mp m)f$ respectively, according as $k = \pm m^{-1}$; and in either case their radii $R$, $R'$ obey the relation

$$R' = |m|R. \quad (63.14)$$

The sharp imaging of the two alternative object spheres can only be effected by *distinct* systems.

The last remark requires a careful comment. To this end, consider more generally the conjugate surfaces corresponding to $(k, m)$ having the values $(k_1, m_1)$, say. Reversibility of the system entails that if one takes as new object surface the 'reflection' of the previous image surface in $\mathscr{C}$, then the system will produce a sharp image of this, namely, the 'reflection' of the original object surface in $\mathscr{C}$. In other words, the conjugate surfaces corresponding to

$$(k, m) = (1/k_1, 1/m_1)$$

are conjugate with respect to the *same* system. (63.12) and (63.13) are indeed consistent with this elementary conclusion, since these equations in effect interchange place when $k$ and $m$ are replaced by $1/k$ and $1/m$ respectively. Again, contemplating the usual angle characteristic, the sharp imagery certainly requires that the coefficients of primary spherical aberration and coma vanish, i.e. $p_1 = p_2 = 0$ for given $(k, m)$. Then, however, $\hat{p}_1 = \hat{p}_2 = 0$ must also hold for $(1/k, 1/m)$, and one confirms by means of (45.9), with $p = m+1$, that this is the case in virtue of the relations (56.4) and (56.10). Now return to the two pairs of spheres considered above. To the first pair, the magnification being $m$, there corresponds another pair at magnification $1/m$. The centres of the object spheres are, as we have seen, at $x = -(m^{-1}-1)f$ and $x = -(m-1)f$, respectively. As regards the second pair, one has again to consider two object spheres, namely at $x = -(m^{-1}+1)f$ and at $x = -(m+1)f$. If now the given system were to form a sharp image of all four spheres, the coefficients of primary coma would have to vanish for *four* distinct values of the magnification. This, however, is impossible, since, according to (45.9), the equation $\hat{p}_2 = 0$ is *cubic* in $p$, bearing in mind that $b \neq 1$.

## B. Concentric systems

### 64. Definition of the concentric system

A system $K$ will be called *concentric* if there exists a point $C$ such that an arbitrary, sufficiently small, rotation of $K$ about $C$ results in a system indistinguishable from $K$. It goes without saying that, for the purposes of this definition, neither the pupil planes nor the object and image planes are to be reckoned as 'parts of $K$'. We choose any line through $C$ as the axis $\mathscr{A}$ of $K$, and to this $\mathscr{I}$ and $\mathscr{I}'$ are normal. Obviously $K$ is symmetric in the usual sense with respect to $\mathscr{A}$. $C$ will be called the *centre* of $K$, and the normal plane $\mathscr{C}$ through it the *central plane*. In practice one usually only utilizes, indeed only constructs, that part of $K$ which, roughly speaking, is contained within a circular cylinder of which $\mathscr{A}$ is the axis; but this is a feature which need not concern us. One need only reflect that if $\mathscr{R}$ is some ray, not too distant from $\mathscr{A}$, which passes through $K$, then $\mathscr{R}$ 'does not know' whether it is in fact passing through $K$ or through the rotated system $\tilde{K}$, provided the rotation was not too large. At any rate, the concentric system has the highest degree of symmetry any system can possess; as a consequence of which the characteristic function is determined to within one unknown function of a single argument, as we shall see.

### 65. Generic form of the angle characteristic

We shall generally proceed in terms of the angle characteristic $T$, granted that $K$ is non-telescopic (see, however, Section 70). Then the most general form of $T$ consistent with the central symmetry of $K$ is to be determined. This problem will be solved first by a relatively long and somewhat formal argument. Once the result is at hand we shall recognize how it can be obtained in a rather more elegant fashion.

Let the base-points $B, B'$ be arbitrarily selected for the time being, and let them be situated at distances $q$ and $q'$ to the left and right of $C$ respectively. The coordinate axes are disposed in the usual way. Then let $\mathscr{A}$ be rotated through an angle $\psi$ about a line through $C$, normal to the meridional plane; this rotation being equivalent to a rotation of $K$ through the angle $-\psi$. As a consequence, $B$, $B'$ go into $\tilde{B}$, $\tilde{B}'$, and the direction cosines $\boldsymbol{\beta}'$, $\boldsymbol{\beta}$ of a given ray $\mathscr{R}$ into $\tilde{\boldsymbol{\beta}}'$, $\tilde{\boldsymbol{\beta}}$. If $T$ is the angle characteristic referred to $B$ and $B'$, and $\tilde{T}$ that referred to $\tilde{B}$ and $\tilde{B}'$,

$$\tilde{T}(\tilde{\boldsymbol{\beta}}', \tilde{\boldsymbol{\beta}}) = T(\tilde{\boldsymbol{\beta}}', \tilde{\boldsymbol{\beta}}), \tag{65.1}$$

on account of the central symmetry of $K$. In other words, the *functional forms* of the old and new angle characteristics (referred to the appropriate base-points) are the same, since there is nothing to distinguish $K$ (with $B$, $B'$ as base-points) from $\tilde{K}$ (with $\tilde{B}$, $\tilde{B}'$ as base-points). Note that

$$\tilde{\beta} = \beta \cos\psi - \alpha \sin\psi, \quad \tilde{\gamma} = \gamma, \tag{65.2}$$

with analogous equations for the primed direction cosines. If $Q$, $Q'$ are the feet of the normals drawn from $B$, $B'$ on to $\mathscr{R}$ (see Fig. 2.3), one has corresponding points $\tilde{Q}$, $\tilde{Q}'$; and one then has the relation

$$\tilde{T} = T + Q'\tilde{Q}' + \tilde{Q}Q, \tag{65.3}$$

regarded as being between optical distances. It is not difficult to confirm that

$$\left.\begin{aligned} \tilde{Q}Q &= q[\beta \sin\psi + \alpha(\cos\psi - 1)], \\ Q'\tilde{Q}' &= q'[\beta' \sin\psi + \alpha'(\cos\psi - 1)]. \end{aligned}\right\} \tag{65.4}$$

There follows the *identity*

$$\begin{aligned} T(\beta' \cos\psi - \alpha' \sin\psi, \gamma', \beta\cos\psi - \alpha\sin\psi, \gamma) &= T(\beta', \gamma', \beta, \gamma) \\ + q'[\beta' \sin\psi + \alpha'(\cos\psi - 1)] &+ q[\beta \sin\psi + \alpha(\cos\psi - 1)], \end{aligned} \tag{65.5}$$

valid for all values of $\psi$. This must hold, in particular, for infinitesimal values of $\psi$:

$$T(\beta' - \psi\alpha', \gamma', \beta - \psi\alpha, \gamma) \equiv T(\beta', \gamma', \beta, \gamma) + (q'\beta' + q\beta)\psi,$$

whence

$$-\alpha' \frac{\partial T}{\partial \beta'} - \alpha \frac{\partial T}{\partial \beta} = q'\beta' + q\beta. \tag{65.6}$$

It will have been noticed that, whereas we might have considered *arbitrary* rotations about $C$, we confined ourselves to rotations about a line through $C$. This is, in fact sufficient, as long as we also make use of the symmetry of $K$ (relative to $\mathscr{A}$). This we do now, by taking $T$ to depend on $\boldsymbol{\beta}'$, $\boldsymbol{\beta}$ only through the usual rotational invariants $\bar{\xi}, \bar{\eta}, \bar{\zeta}$:

$$T = T(\bar{\xi}, \bar{\eta}, \bar{\zeta}), \tag{65.7}$$

and then (65.6) leads to the two equations

$$2\alpha' \frac{\partial T}{\partial \bar{\xi}} + \alpha \frac{\partial T}{\partial \bar{\eta}} = -q', \quad \alpha' \frac{\partial T}{\partial \bar{\eta}} + 2\alpha \frac{\partial T}{\partial \bar{\zeta}} = -q, \tag{65.8}$$

where, of course, $\alpha' = (1-\bar{\xi})^{\frac{1}{2}}, \quad \alpha = (1-\bar{\zeta})^{\frac{1}{2}}.$ (65.9)

The equations (65.8) are most easily solved by introducing in place of $\bar{\eta}$ the new independent variable

$$\chi = 1 - (1-\bar{\xi})^{\frac{1}{2}}(1-\bar{\zeta})^{\frac{1}{2}} - \bar{\eta} \quad (\equiv 1 - (\alpha\alpha' + \beta\beta' + \gamma\gamma')). \tag{65.10}$$

Then $$T = G(\chi) + q'(1-\bar{\xi})^{\frac{1}{2}} + q(1-\bar{\zeta})^{\frac{1}{2}}, \tag{65.11}$$

and here only a single function of one argument remains unknown; so that only one new constant enters into the aberration coefficients of order $2n-1$, given those of the lower orders.

Now that we have obtained the required result, a short-cut to its derivation is plain. Thus initially one takes both base-points to coincide with $C$. Then the corresponding angle characteristic $'T(\boldsymbol{\beta}', \boldsymbol{\beta})$ must be invariant under arbitrary rotations. The only invariant function of $\alpha', \beta', \gamma', \alpha, \beta, \gamma$ (which transform like the Cartesian coordinates themselves) is $\alpha\alpha' + \beta\beta' + \gamma\gamma'$, bearing in mind that $\alpha^2 + \beta^2 + \gamma^2 = \alpha'^2 + \beta'^2 + \gamma'^2 = 1$, always. Thus $'T$ is merely a function of $\chi$, and going over to the base-points to which $T$ refers, (65.11) follows at once.

We note in passing that

$$\chi = \tfrac{1}{2}(\bar{\xi} - 2\bar{\eta} + \bar{\zeta}) + O(4). \tag{65.12}$$

If we now take $B$, $B'$ to coincide with $O_0$, $O_0'$ as usual, we have

$$q' = 1 - m, \quad q = 1 - 1/m, \tag{65.13}$$

since the central plane and both (paraxial) pupils coincide when $s = 1$. Writing

$$G(\chi) = G_0 + G_1\chi + G_2\chi^2 + \dots, \tag{65.14}$$

(65.11) is in harmony with (14.35) provided

$$G_1 = 1. \tag{65.15}$$

We note in passing that when the base-points are equidistant from $C$ the angle characteristic satisfies the condition of reversibility; but $K$ need not be reversible.

## 66. The third- and higher-order aberrations

It is instructive to write down the values of the primary aberration coefficients. Replacing $2G_2$ by the more convenient symbol $k$ we find directly from (65.11), together with (65.10) and (65.14), that

$$t^{(2)} = \tfrac{1}{8}[k(\bar{\xi} - 2\bar{\eta} + \bar{\zeta})^2 + m\bar{\xi}^2 - 2\bar{\xi}\bar{\zeta} + m^{-1}\bar{\zeta}^2]. \tag{66.1}$$

On the right we now introduce $\xi$, $\eta$, $\zeta$ according to (15.12), and a little manipulation then gives, with $c = \frac{1}{8}(s-m)^{-4}$,

$$\left.\begin{aligned}
p_1 &= c(1-m)^2\,[k(1-m)^2 + m],\\
p_2 &= -4c(1-m)(1-s)\,[k(1-m)^2 + m],\\
p_3 &= 2c(1-m)\,[k(1-m)(1-s)^2 + (m-s^2)],\\
p_4 &= 4c(1-s)^2\,[k(1-m)^2 + m],\\
p_5 &= -4c(1-s)\,[k(1-m)(1-s)^2 + (m-s^2)],\\
p_6 &= c[k(1-s)^4 + (m-s^2)^2].
\end{aligned}\right\} \tag{66.2}$$

From these a number of interesting conclusions may be drawn straight away. Thus (i) for a central stop ($s = 1$), coma, astigmatism and distortion vanish; (ii) when, with $m \neq 1$, the spherical aberration vanishes, so do coma and astigmatism; (iii) if spherical aberration is corrected for magnification $m$ it is also corrected at magnification $1/m$; (iv) the invariant $\alpha_3 = 2(s-m)^2(2p_3 - p_4)$ of Section 50 has the value $-1$ (i.e. $-\alpha_3$ = focal length of $K$). In the primary domain, where the aberrations associated with the pupil planes may be ignored as far as the actual displacement is concerned, (i) is obvious enough, for every ray through $C$ passes through $K$ undeviated. (ii) implies that $K$ will, to this order, produce a sharp image on a surface $\mathscr{I}^{*\prime}$ whose curvature is, in view of (19.15),

$$4d'(s-m)p_3 = -1. \tag{66.3}$$

(Recall that (19.15) related to a unit of length such that $d' = 1$;

and the factor $(s-m)^{-1}$ on the right of (19.24) is not included in $t^{(2)}$ here.) (66.3) reflects the fact that in the absence of spherical aberration, $K$ produces a sharp image of a spherical object surface $\mathscr{I}^*$ of radius $q$, the image surface $\mathscr{I}^{*\prime}$ being a sphere of radius $-q'$, the minus sign meaning that it is concave towards $C$. This conclusion is obvious, since the central symmetry of $K$ now extends also to $\mathscr{I}^*$ and $\mathscr{I}^{*\prime}$. We shall shortly consider questions of sharp imagery, as well as generalizations to all orders of the various results just arrived at.

For the sake of simplicity we shall henceforth usually take the stop to be central ($s = 1$). This choice corresponds, in any event, to the situation most often encountered in practice. Other values of $s$ may, if desired, then be accommodated by using the results of Section 44. From (65.11),

$$T = \sum_{n=0}^{\infty}\left[G_n\chi^n + \binom{\frac{1}{2}}{n}(q'\bar{\xi}^n + q\bar{\zeta}^n)\right], \tag{66.4}$$

where $\bar{\xi}$, $\bar{\eta}$, $\bar{\zeta}$ are yet to be replaced by $\xi$, $\eta$, $\zeta$, using (15.12), with $s = 1$. Write

$$\chi = \tau + \sum_{n=2}^{\infty}\phi^{(n)}(\bar{\xi}, \bar{\zeta}), \tag{66.5}$$

where $\tau = \frac{1}{2}(\bar{\xi} - 2\bar{\eta} + \bar{\zeta})$, and $\phi^{(n)}$ is homogeneous of degree $n$ in $\bar{\xi}$, $\bar{\zeta}$. Then

$$\left.\begin{aligned} t^{(3)} &= G_3\tau^3 + 2G_2\tau\phi^{(2)} + \phi^{(3)} - \tfrac{1}{16}(q'\bar{\xi}^3 + q\bar{\zeta}^3), \\ t^{(4)} &= G_4\tau^4 + 3G_3\tau^2\phi^{(2)} + G_2[2\tau\phi^{(3)} + (\phi^{(2)})^2] \\ &\quad + \phi^{(4)} - \tfrac{5}{128}(q'\bar{\xi}^4 + q\bar{\zeta}^4), \end{aligned}\right\} \tag{66.6}$$

and so on.

We consider the special case $m = -1$. Then, in terms of $\xi, \eta, \zeta$,

$$\left.\begin{aligned} \tau = \tfrac{1}{2}\xi, \quad \phi^{(2)} = \tfrac{1}{8}\eta^2, \quad \phi^{(3)} = \tfrac{1}{32}(\xi+\zeta)\eta^2, \\ \phi^{(4)} = \tfrac{1}{128}(\xi^2 + 2\xi\zeta + \eta^2 + \zeta^2)\eta^2, \end{aligned}\right\} \tag{66.7}$$

and so on. As a matter of fact it is possible to show that

$$\begin{aligned} (n+1)\,\phi^{(n+1)} &= \tfrac{1}{4}(2n-1)\,(\xi+\zeta)\,\phi^{(n)} \\ &\quad - \tfrac{1}{16}(n-2)\,(\xi^2 + 2\xi\zeta - 4\eta^2 + \zeta^2)\,\phi^{(n-1)}, \end{aligned} \tag{66.8}$$

from which it follows incidentally that $\phi^{(n)}$ has $\eta^2$ as a factor for every $n$, since $\phi^{(2)}$ and $\phi^{(3)}$ have this factor. Bearing in mind that

$m = -1$ it follows that $t^{(n)}$ is an even function of $\eta$, consistently with the fact that $K$ satisfies the condition of reversibility. In particular (59.5) will here also be valid. From (66.6) and (66.7) one has, in particular,

$$t^{(3)} = \tfrac{1}{8}G_3\xi^3 + \tfrac{1}{8}G_2\xi\eta^2 - \tfrac{1}{256}(\xi^3 + 3\xi^2\zeta + 4\xi\eta^2 + 3\xi\zeta^2 + 4\eta^2\zeta + \zeta^3). \tag{66.9}$$

According to (66.2) the coefficient of primary spherical aberration vanishes when $G_2 = \frac{1}{8}$; we now see that the secondary coefficient vanishes when $G_3 = \frac{1}{32}$. These results may be very easily generalized to all orders. Under the present circumstances we need only set $\bar{\xi} \to \frac{1}{4}\xi, \bar{\eta} \to \frac{1}{4}\xi, \bar{\zeta} \to \frac{1}{4}\xi$, since only the coefficient of $\xi^n$ is required, and then

$$T(\xi, 0, 0) = G(\tfrac{1}{2}\xi) + 4(1 - \tfrac{1}{4}\xi)^{\frac{1}{2}}. \tag{66.10}$$

Thus, using the notation of equation (45.12), $S(\xi)$ will be zero provided

$$G(\chi) = \text{const.} - 2(4 - 2\chi)^{\frac{1}{2}}. \tag{66.11}$$

## 67. The displacement

We proceed to consider the displacement $\boldsymbol{\epsilon}'$, reverting to the variables $\bar{\xi}$, $\bar{\eta}$, $\bar{\zeta}$ for the time being. Write

$$\dot{G}(\chi) = dG/d\chi, \quad \boldsymbol{\Gamma} = (\alpha'\boldsymbol{\beta} - \alpha\boldsymbol{\beta}')\dot{G}. \tag{67.1}$$

Then, from (65.11),

$$\mathbf{y}' = (\boldsymbol{\Gamma} + q'\boldsymbol{\beta}')/\alpha', \quad \mathbf{y} = (\boldsymbol{\Gamma} - q\boldsymbol{\beta})/\alpha. \tag{67.2}$$

The coordinates $\mathbf{y}'_e$, $\mathbf{y}_e$ of the points of intersection of rays with the pupil planes follow from this by setting $q = q' = 0$; e.g. in the exit pupil

$$\mathbf{y}'_e = \boldsymbol{\Gamma}/\alpha'. \tag{67.3}$$

From (67.2) $\quad \boldsymbol{\epsilon}' = [(\boldsymbol{\beta}'/\alpha') - (\boldsymbol{\beta}/\alpha)]\,[(m\alpha' - \alpha)\dot{G} + q'], \quad$ (67.4)

since $q' = -mq$. Taking (67.3) into account, one concludes that

$$\epsilon'_y/\epsilon'_z = y'_e/z'_e \tag{67.5}$$

for all rays. It follows immediately that (when the stop is central) *all barred effective aberration coefficients vanish* (cf. equation (23.1)),

$$\bar{u}^{(n)}_{\mu\nu} = 0; \tag{67.6}$$

so that *the total sagittal comatic asymmetry* $K_s$ *vanishes* (cf. equation (19.22)). Further, *an ideal (i.e. a plane, undistorted) image is unattainable*, since otherwise one would have to have

$$\dot{G} = (1-m)/(\alpha - m\alpha') \tag{67.7}$$

for all rays. This means that $\dot{G}$ would have to be independent of $\bar{\eta}$, which is impossible outside the paraxial region.

## 68. Spherical aberration

Let it be assumed for the time being that $m \neq 0$. Then axial rays are distinguished by the property that $\mathbf{y} = 0$. Using the second member of (67.2), (67.4) then reduces to

$$\boldsymbol{\epsilon}' = q(\boldsymbol{\beta} - m\boldsymbol{\beta}')/\alpha'. \tag{68.1}$$

If spherical aberration of all orders be absent, one must have

$$\boldsymbol{\beta} = m\boldsymbol{\beta}', \tag{68.2}$$

which means that *the sine condition is automatically fulfilled*. Thus absence of spherical aberration *entails* that the total circular comatic displacement be zero. This must, of course, be so, in view of the result stated just after (67.6), bearing Section 31 in mind.

(66.11) may now be generalized to arbitrary values of $m\,(\neq 1)$. When $S(\xi) \equiv 0$, equation (67.7) must hold for axial rays. For these one may set $\gamma = \gamma' = 0$, so that the following equation results:

$$\dot{G}(1 - \alpha\alpha' - \beta\beta') = (1-m)/(\alpha - m\alpha'), \tag{68.3}$$

where

$$\alpha = [1 - m^2(1-\alpha'^2)]^{\frac{1}{2}}, \quad \beta = (1-\alpha^2)^{\frac{1}{2}}, \quad \beta' = (1-\alpha'^2)^{\frac{1}{2}}. \tag{68.4}$$

If one writes $\chi$ for the argument of the function $\dot{G}$, the right-hand member of (68.3) may be expressed as a function of $\chi$, and this leads to the equation

$$\dot{G}(\chi) = \pm(1-m)\,[(1-m)^2 + 2m\chi]^{-\frac{1}{2}}. \tag{68.5}$$

Here the upper and lower sign are to be chosen according as $m < 1$ or $m > 1$; for

$$\alpha' = (1 - m - \chi)\,[(1-m)^2 + 2m\chi]^{-\frac{1}{2}}, \tag{68.6}$$

and the condition that this must be positive determines whether one has to choose the positive or the negative square root here. In (68.5) the choice of the positive square root is understood. Thus now

$$G(\chi) = \pm m^{-1}(1-m)\,[(1-m)^2+2m\chi]^{\frac{1}{2}}. \tag{68.7}$$

When $m = -1$ this reduces correctly to (66.11). When $m = 0$ some of the preceding results require modification. However, as $m \to 0$ (68.7) breaks down solely on account of the coordinate independent term $(1-m)^2/m$. We simply subtract it out, for this process corresponds merely to the choice of a finitely situated anterior base-point. Then (68.7) reduces to

$$G(\chi) = \chi \tag{68.8}$$

in the limit $m = 0$.

## 69. The displacement when spherical aberration is absent

In the absence of spherical aberration the angle characteristic is fully known, on account of (65.11) and (68.7). The actual displacement $\boldsymbol{\epsilon}'$ is thus determined. We consider it only in the special, but important, case of the object being at infinity, so that $m = 0$; and then one has a situation of great simplicity. To begin with, taking the anterior base-point at $E$, so that $q = 0$, we have

$$T = \text{const.} + \alpha' - (\alpha\alpha' + \beta\beta' + \gamma\gamma') \tag{69.1}$$

then, from (67.4) and (68.8),

$$\boldsymbol{\epsilon}' = [(\boldsymbol{\beta}'/\alpha') - (\boldsymbol{\beta}/\alpha)]\,(1-\alpha). \tag{69.2}$$

Again, from (67.3), $\qquad \mathbf{y}'_e = (\alpha'\boldsymbol{\beta} - \alpha\boldsymbol{\beta}')/\alpha', \qquad (69.3)$

so that (69.2) becomes $\qquad \boldsymbol{\epsilon}' = \mathbf{y}'_e\left(1 - \dfrac{1}{\alpha}\right). \qquad (69.4)$

Since $\beta/\alpha = h'$, $\gamma/\alpha = 0$, one thus has, on introducing the usual polar coordinates in the exit pupil,

$$\boldsymbol{\epsilon}' = \rho[1-(1+h'^2)^{\frac{1}{2}}] \times \begin{cases}\cos\theta \\ \sin\theta.\end{cases} \tag{69.5}$$

(69.5) represents one of those rare occasions on which the exact displacement can be exhibited in closed form, as a function of the variables which enter into the series (23.1) defining the effective aberration coefficients. At any rate, *the effective coefficients of all types of coma of all orders are zero.*

It will not come amiss to contemplate for the moment the consequences which would arise if instead of defining effective coefficients relative to $\mathbf{y}'_e$ and $h'$ we were to define them relative to $\mathbf{y}_e$ and $h'$; so that in place of $\rho$ and $\theta$ we have polar coordinates, $\rho_1$, $\theta_1$, say, in the *entrance* pupil. Then one has

$$\boldsymbol{\epsilon}' = \mathbf{y}_e(\alpha - 1)/\alpha', \tag{69.6}$$

in place of (69.4). Evidently we need to obtain an expression for $\alpha/\alpha'$ in terms of $\mathbf{y}_e$ and $h'$, and this can be done starting from

$$\alpha'\boldsymbol{\beta} - \alpha\boldsymbol{\beta}' = \alpha\mathbf{y}_e. \tag{69.7}$$

Eventually there comes

$$\boldsymbol{\epsilon}' = \rho_1[1-(1+h'^2)^{-\frac{1}{2}}](1-\rho_1^2)^{-1}\{\rho_1 h' \cos\theta_1 - [(1-\rho_1^2)(1+h'^2)+\rho_1^2 h'^2\cos^2\theta_1]^{\frac{1}{2}}\} \times \begin{cases}\cos\theta_1 \\ \sin\theta_1.\end{cases} \tag{69.8}$$

The *new* effective comatic coefficients are now not zero. Indeed, the comatic displacement is

$$(\boldsymbol{\epsilon}')_{\text{com}} = \tfrac{1}{2}\rho_1^2(1-\rho_1^2)^{-1}h'[1-(1+h'^2)^{-\frac{1}{2}}] \times \begin{cases}1+\cos 2\theta_1 \\ \sin 2\theta_1 \;.\end{cases} \tag{69.9}$$

The dependence upon $\theta_1$ is the same for all orders, and the image patch is circular. (The expression on the right of (69.9) is of course $O(5)$.) Going over from $\rho$ to $\rho_1 (= \alpha\rho/\alpha')$, the set of rays selected by the condition $\rho_1 =$ constant is different from the set which has $\rho$ constant. In short, one must guard against attaching basic significance to results of a kind which depend fundamentally upon how one happens to split up the family of rays from a given point of the object into sets corresponding to this or that variable having a constant value. In the end it is only the sum total of these sets that counts, and it is relevant to recall that even in the absence of vignetting the family of rays whose passage through $K$ is assured is characterized *neither* by $\rho = \rho_{\text{max}}$, *nor* by $\rho_1 = (\rho_1)_{\text{max}}$, but rather by $\rho_s = (\rho_s)_{\text{max}}$, where $\rho_s$ is a radial coordinate in the plane of the stop.

## 70. On the existence of curved conjugate surfaces

It was already noted in Section 67 that an ideal image is unattainable. However, we may inquire whether there may not exist pairs of conjugate surfaces. To have such a pair at magnification $m$, spherical aberration must certainly be absent for this magnification; so that $G(\chi)$ must have the form given by (68.7). When this is the case, it follows from the central symmetry of $K$ that the spherical surface $\mathscr{I}^*$ of radius $|q|$, will have as conjugate image surface $\mathscr{I}^{*\prime}$ the spherical surface of radius $|q'|$, both having $C$ as centre.

One might at first sight be inclined to conclude that this is the only possibility. This is, however, not so, as reference to the condition of reversibility shows at once. We already saw in Section 63 that *if* a reversible system has a pair of conjugate surfaces at magnification $m$ then it has another such pair at magnification $1/m$. This conclusion evidently rests only upon the requirement that the *condition* of reversibility be satisfied—whether the system is in fact reversible or not is irrelevant. However, *every* concentric system satisfies the condition of reversibility; from which we infer that there is another pair of conjugate spheres, as described above, but with $q$ and $q'$ calculated for magnification $1/m$.

If this conclusion is not to be in conflict with the work of Section 68, $G(\chi)$ must remain unaffected when $m$ is replaced by $1/m$, since otherwise spherical aberration would be present at magnification $1/m$, and one could not attain a sharp image. It is easily confirmed that $G(\chi)$ has, indeed, the required property. Further, reference to the discussion following upon equation (63.14) shows that all possibilities as regards pairs of conjugate surfaces are now exhausted, bearing in mind that $m \to 1/m$ is the only substitution which leaves $G$ invariant.

In the course of the preceding investigation it was understood throughout that $K$ satisfied the condition of not being telescopic. In the contrary case the situation is entirely different, for there then exists the possibility that *every* point of the object space has a sharp image. Let it be supposed, then, that $K$ is telescopic.

We adopt $W_2(\beta', \gamma', x, y, z)$ as the most appropriate characteristic function. It will be noted that its dependence on $x$ is now being explicitly contemplated. We further refer all points to a coordinate basis both origins of which coincide with the centre $C$; and it is convenient to take this point to be also the location of the posterior base-point $B'$. Because of our general assumption, every point $O$ of the normal plane $\mathscr{I}$ through $O_0(x, 0, 0)$ must have a sharp image, the coordinates of which shall be $X'$, $\mathbf{Y}'$. Thus there must exist functions $K(\zeta, x)$, $D(\zeta, x)$, $x'(x)$ such that

$$X' = x' + K, \quad \mathbf{Y}' = \mathbf{y}D, \tag{70.1}$$

$x'$ being the distance between $B'$ and the point conjugate to $O_0$. (In the notation of Section 65, $x = -q$, $x' = q'$.) We therefore must have

$$\frac{\partial W_2}{\partial \boldsymbol{\beta}'} + (x' + K)\frac{\boldsymbol{\beta}'}{\alpha'} = \mathbf{y}D,$$

which splits up into the two equations

$$2\alpha' W_{2\xi} = x' + K, \quad W_{2\eta} = -D. \tag{70.2}$$

However, since the system is telescopic, $\boldsymbol{\beta}'$ must vanish when $\boldsymbol{\beta}$ does so, which implies that $W_2$ must be independent of $\zeta$. (70.2) then shows that the same is true of $D$ and $K$. Since $K(0, x) = 0$ we see that $K$ is zero, whilst $D$ can depend at most upon $x$:

$$D(\zeta, x) = m(x), \tag{70.3}$$

say. It further follows from (70.2) that

$$W_2 = -\alpha' x' - m\eta + k, \tag{70.4}$$

where $k$ can depend at most upon $x$. Pursuing now an argument which runs parallel to that following upon equation (65.11), central symmetry entails that $W_2$ can only be a function of

$$\alpha' x + \beta' y + \gamma' z \quad \text{and} \quad x^2 + y^2 + z^2,$$

these being invariant under arbitrary rotations about $C$. Considering (70.4) in the light of this conclusion we infer that

$$x' = mx. \tag{70.5}$$

At the same time consistency of this result with the linearity of the relation (14.36) requires that $m$ be a constant. The last step in the argument uses the differential equation

$$\left(\frac{\partial W_2}{\partial x}\right)^2 + \left(\frac{\partial W_2}{\partial y}\right)^2 + \left(\frac{\partial W_2}{\partial z}\right)^2 = 1. \tag{70.6}$$

Inserting (70.4) (with (70.5)) in this, there comes

$$k = \text{const.}, \quad m^2 = 1. \tag{70.7}$$

Thus finally, $W_2 = \pm(\alpha' x + \beta' y + \gamma' z) + \text{const.}$ (70.8)

Since $x$ can be chosen at will, we conclude that every point of the object space has a well-defined image point. Moreover, bearing in mind that $|x| = |x'|$, it is possible for any part of the object space to have a sharply defined image geometrically similar to it. However, such imagery is somewhat trivial in as far as the (reduced) magnification has of necessity the value unity. The universal validity of the last statement may be established under quite general circumstances.

## 71. On the absolutely invariant aberrations

The angle characteristic of a concentric system being known to within a single function of one argument, it is of interest to examine the absolute invariants $\alpha_{4N-1}$ considered in general circumstances in Sections 49 and 50. However, since it is convenient to retain the variables $\bar{\xi}, \bar{\eta}, \bar{\zeta}$, we consider, in place of (49.6), the equivalent relation

$$\alpha_{4N-1} = \bar{A}^N G^{(2N)}(\chi). \tag{71.1}$$

$G^{(n)}(\chi)$ is the $(2n-1)$th-order part of $G(\chi)$; and the remaining terms of $T$ are in any event annihilated by $\bar{A}$. Evidently we may replace $G^{(2N)}(\chi)$ by $G(\chi)$ in (71.1), provided all derivatives are evaluated at $\bar{\xi} = \bar{\eta} = \bar{\zeta} = 0$. Upon recalling (65.10) we see that

$$\chi^n = \sum_{m=0}^{n} \binom{n}{m} (-\bar{\eta})^{n-m} [1 - (1-\bar{\xi})^{\frac{1}{2}} (1-\bar{\zeta})^{\frac{1}{2}}]^m. \tag{71.2}$$

Write $\beta_{pm}$ for the coefficient of $(\bar{\xi}\bar{\zeta})^p$ in the expansion of the last factor on the right, so that incidentally

$$2p \geqslant m \tag{71.3}$$

for all $\beta_{pm}$ which do not vanish identically. By straightforward use of the binomial expansion one finds that

$$\beta_{pm} = \sum_{j=0}^{m} (-1)^j \binom{m}{j} \left[\binom{\frac{1}{2}j}{p}\right]^2, \tag{71.4}$$

which is somewhat clumsy, in as far as the validity of (71.3) is not obvious by inspection. At any rate, from (71.4) or otherwise, it is not difficult to construct Table 6.1 for the values of $\beta_{pm}$, for selected small values of $p$ and $m$.

TABLE 6.1

| $p$ | $m$: 0 | 1 | 2 | 3 | 4 | 5 | 6 |
|---|---|---|---|---|---|---|---|
| 0 | 1 | 0 | 0 | 0 | 0 | 0 | 0 |
| 1 | 0 | $-\frac{1}{4}$ | $\frac{1}{2}$ | 0 | 0 | 0 | 0 |
| 2 | 0 | $-\frac{1}{64}$ | $-\frac{1}{32}$ | $-\frac{3}{16}$ | $\frac{3}{8}$ | 0 | 0 |
| 3 | 0 | $-\frac{1}{256}$ | $-\frac{1}{128}$ | $-\frac{1}{64}$ | $-\frac{1}{32}$ | $-\frac{5}{32}$ | $\frac{5}{16}$ |

(71.2) now becomes

$$\chi^n = \sum_m \sum_p \binom{n}{m} \beta_{pm} \bar{\eta}^{n-m} (\xi\zeta)^p + \dots \quad (n-m \text{ even}), \tag{71.5}$$

the dots indicating all terms which are eventually annihilated by $\bar{A}^N$, $(\bar{\xi} = \bar{\eta} = \bar{\zeta} = 0)$. The remaining steps are quite similar to those of Section 50, and one finally arrives at the result

$$\alpha_{4N-1} = N! \sum_{r=0}^{N} \sum_{n=2r}^{2N} (-1)^r \, 2^{2N-2r} \frac{n!\,(N-r)!}{r!\,(n-2r)!} \beta_{N-r,n-2r} G_n. \tag{71.6}$$

Table 6.1 is sufficiently extensive to evaluate $\alpha_3$, $\alpha_7$ and $\alpha_{11}$ from the last equation, and it turns out that

$$\alpha_3 = -1, \quad \alpha_7 = -(1+2G_2), \quad \alpha_{11} = -9(1+2G_2+2G_3). \tag{71.7}$$

Remarkably, the secondary and tertiary coefficients $G_3$, $G_4$ are absent from $\alpha_7$, and the coefficients $G_4$, $G_5$, $G_6$, which relate to orders 7, 9 and 11, do not enter into the eleventh-order invariant $\alpha_{11}$. Nevertheless, the fact remains that the $G_n$ *do* enter into the $\alpha_{4N-1}$. If desired, they may be expressed in terms of the coefficients of spherical aberration; e.g. when $s = 1$ and $m = 0$, one has just

$$G_2 = \tfrac{1}{2}k = 4p_1 = \sigma_1, \tag{71.8}$$

in view of (66.2) and (19.6). The effect of the presence of primary spherical aberration on the value of $\alpha_7$ need therefore be by no means negligible. However, when spherical aberration of all orders is absent, $\alpha_{4N-1}$ reduces to

$$\alpha_{4N-1} = -[(2N-3)(2N-5)\dots 5\cdot 3\cdot 1]^2. \tag{71.9}$$

## Problems

**P.6 (i).** Can a reversible system form a sharp image of a paraboloid? If so, what is the nature of the image?

**P.6 (ii).** A reversible system transforms an hyperboloid into an hyperboloid. Their meriodonal sections have asymptotes the angles between which are $2\psi$ and $2\psi'$ respectively. Show that

$$\tan\psi = \sin\psi'(m^2 - \sin^2\psi')^{-\frac{1}{2}}.$$

**P.6 (iii).** The stop of a reversible system is so situated that $s = 1/m$. Consider the condition of reversibility from the point of view of the point characteristic. In particular, recover equation (56.12) exactly.

**P.6 (iv).** The form of $V_1$ given by equation (63.11) certainly must entail the absence of spherical aberration. This function does not, however, have the familiar form $-(1+\xi)^{\frac{1}{2}}$ when $\eta = \zeta = 0$, apart from an irrelevant additive constant. Resolve the apparent contradiction.

**P.6 (v).** One often requires the homogeneous polynomials $\phi_n(\xi, \zeta)$ defined by

$$(1-\xi)^{\frac{1}{2}}(1-\zeta)^{\frac{1}{2}} = \sum_{n=0}^{\infty} \phi_n(\xi, \zeta).$$

Show that they satisfy the recurrence relation

$$(n+1)\,\phi_{n+1} = (n-\tfrac{1}{2})(\xi+\zeta)\,\phi_n - (n-2)\,\xi\zeta\phi_{n-1} \quad (n > 1).$$

CHAPTER 7

# THE SEMI-SYMMETRIC SYSTEM

## 72. Definition of the semi-symmetric system. Generic form of the characteristic functions

Having considered the symmetric system at great length, ending up with the highest possible degree of symmetry, as represented by the concentric system, we move now in the other direction, as it were, and drop the second condition which formed part of the definition of the general symmetric system in Section 12. Accordingly, a system is called *semi-symmetric* (more precisely, *r*-semi-symmetric) if it has an *axis* of symmetry, but *no* plane of symmetry containing the axis. In other words $K$ will now have a built-in screw-sense; see also the end of this section.

In the optical field a physical realization of semi-symmetry is likely to be rather artificial. We may, however, with advantage include in the term 'optical system' any image-forming device, no matter whether the image is formed by means of light or of particles. In that case an excellent example of a semi-symmetric system is furnished by the electron microscope, provided focusing is achieved, at least in part, by magnetic fields.

Choosing a coordinate basis *for the time being* exactly as in the symmetric case, and assuming regularity of $K$ with respect to it, we know already from the discussion of Section 13 that the point characteristic $V$ must now depend explicitly upon the skew-symmetric rotational invariant $\tau = z'y - zy'$, that is to say, $V$ is to be written as a power series in the four variables $\xi$, $\eta$, $\zeta$, $\tau$, rather than merely $\xi$, $\eta$, $\zeta$. On the other hand, wherever $\tau^2$ occurs it can be removed by replacing it by $\xi\zeta - \eta^2$. Thus the point characteristic of a regular semi-symmetric system has the generic form

$$V(\xi, \eta, \zeta, \tau) = V^{\#}(\xi, \eta, \zeta) + \tau \tilde{V}^{\#}(\xi, \eta, \zeta), \tag{72.1}$$

where $V^{\#}$ and $\tilde{V}^{\#}$ are power series in $\xi$, $\eta$, $\zeta$.

An analogous situation prevails with respect to the other characteristic functions. In the case of the angle characteristic, for instance,

$$T(\xi, \eta, \zeta, \tau) = T^{\#}(\xi, \eta, \zeta) + \tau \tilde{T}^{\#}(\xi, \eta, \zeta), \tag{72.2}$$

where $\xi$, $\eta$, $\zeta$ are the rotational invariants (15.10), whilst the skew-symmetric invariant $\tau$ is defined by

$$\tau = (s\gamma' - \gamma)(m\beta' - \beta) - (m\gamma' - \gamma)(s\beta' - \beta)$$

$$= \sigma_z \mu_y - \sigma_y \mu_z. \quad (72.3)$$

$s$ and $m$ have their usual significance, for we shall see that the paraxial behaviour of the semi-symmetric system is entirely equivalent to that of the symmetric system. When $\bar{\xi}$, $\bar{\eta}$, $\bar{\zeta}$ occur as arguments of $T$, one replaces $\tau$ at the same time by $\bar{\tau}$, where

$$\bar{\tau} = \beta'\gamma - \beta\gamma', \quad (72.4)$$

and then (72.2) can be taken over by merely supplying all the variables with bars. However, note that, since

$$\tau = (s - m)\bar{\tau}, \quad (72.5)$$

the functions $\tilde{T}^{\#}(\xi, \eta, \zeta)$ and $\tilde{T}^{\#}(\bar{\xi}, \bar{\eta}, \bar{\zeta})$ for which one should really use distinct functional symbols, differ by a factor $s - m$.

It must be remarked that the 'optical medium' of any semi-symmetric system must be anisotropic; that is to say, somewhere within $K$ the refractive index at any point must be dependent upon the direction of the ray—passing through this point—which is being considered. To see this, one need only reflect that, were the medium isotropic everywhere, a ray initially meridional would certainly have to be meridional throughout $K$; a conclusion which follows from the elementary laws of refraction in isotropic systems. The fact that the anisotropy here reveals itself only in the explicit dependence of the characteristic function upon the skew-symmetric rotational invariant $\tau$ is due to our continued adherence to the agreement that in the object space and in the image space—especially in the latter—the refractive index is constant in the kind of system under consideration. This entails that the formal complications arising from the need to distinguish between the direction of the wave-normal on the one hand, and the ray (i.e. the direction of energy transport) on the other, are absent in the region in which the image is being considered. Further details appear in Chapter 11, particularly in Section 114.

## 73. The paraxial region. Adapted coordinate basis

Let the point characteristic $V$ be referred to base-points $O_0$ and $B'$, where $B'$ may, in the first instance, be chosen arbitrarily, so long as it is not conjugate to $O_0$. Paraxially we write, in analogy with (14.1),

$$V = a_0 + \tfrac{1}{2}k_1\xi + k_2\eta + \tfrac{1}{2}k_3\zeta + k_4\tau, \tag{73.1}$$

so that, in principle, we now have four paraxial constants, instead of three as previously. We shall, however, see in a moment that $k_4$ is effectively redundant.

To study the consequences of (73.1) as it stands, it is advisable somewhat to extend the notation first introduced just after equations (14.4) in the following way. Given any *two*-component quantity, say $\mathbf{U}\,(\equiv (U_y, U_z))$ one adjoins to it the quantity

$$\mathbf{U}^{\#} = (-U_z, U_y).$$

If $\mathbf{U}$ be regarded as a Euclidean two-vector, $\mathbf{U}^{\#}$ has the same length as $\mathbf{U}$, but is orthogonal to it. Then, from (6.1),

$$\boldsymbol{\beta}' = k_1\mathbf{y}' + k_2\mathbf{y} + k_4\mathbf{y}^{\#}, \quad \boldsymbol{\beta} = -k_2\mathbf{y}' - k_3\mathbf{y} + k_4\mathbf{y}'^{\#}. \tag{73.2}$$

The coordinates of the point of intersection of a ray $\mathscr{R}$ with the normal plane $\bar{x}' = d'$ are therefore

$$\mathbf{Y}' = (1 + d'k_1)\,\mathbf{y}' + d'(k_2\mathbf{y} + k_4\mathbf{y}^{\#}), \tag{73.3}$$

cf. equation (14.8). All rays through $O$ therefore intersect the plane which has $d' = -1/k_1$ in just one point $O'$ the coordinates of which are

$$Y' = \bar{d}'(y\cos\psi - z\sin\psi), \quad Z' = \bar{d}'(y\sin\psi + z\cos\psi), \tag{73.4}$$

with
$$\bar{d}' = d'(k_2^2 + k_4^2)^{\frac{1}{2}}, \quad \tan\psi = k_4/k_2.$$

This means that in the paraxial domain $K$ produces a sharp and undistorted image in the plane $\mathscr{I}'$ given by $d' = -1/k_1$ of an object lying in the normal plane $\mathscr{I}$ through $O_0$. However, this image is now turned as a whole through an angle $\psi$ about the axis of $K$ relative to the object, in the sense of equations (73.4). This conclusion informs us that we have proceeded in a clumsy way, and we go on to rectify this state of affairs.

To this end, having started with the *usual* coordinate basis as above, rotate the axes *in the image space* through an angle $\psi$ according to

$$\bar{\mathbf{y}}' \to \bar{\mathbf{y}}' \cos\psi' + \bar{\mathbf{y}}'^{\#} \sin\psi. \tag{73.5}$$

As a result of this

$$\xi \to \xi, \quad \eta \to \eta\cos\psi - \tau\sin\psi, \quad \zeta \to \zeta, \quad \tau \to \eta\sin\psi + \tau\cos\psi,$$

whence

$$V \to a_0 + \tfrac{1}{2}k_1\xi + (k_2\cos\psi + k_4\sin\psi)\,\eta + \tfrac{1}{2}k_3\zeta + (k_4\cos\psi - k_2\sin\psi)\,\tau.$$

If we now choose $\psi$ so that $\tan\psi = k_4/k_2$, this reduces to

$$V = a_0 + \tfrac{1}{2}k_1\xi + k_2\eta + \tfrac{1}{2}k_3\zeta, \tag{73.6}$$

where we have written $k_2$ in place of $(k_2^2 + k_4^2)^{\frac{1}{2}}$, with the understanding that in (73.6) $k_2$ is the coefficient of $\eta$ when $\psi$ has just that value which causes $\tau$ to disappear from the paraxial point characteristic; consistently with the fact that $\xi, \eta, \zeta$ in (73.6) of course also refer to this situation.

In short, we can always choose a coordinate basis (distinguished from that used hitherto only by a rotation of the $\bar{y}'$-, $\bar{z}'$-axes) such that in the paraxial region $V$ takes the usual form (73.6); and we call it the *adapted coordinate basis*. Alternatively we shall often simply speak of adapted coordinates. In the following investigation of semi-symmetric systems the use of adapted coordinates will be understood throughout.

The manifold results of Section 14 can now happily be taken over as they stand. One small point of terminology, however, remains to be examined. In the context of the symmetric system the coordinate basis was always such that the planes $\bar{z} = 0$ and $\bar{z}' = 0$ were in fact one and the same plane; and we called it the meridional plane. Now, however, unless $\psi = 0$ the planes $\bar{z} = 0$ and $\bar{z}' = 0$ are distinct. The best we can do, if we wish to retain something like the usual terminology, is to take 'the meridional plane' to consist of two distinct 'sheets'. They contain a common line, but their normals make an angle $\psi$ with each other. Even so, outside the paraxial region we cannot simply speak of 'meridional rays'. It is true that if the initial ray is meridional the final ray will intersect the axis in the image space, or else is parallel to the axis there. This may be inferred directly from the fact that the invariant $j$, defined by (8.6), and whose existence is guaranteed by the *axial* symmetry of $K$, vanishes under the present conditions. It does *not* follow, however, that the final ray is also meridional.

## 74. Remark on the ideal characteristic function

We have already considered the ideal point characteristic twice, namely in Sections 7 and 15, and it should be clear that (15.6) applies here also. Were we not using adapted coordinates, however, we should have to set

$$\mathbf{Y}' = \mathbf{y}_1 \cos\psi - \mathbf{y}_1^{\#} \sin\psi \tag{74.1}$$

in calculating the optical distance between $D'$ and $O'$ as in Section 15, with the result that we should then have to write

$$u = \xi - 2\eta\cos\psi + \zeta - 2\tau\sin\psi, \tag{74.2}$$

where $\mathbf{y}_1$ now takes the place of $\mathbf{y}$ in $\tau$.

Suppose now that we regard the image as 'ideal' provided it is merely plane and sharp. In the case of the symmetric system this meant that there existed a function $D(\zeta)$ such that $\mathbf{Y}' = \mathbf{y}_1 D(\zeta)$. Now, however, the line $O_0' O'$ need not make the same angle with the meridional plane as $O_0 O$ does. To accommodate this feature we therefore have the weaker condition

$$\mathbf{Y}' = \mathbf{y}_1 D(\zeta) + \mathbf{y}_1^{\#} \tilde{D}(\zeta), \tag{74.3}$$

where $$D(\zeta) = 1 + d_1\zeta + \dots, \quad \tilde{D}(\zeta) = \tilde{d}_1\zeta + \dots. \tag{74.4}$$

Then, taking $d' = 1$ as usual,

$$V_0 = g(\zeta) - [1 + \xi - 2\eta D + \zeta(D^2 + \tilde{D}^2) - 2\tau\tilde{D}]^{\frac{1}{2}}. \tag{74.5}$$

In short, we now have *two* 'distortion functions', i.e. $D$ and $\tilde{D}$.

## 75. The normal and skew aberration functions

The aberration function is, as usual, the difference between the actual and the ideal characteristic functions. We continue to work with the point characteristic, so that the aberration function is $v = V - V_0$. Recalling (72.1), we write

$$v = v^{\#} + \tau\tilde{v}^{\#}, \tag{75.1}$$

and call $v^{\#}$, $\tilde{v}^{\#}$ the *normal* and *skew* aberration functions respectively. The terminology is convenient even though it is $\tau\tilde{v}^{\#}$ rather than $\tilde{v}^{\#}$ which induces the 'skew aberrations'. Order by order,

$$v^{\#} = \sum_{n=2}^{\infty} v^{\#(n)}, \quad \tilde{v}^{\#} = \sum_{n=2}^{\infty} \tilde{v}^{\#(n)}, \tag{75.2}$$

where $v^{\#(n)}$ and $\tilde{v}^{\#(n)}$ are homogeneous polynomials, of degree $n$ and $n-1$ respectively, in $\xi$, $\eta$, $\zeta$. Explicitly,

$$\left.\begin{aligned} v^{\#(n)} &= \sum_{\mu=0}^{n} \sum_{\nu=0}^{\mu} v_{\mu\nu}^{(n)} \xi^{n-\mu} \eta^{\mu-\nu} \zeta^{\nu}, \\ \tilde{v}^{\#(n)} &= \sum_{\mu=0}^{n-1} \sum_{\nu=0}^{\mu} \tilde{v}_{\mu\nu}^{(n)} \xi^{n-\mu-1} \eta^{\mu-\nu} \zeta^{\nu}, \end{aligned}\right\} \tag{75.3}$$

and these define the (characteristic) *normal and skew aberration coefficients*. There is no need to attach the superscript # to these, for the $v_{\mu\nu}^{(n)}$ cannot possibly refer to a formal power series for $v^{(n)}$ itself: the latter depends on four variables, and so its power series has coefficients which, in principle, must have *three* subscripts.

$v_{nn}^{(n)}$ does not enter into the aberrations associated with the conjugate planes $\mathscr{I}$ and $\mathscr{I}'$. One must, however, not fall into the error of supposing—from force of habit—that the same is true of the corresponding coefficient $\tilde{v}_{n-1,n-1}^{(n)}$. Accordingly one has in $(2n-1)$th-order $\frac{1}{2}n(n+1)$ skew coefficients in addition to the $\frac{1}{2}n(n+3)$ normal coefficients, or $n(n+2)$ in all.

## 76. The third-order displacement

We shall now investigate the third-order displacement at some length, since it already brings out very clearly the effects of the presence of the skew aberration function. The quadratic polynomial (19.1) represents $v^{\#(2)}$, and in addition we have

$$\tilde{v}^{\#(2)} = \tilde{p}_1\xi + \tilde{p}_2\eta + \tilde{p}_3\zeta. \tag{76.1}$$

We note that if $v$ has no terms of order less than $2n-1$, then

$$\boldsymbol{\epsilon}'_{2n-1} = 2v_{\xi}^{(n)}\mathbf{y}' + v_{\eta}^{(n)}\mathbf{y}_1 + v_{\tau}^{(n)}\mathbf{y}_1^{\#}, \tag{76.2}$$

which reduces to (17.7) when $\tilde{v}^{\#} = 0$. Thus, when $n = 2$,

$$\boldsymbol{\epsilon}'_3 = (4p_1\xi + 2p_2\eta + 2p_3\zeta + 2\tilde{p}_1\tau)\mathbf{y}' + (p_2\xi + 2p_4\eta + p_5\zeta + \tilde{p}_2\tau)\mathbf{y}_1 + (\tilde{p}_1\xi + \tilde{p}_2\eta + \tilde{p}_3\zeta)\mathbf{y}_1^{\#}. \tag{76.3}$$

Since $K$ is axially symmetric we can always arrange $z_1$ to be zero, as in Section 19. Accordingly we can use (19.3) and (19.4) in (76.3), whilst

$$\tau = \rho h' \sin\theta. \tag{76.4}$$

Further, we retain (19.6) since the skew aberration function merely causes the addition of new terms to those previously encountered in equations (19.7). For the sake of uniformity we also define

$$\tilde{\sigma}_2 = \tilde{p}_1, \quad \tilde{\sigma}_3 = \tilde{p}_2, \quad \tilde{\sigma}_5 = \tilde{p}_3. \tag{76.5}$$

(76.3) then becomes

$$\left.\begin{aligned} \epsilon'_{3y} &= \sigma_1\rho^3\cos\theta + \rho^2 h'[\sigma_2(2+\cos 2\theta) + \tilde{\sigma}_2\sin 2\theta] \\ &\qquad + \rho h'^2[(3\sigma_3+\sigma_4)\cos\theta + \tilde{\sigma}_3\sin\theta] + \sigma_5 h'^3, \\ \epsilon'_{3z} &= \sigma_1\rho^3\sin\theta + \rho^2 h'[\sigma_2\sin 2\theta + \tilde{\sigma}_2(2-\cos 2\theta)] \\ &\qquad + \rho h'^2[(\sigma_3+\sigma_4)\sin\theta + \tilde{\sigma}_3\cos\theta] + \tilde{\sigma}_5 h'^3. \end{aligned}\right\} \tag{76.6}$$

The number of *types* of aberrations is not affected by the presence of $\tilde{v}^{\#}$, since types are essentially characterized by their dependence upon $\rho$ and $h'$. Nevertheless, each type is now subdivided into two, of which one is normal, the other skew. They may be considered separately or jointly: here we pursue the second course.

The terms varying as $\rho^3$ evidently represent spherical aberration, and since they are not affected by the presence of $\tilde{v}^{\#(2)}$ they need not be discussed again. The partial displacement which varies as $\rho^2 h'$ may be written

$$\left.\begin{aligned} \hat{y} &= \mathfrak{a}[(2+\cos 2\theta) + k\sin 2\theta], \\ \hat{z} &= \mathfrak{a}[\sin 2\theta + k(2-\cos 2\theta)]. \end{aligned}\right\} \tag{76.7}$$

This corresponds exactly to (19.12), whilst the additional parameter $k = \tilde{\sigma}_2/\sigma_2$. To begin with, we observe that when $\sigma_2 = 0$ we have exactly the usual primary comatic flare, *except* that its axis of symmetry lies along the $\hat{z}$-axis instead of pointing along the $\hat{y}$-axis, as it does when $\tilde{\sigma}_2 = 0$. Apart from this rotation of the flare as a whole (which must not be confused with the rotation of the normal flare accounted for by the rotation of the ideal image relative to the object) normal and skew primary coma therefore have essentially the same character. As a matter of fact it is not difficult to show that, to within a rotation, one has exactly the usual comatic flare when normal and skew coma are taken jointly. To this end, rotate the local $\hat{y}$-, $\hat{z}$-axes through an angle $\bar{\psi} = \arctan k$:

$$'\hat{\mathbf{y}} = \hat{\mathbf{y}}\cos\bar{\psi} - \hat{\mathbf{y}}^{\#}\sin\bar{\psi}; \tag{76.8}$$

and thereafter replace the angle $\theta$ by $'\theta + \frac{1}{2}\bar{\psi}$. Then

$$\left.\begin{aligned} '\hat{y} &= \rho^2 h'(\sigma_2^2 + \tilde{\sigma}_2^2)^{\frac{1}{2}}(2 + \cos 2'\theta), \\ '\hat{z} &= \rho^2 h'(\sigma_2^2 + \tilde{\sigma}_2^2)^{\frac{1}{2}}\sin 2'\theta. \end{aligned}\right\} \tag{76.9}$$

The state of affairs represented by (76.9) is illustrated in Fig. 7.1, in which the comatic displacement is of course very greatly exaggerated.

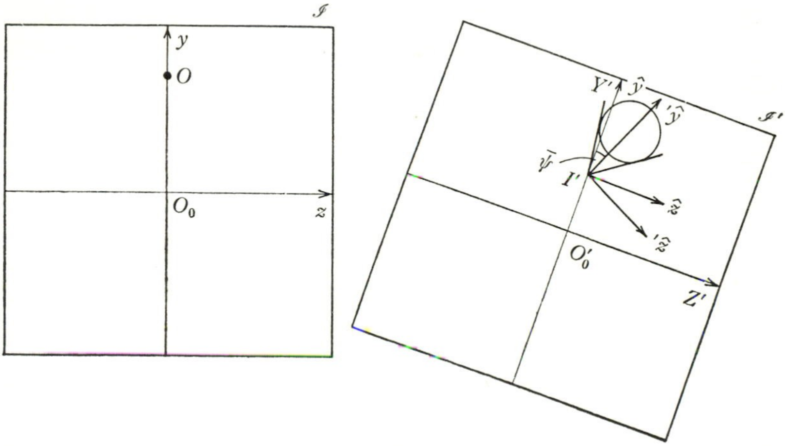


Fig. 7.1

Next we come to the terms of (76.6) varying as $\rho h'^2$. The partial displacement may be considered in an out-of-focus plane shifted relatively to $\mathscr{I}'$ by an amount $\chi = kh'^2$, exactly as in Section 19, so that

$$\left.\begin{aligned} \hat{y} &= [(3\sigma_3 + \sigma_4 - k)\cos\theta + \tilde{\sigma}_3\sin\theta]\rho h'^2, \\ \hat{z} &= [(\sigma_3 + \sigma_4 - k)\sin\theta + \tilde{\sigma}_3\cos\theta]\rho h'^2. \end{aligned}\right\} \tag{76.10}$$

The displacement is symmetric with respect to $I'$ and has essentially the usual character of curvature of field. The image is in general elliptical, but the axes of the ellipse do not lie along the $\hat{y}$-, $\hat{z}$-axes, being rotated through an angle $\frac{1}{2}\arctan[\tilde{\sigma}_3/(2\sigma_3 + \sigma_4 - k)]$ relative to them. Again, two mutually perpendicular focal lines exist, namely in the image planes for which $k$ has one or other of the values

$$k = 2(\sigma_3 + \sigma_4) \pm (\sigma_3^2 + \tilde{\sigma}_3^2)^{\frac{1}{2}}. \tag{76.11}$$

The skew coefficient $\tilde{\sigma}_3$ evidently represents a form of astigmatism, since its presence leads to focal lines even when $\sigma_3 = \sigma_4 = 0$. The image patch is circular when $k = 2\sigma_3 + \sigma_4$, irrespectively of the value of $\tilde{\sigma}_3$.

Finally we come to distortion, for which the partial displacement is

$$\hat{y} = \sigma_5 h'^3, \quad \hat{z} = \tilde{\sigma}_5 h'^3. \tag{76.12}$$

The presence of two coefficients here reflects just the presence of the two distortion functions in the point characteristic of equation (74.5). From the latter one verifies easily that $\sigma_5 = d_1$, $\tilde{\sigma}_5 = \tilde{d}_1$, so that (76.12) is of course consistent with the more general equation (74.3).

## 77. Higher-order displacements

We consider very briefly the displacement of order $2n-1$ induced by the aberration function of that order. The situation is clearly very similar to that dealt with in Section 21. In view of this, a few general remarks will suffice. Accordingly we make the transition to the variables $\rho, \theta, h'$ in (75.3). Equations (21.2) may be retained, but now

$$\phi_\lambda^{(n)} = \Sigma\,(v_{\mu\nu}^{(n)} \cos\theta + \tilde{v}_{\mu-1,\nu}^{(n)} \sin\theta) \cos^{\mu-\nu-1}\theta, \tag{77.1}$$

the summation going over values of $\mu$ and $\nu$ such that $\mu + \nu = \lambda$, with the understanding that when the value of the second subscript of any coefficient exceeds that of the first, the coefficient in question is to be taken as zero. From this the partial displacements follow by means of (21.4). For instance, with regard to circular coma, the position is very much the same as that encountered in third order. In fact, one has exactly the $(2n-1)$th-order flare familiar from the symmetric case (cf. equations (21.9)), except that it is rotated bodily through an angle $\overline{\psi}^{(n)} = \arctan(\tilde{v}_{00}^{(n)}/v_{10}^{(n)})$ relative to the $\hat{y}$-axis. When several orders are superposed one has a similar state of affairs with regard to any particular zone. However, one must bear in mind that $\overline{\psi}^{(n)}$ depends upon $n$. Therefore, under conditions such that a zonal family of circles possesses an envelope, this will, in general, no longer be a pair of straight lines, but two curves of a more or less complicated character. Again, spherical aberration is unaffected by the presence of the skew aberration function, whilst the effects of the latter on curvature of field and distortion are amply illustrated by (76.10) and (76.12).

The additional types which enter as one goes from order three to order five are, of course, oblique spherical aberration and elliptical coma. The first of these now involves three, rather than two, coefficients, so that the generic shapes of the usual zonal curves form a two-parameter family. It is therefore hardly feasible, nor is it useful, to display representative examples after the fashion of Fig. 4.4. Elliptical coma is, in a sense, even more complicated, in as far as it is jointly governed by four coefficients, of which two are normal and two are skew. The only truly simple feature is here that every zonal curve is an ellipse, though in special circumstances this may degenerate into a straight line. In fact, absorbing a factor $\rho^2 h'^3$ in each of the aberration coefficients, set

$$v^{(n)}_{n,n-3} = \tfrac{2}{3}(\mathfrak{a}-\mathfrak{b}), \quad v^{(n)}_{n-1,n-2} = \mathfrak{b},$$
$$\tilde{v}^{(n)}_{n-1,n-3} = \tfrac{2}{3}(\tilde{\mathfrak{a}}+\tilde{\mathfrak{b}}), \quad \tilde{v}^{(n)}_{n-2,n-2} = \tfrac{1}{3}(\tilde{\mathfrak{a}}-2\tilde{\mathfrak{b}}).$$

Then

$$\left.\begin{aligned} \hat{\mathfrak{y}} &= \mathfrak{a}\cos 2\theta + \tilde{\mathfrak{a}}\sin 2\theta + (\mathfrak{a}+\mathfrak{b}), \\ \hat{\mathfrak{z}} &= \mathfrak{b}\sin 2\theta + \tilde{\mathfrak{b}}\cos 2\theta + (\tilde{\mathfrak{a}}-\tilde{\mathfrak{b}}). \end{aligned}\right\} \tag{77.2}$$

As before, one may have a pair of real tangents to the set of curves generated by varying $\rho$; and when they exist they are a pair of straight lines through $I'$. A typical case is illustrated by Fig. 7.2. The line of centres does not lie along either coordinate axis, nor does it coincide with the line bisecting the angle between the tangents. The image will be triangular when $\mathfrak{a}\mathfrak{b} = \tilde{\mathfrak{a}}\tilde{\mathfrak{b}}$.

From the particular cases we have discussed it will be seen that, qualitatively speaking, the presence of the skew aberration function $\tilde{v}^{\#}$ has two principal effects on the zonal curves, when these are compared with the zonal curves when $\tilde{v}^{\#} = 0$: (i) it displaces their centres away from the $\hat{\mathfrak{y}}$-axis, and (ii) rotates them bodily about their centres.

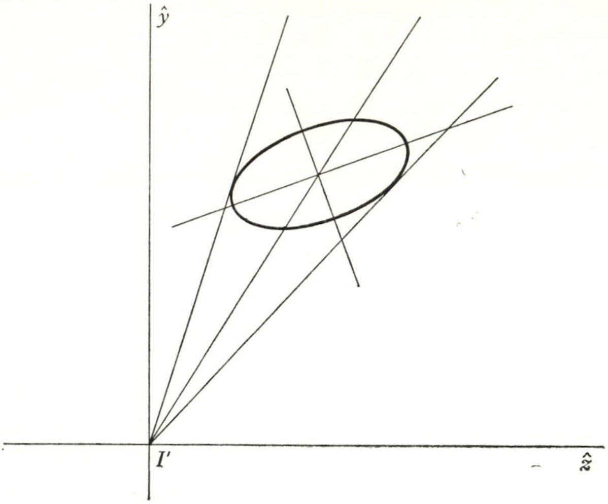


Fig. 7.2

## 78. Effective coefficients. Ancillary functions

The effective aberration coefficients are, as always, defined to be the coefficients of the power series for the *actual* displacement $\boldsymbol{\epsilon}'$, regarded as a function of appropriate variables. These we again take to be $\mathbf{y}'$ and $\mathbf{y}_1$. However, in place of the rather simple polynomial (23.1) for the displacement of order $2n-1$, the corresponding polynomial here is a good deal more complicated, in that it involves *five*, instead of only two, kinds of effective coefficients. Indeed,

$$\boldsymbol{\epsilon}'_{2n-1} = \sum_{\mu=0}^{n-1} \sum_{\nu=0}^{\mu} (u_{\mu\nu}^{(n)}\mathbf{y}' + \bar{u}_{\mu\nu}^{(n)}\mathbf{y}_1 + \tilde{u}_{\mu\nu}^{(n)}\mathbf{y}_1^{\#})\, \xi^{n-\mu-1}\eta^{\mu-\nu}\zeta^{\nu}$$

$$+ \sum_{\mu=0}^{n-2} \sum_{\nu=0}^{\mu} ({}^{\#}u_{\mu\nu}^{(n)}\mathbf{y}' + {}^{\#}\bar{u}_{\mu\nu}^{(n)}\mathbf{y}_1)\, \xi^{n-\mu-2}\eta^{\mu-\nu}\zeta^{\nu}\tau. \quad (78.1)$$

The number of effective coefficients of order $2n-1$ is $n(2n+1)$, so that there must be $n(2n+1)-n(n+2) = n(n-1)$ identities between them (cf. the end of Section 23).

In fifth order the state of affairs is fortunately rather simple, since (in an obvious notation) the $s_\alpha^*$ and $\bar{s}_\alpha^*$ are exactly given by (23.3); in other words, the coefficients $\tilde{p}_\alpha$ do not occur in them. (Note, however, that in seventh order, for instance, they *will* turn up in

$t^*_\alpha$ and $\tilde{t}^*_\alpha$.) We therefore need to write down only the following additional relations

$$\begin{aligned}
\tilde{s}^*_1 &= \tilde{s}_1 + \tfrac{1}{2}\tilde{p}_1, \\
\tilde{s}^*_2 &= \tilde{s}_2 - \tilde{p}_1 + \tfrac{1}{2}\tilde{p}_2, \\
\tilde{s}^*_3 &= \tilde{s}_3 + \tfrac{1}{2}\tilde{p}_1 + \tfrac{1}{2}\tilde{p}_3, \\
\tilde{s}^*_4 &= \tilde{s}_4 - \tilde{p}_2, \\
\tilde{s}^*_5 &= \tilde{s}_5 + \tfrac{1}{2}\tilde{p}_2 - \tfrac{1}{2}\tilde{p}_3, \\
\tilde{s}^*_6 &= \tilde{s}_6 + \tfrac{1}{2}\tilde{p}_3,
\end{aligned}$$

$$\begin{aligned}
{}^\#s^*_1 &= 4\tilde{s}_1 + 3\tilde{p}_1, & {}^\#\bar{s}^*_1 &= \tilde{s}_2 - 3\tilde{p}_1 + \tfrac{1}{2}\tilde{p}_2, \\
{}^\#s^*_2 &= 2\tilde{s}_2 - 4\tilde{p}_1 + 2\tilde{p}_2, & {}^\#\bar{s}^*_2 &= 2\tilde{s}_4 + 2\tilde{p}_1 - 3\tilde{p}_2, \\
{}^\#s^*_3 &= 2\tilde{s}_3 + \tilde{p}_1 - \tilde{p}_2 + \tilde{p}_3, & {}^\#\bar{s}^*_3 &= \tilde{s}_5 + \tfrac{3}{2}\tilde{p}_2 - \tilde{p}_3.
\end{aligned} \tag{78.2}$$

One could go on like this to higher orders, but one would evidently be confronted with a confusing and rather lengthy array of relations. It is preferable to restore some semblance of simplicity by going over to one of the ancillary functions previously introduced in Section 37, i.e. to the spherical point characteristic, or better still, the modified spherical point characteristic. Owing to the use of adapted coordinates the basic equations relating to $V^\dagger$ are unaffected by the presence of the skew aberrations. Consequently, let the $v^{(n)}_{\mu\nu}$ and $\tilde{v}^{(n)}_{\mu\nu}$ be understood to be the coefficients of the modified spherical point characteristic aberration function, and further, let the $u^{(n)}_{\mu\nu}$, $\bar{u}^{(n)}_{\mu\nu}$, ... be interpreted in the sense of equation (78.1), but with $\mathbf{y}'$ referring to the ideal wavefront, and not to the plane of the exit pupil. Then the equations

$$\left.\begin{aligned}
u^{(n)}_{\mu\nu} = 2(n-\mu)\,v^{(n)}_{\mu\nu}, \quad \bar{u}^{(n)}_{\mu\nu} = (\mu-\nu+1)\,v^{(n)}_{\mu+1,\nu}, \quad \tilde{u}^{(n)}_{\mu\nu} = \tilde{v}^{(n)}_{\mu\nu}, \\
{}^\#u^{(n)}_{\mu\nu} = (n-\mu+1)\,\tilde{v}^{(n)}_{\mu\nu}, \quad {}^\#\bar{u}^{(n)}_{\mu\nu} = (\mu-\nu+1)\,\tilde{v}^{(n)}_{\mu+1,\nu}
\end{aligned}\right\} \tag{78.3}$$

are exact for $n = 2$ and $n = 3$, and are likely to give a very close approximation to the actual effective coefficients when $n > 3$.

Finally, it may be noted that the considerations involving the wavefront and the various functions associated with it, as set out in Sections 38 and 39, are not affected by the presence of the skew aberration function.

## 79. Shifts of stop and object

We investigate briefly the effects of shifts of the object and of the stop. To this end we naturally use the angle characteristic, though, if one is not content with the pseudo-displacement or some analogous approximation to $\epsilon'$, the complex relationship between the effective and the characteristic coefficients always lurks threateningly in the background.

Proceeding exactly as in Section 46, equation (45.3) is the basic relationship from which all the required results flow, $t(\xi, \eta, \zeta, \tau)$ now replacing $t(\xi, \eta, \zeta)$, of course. (This simple state of affairs is due to our use of adapted coordinates, coupled with the fact that the angle $\psi$ in equations (73.4) in no way depends on the values of $s$ or $m$.) In addition to (46.4) we have

$$\hat{\tau} = \frac{1-pq}{(1-p)(1-q)}\tau. \tag{79.1}$$

Now observe that $\xi, \eta, \zeta$ on the one hand and $\tau$ on the other transform independently of each other, from which it follows that (45.3) splits up into two separate identities. Thus we have an equation exactly like (45.3) for the *normal* aberration function $t^{\#}$, whilst the skew aberration function $\tilde{t}^{\#}$ enters only into the identity

$$\hat{\tau}\hat{\tilde{t}}^{\#}(\hat{\xi}, \hat{\eta}, \hat{\zeta}) = \tau\tilde{t}^{\#}(\xi, \eta, \zeta),$$

or, on account of (79.1),

$$\hat{\tilde{t}}^{\#}(\hat{\xi}, \hat{\eta}, \hat{\zeta}) = \frac{(1-p)(1-q)}{1-pq}\tilde{t}^{\#}(\xi, \eta, \zeta). \tag{79.2}$$

It is convenient to write $1/\bar{d}$ for the initial factor on the right; see equations (46.1). Then one finds very easily that the new primary skew coefficients are given by

$$\left.\begin{aligned}\hat{\tilde{p}}_1 &= \bar{d}^{-1}c^2(\tilde{p}_1 + p\tilde{p}_2 + p^2\tilde{p}_3),\\ \hat{\tilde{p}}_2 &= \bar{d}^{-1}bc[2q\tilde{p}_1 + (1+pq)\tilde{p}_2 + 2p\tilde{p}_3],\\ \hat{\tilde{p}}_3 &= \bar{d}^{-1}b^2(q^2\tilde{p}_1 + q\tilde{p}_2 + \tilde{p}_3).\end{aligned}\right\} \tag{79.3}$$

For the normal coefficients one has, of course, (46.5) as before.

We do not need to write down the equations for the secondary skew coefficients, for a moment's reflection shows that they result from equations (46.5) by the following formal prescription: (i) set

$j_1 = 0$; (ii) replace every $p_\alpha$ by $\tilde{s}_\alpha$; and (iii) replace every $\hat{p}_\alpha$ by $\bar{d}\hat{\tilde{s}}_\alpha$. A similar situation prevails with respect to the tertiary skew coefficients, for which the required equations are to be read off from (46.9) using the formal prescriptions (i) $j_2 \to 0$, (ii) $s_\alpha \to \tilde{t}_\alpha$, (iii) $\hat{s}_\alpha \to \bar{d}\hat{\tilde{t}}_\alpha$. Evidently our patience in writing out many long sets of equations in the context of the symmetric system is paying dividends now.

## 80. On invariant aberrations

The ideas of Sections 46–52 remain entirely relevant to the semi-symmetric system. Indeed everything that was said before applies quite unchanged to the normal coefficients. It therefore only remains to investigate invariant and semi-invariant skew aberrations. Equations (79.3) already show at once that, in the third-order domain, there are two semi-invariants, but no absolute invariants; the former being the $s$-invariant $(s-m)^3\,\tilde{p}_1$ and the $m$-invariant $(s-m)^3\,\tilde{p}_3$. The interesting feature which emerges here is that the skew semi-invariants are of comatic type, whereas the normal semi-invariants were of astigmatic type. In particular when primary coma is entirely skew it cannot be removed by a shift of the stop (recall that now $\tilde{\sigma}_2 = (s-m)\,\tilde{p}_1$).

Again, one sees immediately from (46.7) that the expression $(s-m)^3(2\tilde{s}_3 - \tilde{s}_4)$ will be an absolute invariant; and this relates to secondary elliptical coma. Now suppose that $2\tilde{s}_3 - \tilde{s}_4$ vanishes for some position of the object and of the stop. This implies just that the quantity $\mathfrak{b}$ in (77.2)—with $n = 3$—is zero. If therefore we suppose, for the sake of illustration, that the normal aberrations have been removed, then the flare due to secondary skew elliptical coma will be triangular, with its base parallel to the $\hat{y}$-axis, the angle at $I'$ being a right angle; and it will have this shape for all values of $s$ and $m$.

The method of generators of Section 49 may be extended—somewhat trivially perhaps—by adjoining to the previous set (49.1–3) the generator T and its adjoint $\overline{\mathrm{T}}$, defined by

$$\mathrm{T} = (s-m)\frac{\partial}{\partial \tau}, \quad \overline{\mathrm{T}} = \frac{\partial}{\partial \overline{\tau}}, \tag{80.1}$$

so that

$$\mathrm{T} = \overline{\mathrm{T}}. \tag{80.2}$$

Then the various skew invariants and semi-invariants result from the application to $t$ of operators which are products of $\tau$ and appropriate powers of $A$, $S$ and $M$. In particular, there is one absolute skew invariant of order $4N+1$ ($N = 1, 2, \ldots$), and it is given by

$$\tilde{\alpha}_{4N+1} = \mathrm{T}A^N t^{(2N+1)} \equiv (s-m)\,A^N \tilde{t}^{\#(2N+1)}. \qquad (80.3)$$

The semi-invariants and conditional invariants are likewise given by expressions closely resembling those obtained previously.

## 81. Sine-relation and cosine-conditions

Since the ideal point characteristic $V_0$ of the semi-symmetric system $K$ is formally the same as that of the symmetric system, relative to an adapted coordinate basis, it follows at once that when there exists a pair of perfect conjugate planes $\mathscr{I}$ and $\mathscr{I}'$ all rays through the axial points of these satisfy the sine-relation (27.3). On the other hand the situation is more complex with regard to the analogue of the sine-*condition* of Section 28, as will now be set out.

In the first place, the aberration function (26.1)—which included no terms varying non-linearly with $h'$—must now be supplemented by a skew term, i.e. here

$$v = S(\xi) + \eta C(\xi) + \tau \tilde{C}(\xi). \qquad (81.1)$$

Then it is easily confirmed that for axial rays ($\mathbf{y}_1 = 0$)

$$\boldsymbol{\beta}' - m^{-1}\boldsymbol{\beta} = (2\dot{S} + C)\,\mathbf{y}' - \tilde{C}\mathbf{y}'^{\#}. \qquad (81.2)$$

Now, even when $\gamma = 0$, $\gamma'$ will in general fail to vanish, and it is not good enough to retain the sine-condition in the form

$$\sin\phi' - m^{-1}\sin\phi = 0,$$

i.e.

$$m^2(\beta'^2 + \gamma'^2) = \beta^2 + \gamma^2. \qquad (81.3)$$

Taking spherical aberration to be absent for the sake of orientation, (81.3) merely implies

$$C^2 + \tilde{C}^2 + 2C/w = 0, \qquad (81.4)$$

a result which is unnecessarily weak. Evidently the sine-condition should now be expanded into the *cosine-conditions* even in the context of axial rays (cf. Section 34), i.e.

$$\boldsymbol{\beta}' - m^{-1}\boldsymbol{\beta} = 0, \qquad (81.5)$$

for when $\dot{S} = 0$, these entail

$$C = 0 \quad and \quad \tilde{C} = 0. \tag{81.6}$$

In the absence of spherical aberration, satisfaction of the cosine-conditions therefore implies the absence of both normal and skew circular coma.

In the presence of spherical aberration one naturally takes in place of (81.5) the *modified cosine-conditions*

$$\bar{\boldsymbol{\beta}}' - m^{-1}\boldsymbol{\beta} = 0, \tag{81.7}$$

where, in the notation of Section 29, $\bar{\beta}', \bar{\gamma}'$ are the cosines of the angles which the line $D'O_0'$ makes with the $\bar{y}'$- and $\bar{z}'$-axes.

## 82. The two offences against the cosine-conditions

It is of considerable interest to examine whether here again there exists a close relationship between the exact total circular comatic displacement of certain rays and the extent to which the cosine-conditions are not satisfied by axial rays of the same aperture. To do so we first require the appropriate generalization of equation (26.5), to include the effects of the presence of the term $\tau\tilde{C}$ in (81.1). To the various quantities defined by equations (26.3) we naturally adjoin

$$\tilde{Q} = -2w^3\dot{\tilde{C}}, \quad \tilde{R} = w\tilde{C}; \tag{82.1}$$

and then one finds after some rather tedious manipulations that

$$\begin{aligned}\boldsymbol{\epsilon}' = \{(1-JP)+J^3[(P^2R-Q)\eta+(P^2\tilde{R}-\tilde{Q})\tau]\}\,\mathbf{y}' \\ -(1-JR)\,\mathbf{y}_1+J\tilde{R}\mathbf{y}_1^{\#}.\end{aligned} \tag{82.2}$$

For the transverse and longitudinal spherical aberration one has

$$\sigma = \rho(1-JP), \quad \delta = -1+1/JP, \tag{82.3}$$

as before. On the other hand, upon defining meridional and sagittal rays to be rays through $O$—itself in the meridional plane—which have $\theta = 0$ or $\pi$, and $\theta = \pm\frac{1}{2}\pi$ respectively, we see that $\epsilon_z'$ in general does not vanish for tangential rays. We thus have no option but to contemplate, in place of $\kappa_s$, *two* quantities, $\kappa$ and $\tilde{\kappa}$ say, defined as

$$\kappa = \epsilon_y'(\theta = \tfrac{1}{2}\pi), \quad \tilde{\kappa} = \epsilon_z'(\theta = 0). \tag{82.4}$$

At the same time we write $K = \kappa/h'$ and $\tilde{K} = \tilde{\kappa}/h'$, so that

$$K = JR - 1, \quad \tilde{K} = J\tilde{R}. \tag{82.5}$$

Now $$w\boldsymbol{\beta}/m = -R\mathbf{y}' + \tilde{R}\mathbf{y}'^{\#}, \quad w\boldsymbol{\beta}' = -P\mathbf{y}',$$

whence $$P\boldsymbol{\beta}/m = R\boldsymbol{\beta}' - \tilde{R}\boldsymbol{\beta}'^{\#}. \tag{82.6}$$

From the last pair of equations it follows that

$$\frac{mR}{P} = \frac{\beta\beta' + \gamma\gamma'}{\beta'^2 + \gamma'^2}, \quad \frac{m\tilde{R}}{P} = \frac{\beta\gamma' - \gamma\beta'}{\beta'^2 + \gamma'^2}. \tag{82.7}$$

It is convenient to introduce the abbreviation

$$M = m(1 + \delta)(\beta'^2 + \gamma'^2),$$

and then (82.3), (82.5) and (82.7) finally yield

$$K = -1 + (\beta\beta' + \gamma\gamma')/M, \quad \tilde{K} = (\beta\gamma' - \gamma\beta')/M. \tag{82.8}$$

The direction cosines on the right may be taken as relating to an arbitrary axial ray; indeed, $K$ and $\tilde{K}$ are obviously invariant under rotations. The closest resemblance to (31.5) is obtained by choosing it to be sagittal, for then $\gamma' = 0$, and therefore

$$K = -1 + \frac{\beta}{m(1+\delta)\beta'}, \quad \tilde{K} = -\frac{\gamma}{m(1+\delta)\beta'}. \tag{82.9}$$

It is natural to call $K$ and $\tilde{K}$ the *normal and skew offence against the cosine-conditions* respectively. When both vanish simultaneously the comatic zonal circle, corresponding to the value of $\rho$ in question, passes through $I'$.

## 83. Reversible systems

The definition of reversibility given in Section 53 in the context of the symmetric system may be retained for the semi-symmetric system. One must, however, be clear about the implications of this as regards the screw-sense. Thus, consider some ray $\mathscr{R}$ through $K$. The 'reflected' initial and final rays define a certain ray which will be the reflection of $\mathscr{R}$ as a whole only if the screw-sense of the reflected system $K^*$ is the reverse of that of $K$. For example, when $\mathscr{R}$ is, within $K$, a right-handed spiral, then its reflection $\mathscr{R}^*$ will be a left-handed spiral, viewed in each case from the object space;

and realizability of this by $K^*$ requires that the preferential direction implied, at each point of the ray, by the screw-sense must have opposite signs for $K$ and $K^*$ respectively. Physically, in the case of an electron–optical system, the direction of the magnetic field must be reversed.

The reversal of the screw-sense implies that if $T^* = f(\bar{\xi}, \bar{\eta}, \bar{\zeta}, \bar{\tau})$ is the angle characteristic of $K$ referred to base-points $O_0^*$, $O_0'$, as before, then that of $K^*$ is $T^* = f(\bar{\xi}, \bar{\eta}, \bar{\zeta}, -\bar{\tau})$. However, on going over from $\mathscr{R}$ to $\mathscr{R}^*$, $\bar{\xi}, \bar{\eta}, \bar{\zeta}, \bar{\tau}$ go into $\bar{\zeta}, \bar{\eta}, \bar{\xi}, -\bar{\tau}$; so that we now have the basic identity

$$T^*(\bar{\xi}, \bar{\eta}, \bar{\zeta}, \bar{\tau}) = T^*(\bar{\zeta}, \bar{\eta}, \bar{\xi}, \bar{\tau}) \tag{83.1}$$

as the generalization of (53.1). The fact that adapted coordinates are being contemplated need cause no concern on account of the axial symmetry of $K$.

It is immediately clear that (83.1) breaks up into two separate identities. The first of these involves only $T^\#$; and it is just (54.2) with $T$ in this replaced by $T^\#$. It follows that the work of Sections 54–60 may be retained in its entirety, provided it is understood to concern the normal aberration function only. In particular we have the relations (56.5), (57.1), (59.5) governing the normal aberration coefficients of the various orders.

It remains to investigate the effects of reversibility on the skew coefficients, which are implied by the identity

$$\tilde{t}^\#(\xi, \eta, \zeta) = \tilde{t}^\#(\hat{\xi}, \hat{\eta}, \hat{\zeta}). \tag{83.2}$$

Since one does not have a problem here analogous to that of eliminating the unwanted normal coefficients $t_{nn}^{(n)}$, the device of introducing the auxiliary variables $L$, $M$, $N$, defined by (55.1), is, strictly speaking, superfluous. However, as they are already available, we may as well use them, especially as they endow the work with a certain additional simplicity. Consequently the factors multiplying

$$L^{n-\mu-1}M^{\mu-\nu}N^{\nu} \quad \text{and} \quad L^{\nu}M^{\mu-\nu}N^{n-\mu-1}$$

in the polynomial $t^{\#(n)}(u^2L + 2uvM + v^2N,\ w(uL+vM),\ w^2L)$ must be equal to each other. We note here that all relations between the skew coefficients will be homogeneous.

When $n = 2$ one gets just one relation, namely

$$(u^2 - v^2)\,\tilde{p}_1 + wu\tilde{p}_2 + w^2\tilde{p}_3 = 0, \tag{83.3}$$

where we suppose for the time being that $m^2 \neq 1$. In terms of the coefficients of equations (76.5) therefore

$$(1-s^2)\,\tilde{\sigma}_2+(1-sm)\,\tilde{\sigma}_3+(1-m^2)\,\tilde{\sigma}_5 = 0. \qquad (83.4)$$

The two fifth-order relations may be read off from (56.3) and (56.10), since the polynomial to be considered is given by (56.2) if, in this, we set $d^* = 0$ and replace $p_\alpha$ by $\tilde{s}_\alpha$. Thus

$$\left.\begin{aligned} 4(1-s^2)\,u\tilde{s}_1+(3u^2-v^2)\,\tilde{s}_2+2uw(\tilde{s}_3+\tilde{s}_4)+w^2\tilde{s}_5 &= 0,\\ (1-s^2)\,(u^2+v^2)\,\tilde{s}_1+u^3\tilde{s}_2+u^2w(\tilde{s}_3+\tilde{s}_4)+uw^2\tilde{s}_5+w^3\tilde{s}_6 &= 0.\end{aligned}\right\} \qquad (83.5)$$

Since the degree of the polynomial $\tilde{t}^{\#(n)}$ is less by one than that of $t^{\#(n)}$ the number of relations between the skew coefficients of order $2n-1$ may be read off from (58.4), replacing $n$ in this by $n-1$, and increasing the number so obtained by 1, since $\tilde{t}^{(n)}_{n-1,\,n-1}$ need not be eliminated. Thus the number in question is

$$\tfrac{1}{4}(n^2-1)\ \textit{when } n \textit{ is odd}, \quad \text{and} \quad \tfrac{1}{4}n^2\ \textit{when } n \textit{ is even}. \qquad (83.6)$$

Altogether the number of relations between the $n(n+2)$ coefficients of order $2n-1$ is therefore $\tfrac{1}{2}(n-1)(n+2)$, leaving $\tfrac{1}{2}(n+1)(n+2)$ which are mutually independent.

We have yet to consider the case $m^2 = 1$. It will suffice to take $m = -1$, and, at the same time $s = 1$. Proceeding as in Section 59 it follows at once that then

$$\tilde{t}^{(n)}_{\mu\nu} = 0 \quad \textit{when} \quad \mu-\nu \quad \textit{is odd}. \qquad (83.7)$$

In particular, when $n = 2$ or 3,

$$\tilde{p}_2 = 0; \quad \tilde{s}_2 = \tilde{s}_5 = 0. \qquad (83.8)$$

In short, all skew aberration coefficients of the completely reversible semi-symmetric system which are of *astigmatic* type must vanish. This result is to be contrasted with (59.5) which expresses the vanishing of the normal *comatic* types. Note that (83.6) continues to hold when $m^2 = 1$.

Finally, contemplating the possibility of the achievement of sharp imagery by semi-symmetric systems, nothing emerges which goes essentially beyond what was found in Sections 62 and 63 for symmetric systems. The reason for this is quite simply as follows: that to obtain a sharp image all skew aberrations other than distortion must certainly vanish, bearing in mind that skew curvature of field is of astigmatic character. The reversibility condition then requires that the skew distortion function $\tilde{D}$ be constant, and therefore zero; cf. equations (74.3, 4). It follows that the presence of an inherent screw-sense in $K$ must have no effect beyond rotating the image space rigidly through some angle; but such a rotation is irrelevant to surfaces of revolution $\mathscr{I}^{*\prime}$.

## Problems

**P.7 (i).** Show that the line of centres (of the zonal circles) of the circular comatic flare produced by a semi-symmetric system is a straight line if, and only if, the ratio of the comatic functions $C$ and $\tilde{C}$ is a constant. (Spherical aberration of all orders is taken to be absent.)

**P.7 (ii).** Given any two sets of values of the primary skew coefficients, show that if the first relates to $s$ and $m$ then one can always determine $\hat{s}$ and $\hat{m}$ such that for these the primary skew coefficients take the values of the second set.

**P.7 (iii).** Given $T$, show that, with $\boldsymbol{\sigma}$ and $\boldsymbol{\mu}$ defined by (24.2),

$$\boldsymbol{\epsilon}' = (s-m)(2\boldsymbol{\sigma} T_\xi + \boldsymbol{\mu} T_\eta + \boldsymbol{\mu}^\# T_\tau).$$

**P.7 (iv).** Find the explicit form of the absolute skew invariant $\tilde{\alpha}_9$. To what type of aberration does it relate?

CHAPTER 8

# SYSTEMS WITH TRANSLATIONAL SYMMETRY, AND PLANE-SYMMETRIC SYSTEMS

## 84. Definition of translationally symmetric systems with or without additional symmetries

Let a coordinate basis of the kind familiar from the theory of symmetric systems be chosen, so that the $\bar{x}'$- and $\bar{x}$-axes are collinear, and the remaining axes pair-wise parallel. Then a system $K$ is *translationally semi-symmetric* if, upon translating it in a direction parallel to the $\bar{z}$-axis, the resulting system is indistinguishable from $K$. For the purposes of this definition the stop is not to be regarded as part of $K$. On the other hand, the object and image planes may be included, granted that these are to be taken normal to the $\bar{x}$- and $\bar{x}'$-axes. The fact that $K$ cannot, of course, extend to infinity is immaterial. The situation corresponds exactly to that remarked upon in Section 64 in the context of the concentric system: a ray which passes through $K$ both before and after a translation 'does not know' that the system has been displaced.

Consider now the translation of $K$ through a distance $s$. Then, since $z'$ and $z$ are reduced, as usual, $z'$ changes by an amount $N's$, and $z$ by $Ns$. Invariance requires that the relation

$$V(y', z'+N's, y, z+Ns) = V(y', z', y, z)$$

must hold identically for all values of $s$. The generic form of the point characteristic is therefore

$$V = V(y', y, z'-z_1), \tag{84.1}$$

where $z_1 = (N'/N)\,z$; so that $V$ is a function of only three arguments.

By way of example, a system which consists of a number of refracting cylinders having *arbitrary* normal cross-sections, and which are arbitrarily disposed relative to one another except that their generators must be parallel to the $\bar{z}$-axis, is certainly invariant under translations parallel to this axis. However, when $K$ is thus constituted one sees immediately that the positive $\bar{z}'$- and $\bar{z}$-directions are not preferred over the negative $\bar{z}'$- and $\bar{z}$-directions. This will evidently always be so in the absence of anisotropies within $K$. We have a situation analogous to that remarked upon at the end of Section 72: that is to say, unless $K$ was a purely electrostatic electron-optical instrument, the positive and negative $\bar{z}'$- and $\bar{z}$-directions respectively were not equivalent. We call $K$ *translationally symmetric* if it is translationally semi-symmetric *and* if the 'meridional plane' ($\bar{z}' = 0, \bar{z} = 0$) is a plane of symmetry. This means that $V$ must be invariant under the joint reversal of sign of $z'$ and $z_1$. Since we impose the condition of regularity upon $K$ as usual, this means that only even powers of $z' - z_1$ can appear in the power series for $V$; and this situation is reflected formally by writing

$$V = V[y', y, (z' - z_1)^2]. \tag{84.2}$$

If $K$ is translationally symmetric *and*, in addition, the common plane $\bar{y}' = 0$, $\bar{y} = 0$ is a plane of symmetry, then $K$ will be called *c-symmetric*. This case is realized in practice when, in the example discussed previously, the boundaries of the normal cross-sections are symmetric about the axis $\mathscr{A}$ of $K$, i.e. the common line defined by the $\bar{x}'$- and $\bar{x}$-axes; in particular they might be composed of arcs of circles whose centres lie upon $\mathscr{A}$. *c*-Symmetry requires that $V$ be invariant under the simultaneous reversal of sign of $y'$ and $y$, and in place of (84.2) we then write

$$V = V[y'^2, y'y_1, y_1^2, (z' - z_1)^2], \tag{84.3}$$

where $y_1 = my$. Here $m$ is the magnification in the meridional section of $K$; see Section 85. From (84.2) it follows immediately that

$$\gamma' = \nu\gamma \quad (\nu = N/N'), \tag{84.4}$$

which is, once again, a special case of (8.5).

It is natural to put

$$\xi = y'^2, \quad \eta = y'y_1, \quad \zeta = y_1^2, \quad \tau = (z' - z_1)^2, \tag{84.5}$$

although the $\xi$, $\eta$, $\zeta$, $\tau$ here have meanings in no way related to those of the rotational invariants denoted previously by the same symbols. Nevertheless the notation is not inappropriate, since $\xi$, $\zeta$ and $\tau$ may be regarded as the 'elementary reflection invariants' of the *c*-symmetric system. $\eta$ is not elementary, for

$$\eta^2 \equiv \xi\zeta; \tag{84.6}$$

and its inclusion is dictated only by the need easily to accommodate the condition of regularity. Evidently we can write, in complete analogy with (72.1),

$$V(\xi, \eta, \zeta, \tau) = V^{\#}(\xi, \zeta, \tau) + \eta \tilde{V}^{\#}(\xi, \zeta, \tau). \qquad (84.7)$$

However, there does not appear to be any general, interesting property of $K$ which would be reflected in the absence of the aberration function corresponding to $\tilde{V}^{\#}$.

It is possible, in principle, for a translationally semi-symmetric system to have the last symmetry introduced above; in which case we would call it *c*-semi-symmetric. Our purposes will, however, be sufficiently well served if we confine our discussion to the simplest case, i.e. that of the *c*-symmetric system. We may have additional symmetries. Thus $K$ may be *reversible*, i.e. have a normal plane of symmetry. Figure 6.1 may be used as an illustration, if the various curves be interpreted as being the boundaries of the normal cross-sections of the refracting elements. Again, $K$ may be *concentric*, i.e. invariant under rotations about a line through $\mathscr{A}$, normal to the meridional plane. Such additional symmetries entail the use of appropriate formal devices to make them tractable, and we therefore consider them later on. On the other hand, it is hardly appropriate to deal with the *c*-symmetric system in the same detail as we did with the symmetric system; especially as the theory of the latter was given at such great length so that it might serve as a model for the treatment of systems having different basic symmetries.

One final point deserves comment. Suppose that a *c*-symmetric system is also invariant under translations in a direction parallel to the $\bar{y}$-axis. Then, in view of (84.3),

$$V = V[(y' - y_1)^2,\ (z' - z_1)^2]. \qquad (84.8)$$

This is as far as one can go: we conclude that even invariance under translations along two (mutually perpendicular) directions does not imply the absence of anisotropy within $K$; for example, $N$ might depend upon $\gamma^2$. Were $K$ isotropic, $N$ would be constant over every normal plane, and $K$ therefore symmetric; and in (84.8) $V$ would reduce to a function of $(y' - y_1)^2 + (z' - z_1)^2$. In short, when classes of systems are defined generically by the symmetries they possess, one may always have internal anisotropy consistent with these symmetries.

## 85. The paraxial region

In the paraxial domain $V$ reduces to

$$V = \text{const.} + \tfrac{1}{2}k_1\xi + k_2\eta + \tfrac{1}{2}k_3\zeta + \tfrac{1}{2}k_4\tau, \qquad (85.1)$$

the base-points having been chosen arbitrarily for the moment. Then

$$\left.\begin{aligned} \beta' &= k_1 y' + k_2 y_1, & \gamma' &= k_4(z' - z_1), \\ -\beta/m &= k_2 y' + k_3 y_1, & \nu\gamma &= k_4(z' - z_1). \end{aligned}\right\} \qquad (85.2)$$

This is evidently a special case of (10.10), bearing in mind that a $c$-symmetric system has two planes of symmetry containing $\mathscr{A}$. In any such system the $y$- and $z$-directions 'do not interact' paraxially, so to speak. One can therefore always make formal use of the work of Section 14, taking the $y$- and $z$-components of the equations there separately, with the appropriate transcription of the paraxial constants. In the present instance this transcription is

$$k_1, k_2, k_3 \to \begin{cases} k_1, mk_2, m^2k_3 \\ k_4, -k_4/\nu, k_4/\nu^2, \end{cases} \qquad (85.3)$$

where $B$ has now been taken to be $O_0$, whilst $B'$ may be regarded as the axial point of the exit pupil. One then concludes immediately that $K$ is characterized by two distinct focal lengths $f_1$ and $f_2$, where $f_1$ is given exactly by the right-hand member of (14.15), whilst $f_2$ is infinite. The sagittal section of $K$ is therefore telescopic, and we cannot use the angle characteristic. Of course, in the sagittal section the refracting elements are equivalent to a set of plane-parallel slabs. All rays from an object point $O$ will pass through just one point $I'$ of the image plane only if $K$ is such that $k_1 = k_4$, and $d'$ is chosen to have the value $-1/k_1$. $I'$ is evidently virtual (i.e. the rays in the image space meet in $I'$ only upon being produced backwards). If $k_1 \neq k_4$ meridional rays all pass through one point of the plane which has $d' = -1/k_1$, and this point need not be virtual. The complete image, formed by all rays admitted by $K$, will be a line normal to the meridional plane:

$$Y' = y_1, \quad Z' = [(k_1 - k_4)z' + k_4 z_1]/k_1. \qquad (85.4)$$

Such line images will be studied more closely in Section 89.

## 86. The aberration coefficients

Let $\mathscr{I}$ be the object plane as usual, and let $\mathscr{I}'$ be the image plane *meridionally* conjugate to it, so that the location of $\mathscr{I}'$ is given by the value $-1/k_1$ of $d'$. For the present we take the view that we should *like* $K$ to produce a sharp image in $\mathscr{I}'$, the (reduced) magnifications in the $y'$- and $z'$-directions being required to have constant values $m$ and $\nu^{-1}$ respectively. It is quite in order to adopt this definition of ideality even though we know that $K$ cannot possibly produce an image of this kind unless $k_1 = k_4$.

By the usual argument therefore,

$$V = g(\zeta) - (1 + \xi - 2\eta + \zeta + \tau)^{\frac{1}{2}} + v, \tag{86.1}$$

where we have set $d' = 1$, as often before. Note that $g$ cannot depend on $z_1$, since when the imagery is perfect, $\tilde{V}(O, O')$ is a function of $y_1$ and $z_1$ only, but $z_1$ must of necessity occur together with $z'$, as we have seen. If $g(\zeta) = g_0 + \frac{1}{2}g_1\zeta + \ldots$ it follows from (86.1) that

$$V = g_0 - 1 - \tfrac{1}{2}\xi + \eta - \tfrac{1}{2}(1 - g_1)\zeta - \tfrac{1}{2}\tau + v^{(1)} + O(4). \tag{86.2}$$

On the other hand, we already know from the work of Section 85 that

$$V = \text{const.} - \tfrac{1}{2}\xi + \eta - \tfrac{1}{2}[1 + (1/mf_1)]\zeta + \tfrac{1}{2}k_4\tau + O(4), \tag{86.3}$$

by transcription of (14.30), applied to the meridional section of $K$. Comparison of the last two equations shows that $g_1 = -1/mf_1$, and

$$v^{(1)} = \tfrac{1}{2}(k_4 + 1)\tau. \tag{86.4}$$

The presence of the *first*-order aberration function of course simply reflects our previous conclusion that perfect imagery, in the sense understood above, requires that $k_4 = k_1 (= -1)$; see also Section 89.

In harmony with (84.7) we now write

$$v = v^{\#}(\xi, \zeta, \tau) + \eta\tilde{v}^{\#}(\xi, \zeta, \tau). \tag{86.5}$$

The coefficients $v^{(n)}_{\mu\nu}$, $\tilde{v}^{(n)}_{\mu\nu}$ of the power series for $v^{\#}$ and $\tilde{v}^{\#}$ are the aberration coefficients of the $c$-symmetric system. When $n = 1$ there is only the one non-zero coefficient $v^{(1)}_{11} = \frac{1}{2}(k_4 + 1)$, in view of (86.4). As regards the coefficients of orders 3, 5, 7, ... (there are of course none of even order), we have no need to count them, for the present state of affairs is precisely analogous to that encountered in Section 75. Thus, *when* $m \neq 0$ *the number of aberration coefficients of order* $2n-1$ $(n > 1)$ *is* $n(n+2)$. In particular there are *eight* third-order coefficients.

The explicit inclusion here of the condition $m \neq 0$ is necessitated by the fact that one obtains a different result when the object is at infinity, in which case the ideal image must also lie at infinity. This means, in turn, that $K$ must be entirely telescopic, i.e. $f_1 = \infty$ also. We then understand perfect imagery to mean that $K$ transforms every set of mutually parallel initial rays into a set of mutually parallel final rays, i.e.

$$\beta' = \mu\beta, \quad \gamma' = \nu\gamma \tag{86.6}$$

for all rays, where $\mu$ is a constant and $\nu$ has its previous significance. We now go over to the mixed characteristic $W_1$, base-points at $E$ and $E'$, say. According to (84.4) the second member of (86.6) must hold for all translationally symmetric systems, whether perfect or not. Thus $\partial W_1/\partial z' = \nu\gamma$, i.e.

$$W_1 = \nu z'\gamma + A(y', \beta, \gamma), \tag{86.7}$$

where the function $A$ is so far arbitrary. $c$-Invariance requires that its power series be solely in terms of the variables $\xi = y'^2$, $\eta = y'\beta$, $\zeta = \beta^2$, $\tau = \gamma^2$. If the planes at infinity be perfect the first member of (86.6), i.e. $\partial W_1/\partial y' = \mu\beta$, together with (86.7), shows that

$$A = \mu y'\beta + g(\zeta, \tau), \tag{86.8}$$

where $g$ is arbitrary. We are thus led to write

$$W_1 = g(\zeta, \tau) + \mu y'\beta + \nu z'\gamma + w_1, \tag{86.9}$$

where $w_1$ is the aberration function.

To count the number of coefficients in this case, imagine $w_1$ to be written in a form analogous to (86.5). Then there are altogether $\frac{1}{2}(n+1)(n+2)$ coefficients $w_{1\mu\nu}^{(n)}$ of which, however, all those multiplying powers of $\zeta$ *and* $\tau$ only are redundant, and there are $n+1$ of these, so that we are left with $\frac{1}{2}n(n+1)$ coefficients of this type. In addition there are $\frac{1}{2}n(n+1)$ coefficients $\tilde{w}_{1\mu\nu}^{(n)}$. Thus, finally, *when* $m = 0$ *the number of aberration coefficients of order* $2n-1$ *is* $n(n+1)$, i.e. $n$ fewer than when $m \neq 0$. In particular, there are only *six* third-order coefficients. It may be noted that the kind of system to which this result relates occasionally occurs in practice, namely as a periscopic instrument.

## 87. The third-order displacement when $k_1 = k_4$

We first consider the third-order displacement under conditions such that the possibility of ideal imagery (in the sense defined above) exists in principle. This requires that $k_1 = k_4$, and then we naturally contemplate the ideal image plane. In the usual way.

$$\epsilon'_{y3} = 2y'v_\xi + y_1 v_\eta, \quad \epsilon'_{z3} = 2(z'-z_1)\,v_\tau, \tag{87.1}$$

where $v$ is to include only the third-order terms. Although (86.5) was a useful way of writing $v$ in the earlier context, there appears to be little point in adhering to it now, nor does there seem to be a preferential ordering of the various terms of $v^{(2)}$. We therefore write

$$v^{(2)} = p_1\xi^2 + p_2\xi\eta + p_3\xi\zeta + p_4\xi\tau + p_5\eta\zeta + p_6\eta\tau + p_7\zeta\tau + p_8\tau^2. \tag{87.2}$$

Then

$$\left.\begin{aligned} \epsilon'_{y3} &= 2y'(2p_1\xi + p_2\eta + p_3\zeta + p_4\tau) + y_1(p_2\xi + p_5\zeta + p_6\tau), \\ \epsilon'_{z3} &= 2(z'-z_1)(p_4\xi + p_6\eta + p_7\zeta + 2p_8\tau). \end{aligned}\right\} \tag{87.3}$$

If the aperture of the stop is rectangular, one can discuss 'zonal curves' corresponding to $y'$ being kept fixed (see the end of this section).

A stop with a circular aperture will usually lie in the image space. We now definitely suppose this to be the case, and introduce the usual polar coordinates. At the same time, since such a stop does not partake of the translational symmetry of $K$, we cannot simply put $z = 0$. It is convenient, therefore, also to introduce polar coordinates in the ideal image plane. Thus, altogether, we set

$$y' = \rho\cos\theta, \quad z' = \rho\sin\theta, \quad y_1 = h'\cos\psi, \quad z_1 = h'\sin\psi, \tag{87.4}$$

so that $h'$ is the 'ideal image height', i.e. the perpendicular distance of $I'$ from the axis. Of course, $h'$ does not stand in a fixed ratio to the 'object height' $h(=(y^2+z^2)^{\frac{1}{2}})$ unless $\nu m = \pm 1$. (87.3) then gives rise to the rather complicated expressions

$$\left.\begin{aligned}\epsilon'_{y3} = {} & [(4p_1\cos^2\theta+2p_4\sin^2\theta)\cos\theta]\rho^3+[\tfrac{1}{2}(3p_2+p_6)\cos\psi \\ & +\tfrac{1}{2}(3p_2-p_6)\cos\psi\cos 2\theta-2p_4\sin\psi\sin 2\theta]\rho^2h' \\ & +2[(p_3\cos^2\psi+p_4\sin^2\psi)\cos\theta-p_6\sin\psi\cos\psi \\ & \times\sin\theta]\rho h'^2+(p_5\cos^3\psi+p_6\cos\psi\sin^2\psi)h'^3, \\ \epsilon'_{z3} = {} & [(2p_4\cos^2\theta+4p_8\sin^2\theta)\sin\theta]\rho^3+[-(p_4+6p_8) \\ & \times\sin\psi-(p_4-6p_8)\sin\psi\cos 2\theta+p_6\cos\psi\sin 2\theta] \\ & \times\rho^2h'+2[-p_6\cos\psi\sin\psi\cos\theta+(p_7\cos^2\psi \\ & +6p_8\sin^2\psi)\sin\theta]\rho h'^2-(2p_7\cos^2\psi\sin\psi \\ & +4p_8\sin^3\psi)h'^3.\end{aligned}\right\}\qquad(87.5)$$

To the four types of primary aberrations which appear here we attach the usual names and briefly discuss them in turn. Note that any particular coefficient in general now enters into the partial displacement corresponding to *several* types; so that the discussion of the separate partial displacements is inherently even more formal than usual.

(i) *Spherical aberration*

This is the only axial aberration, and it is now jointly governed by *three* coefficients, so that the state of affairs is, in general, far from simple. The image patch is circular only when $2p_1 = p_4 = 2p_8$.

(ii) *Coma*

The partial displacement has the generic form

$$\hat{y} = \mathfrak{a}\cos 2\theta+\mathfrak{b}\sin 2\theta+\mathfrak{a}+\mathfrak{d},\quad \hat{z} = \mathfrak{c}\cos 2\theta+\mathfrak{d}\sin 2\theta+\mathfrak{b}-\mathfrak{c},\qquad(87.6)$$

with an obvious interpretation of $\mathfrak{a}, \ldots$; e.g.

$$\mathfrak{a} = \tfrac{1}{2}\rho^2h'(3p_2-p_6)\cos\psi.$$

For any fixed value of $\psi$ one has a displacement generically of exactly the same kind as that due to higher-order *elliptical* coma in semi-symmetric systems (see (77.2)). One cannot therefore speak of 'circular coma' here, but rather of linear coma, as suggested in

Section 22. A more detailed discussion is superfluous. We merely remark in passing that, for general values of the coefficients, there is always a value of $\psi$ such that the partial displacement, taken by itself, represents a triangular image; but a statement of this kind is somewhat misleading since the partial displacements representing other types will, under these circumstances, certainly not be absent; cf., by contrast, the results of Section 88.

(iii) *Curvature of field*

We consider the image as usual in an out-of-focus plane shifted with respect to $\mathscr{I}'$ through a distance $kh'^2$. The partial displacement is then generically just that which obtained in the case of the ($r$-) semi-symmetric system, with the following formal identifications in (76.10):

$$\begin{aligned}(p_3-p_7)\cos^2\psi+(p_4-6p_8)\sin^2\psi &\to \sigma_3,\\ -(p_3-3p_7)\cos^2\psi-(p_4-18p_8)\sin^2\psi &\to \sigma_4,\\ -2p_6\sin\psi\cos\psi &\to \tilde{\sigma}_3.\end{aligned} \tag{87.7}$$

The discussion of curvature of field given in Section 76 may therefore to a large extent be taken over into the present context. Pretending *for the moment* that the partial displacement can be discussed as if all the other types were absent, we observe that, naturally, the various secondary image surfaces will bear a more complicated character than was the case when axial symmetry obtained. The 'Petzval surface' will serve as an example. Thus, point images result for all values of $\psi$ if $p_6 = 0$, $p_3 = p_7$, and $p_4 = 6p_8$. The surface on which these images are formed therefore has principal curvatures $c_1$ and $c_2$, whose values are

$$c_1 = 4p_3, \quad c_2 = 4p_4 \tag{87.8}$$

in a neighbourhood of the axis. In general one therefore has an ellipsoid or an hyperboloid in a neighbourhood of the axis. Yet, how can that be? If points such that $y_1 = 0$ have sharp images these must surely fall along a straight line, in view of the translational symmetry of $K$, so that the Petzval surface must be cylindrical. The paradox is only apparent, and further emphasizes the artificial nature of this analysis: to have *in fact* a sharp image, spherical aberration at least must be absent, and so certainly $p_4 = 0$; and then $c_2 = 0$, as required.

(iv) *Distortion*

Both $\hat{y}$ and $\hat{z}$ are, in general, non-zero. Once again, this partial displacement has little more than formal significance, unless all other types are absent. Then one is left only with the terms governed by $p_5$, and the displacement in the $z$-direction vanishes, independently of the value of $\psi$.

In conclusion we remark that the occurrence of a particular coefficient (such as $p_4$) in the several partial displacements is essentially a consequence of having a stop whose shape does not conform to the symmetries of $K$. Were the stop a slit (like $K$ as such, unlimited, in principle, in the $z$-direction), $z_1$ could always be arranged to be zero. A zonal curve would then be taken to correspond to a fixed value of $y'$; and, recalling (87.3), the four 'types' of partial displacements would be governed by (i) $p_1, p_4, p_8$, (ii) $p_2, p_6$, (iii) $p_3$, $p_7$, (iv) $p_5$, respectively.

## 88. Digression on plane-symmetric systems

The symmetric and $c$-symmetric systems, as well as those to be considered in some detail in the next chapter, all have one common property, called *double plane-symmetry*, which means the existence of two mutually perpendicular planes of symmetry, whose line of intersection defines a preferred axis $\mathscr{A}$, and so a preferred base-ray. We therefore digress for a moment to consider the doubly plane-symmetric system as such. We do so partly to lend added weight to the point of view that in certain respects the Hamiltonian theory is interesting not so much because it tells us what kind of imagery a given type of system *can* achieve, but rather what kind of imagery it *cannot* achieve. For this reason we prefer to be landed with a slight amount of repetitiveness inherent in separately treating the more highly symmetric systems as we have done here, instead of regarding the systems in question as being doubly plane-symmetric, but with additional symmetries. The confusion likely to be engendered by constantly having to investigate 'special cases' would surely be a disadvantage; and in any event translationally semi-symmetric systems, for example, would be excluded from the outset.

In the context of the point characteristic, one is confronted with the six 'reflection invariants' $\xi_1 = y'^2$, $\eta_1 = y'y_1$, $\zeta_1 = y_1^2$, $\xi_2 = z'^2$, $\eta_2 = z'z_1$, $\zeta_2 = z_1^2$, where $y_1 = m_1 y$, $z_1 = m_2 z$. Here $m_1$ and $m_2$ are the magnifications associated with the pair of conjugate normal planes $\mathscr{I}$ and $\mathscr{I}'$ being contemplated (cf. (10.17)), and it is definitely supposed, for the sake of simplicity, that $K$ does in fact produce a sharp image of a plane object in $\mathscr{I}$; cf. (10.16). Even powers of $\eta_1$ and $\eta_2$ are always to be eliminated by means of the identities

$$\eta_1^2 = \xi_1\zeta_1, \quad \eta_2^2 = \xi_2\zeta_2.$$

The anterior base-point is taken as $O_0$, whilst the posterior base-point $B'$ is chosen arbitrarily. When the stop lies in the image space one will naturally take $B'$ to be its axial point; when it does not, one has the situation that there is, in general, no sharply defined exit pupil, so that one does not then have such a 'natural' choice of the position of $B'$.

The ideal image is taken to be sharp, with constant magnifications in the $y'$- and $z'$-directions. Accordingly, on setting $d' = 1$, as usual,

$$V = g(\zeta_1, \zeta_2) - (1 + \xi_1 - 2\eta_1 + \zeta_1 + \xi_2 - 2\eta_2 + \zeta_2)^{\frac{1}{2}} + v, \quad (88.1)$$

where $v$ is the aberration function. There is no need to go through the paraxial theory once again. Indeed, if desired, one may simply draw upon the general equations of Section 14, adopting the $y$-components as they stand, and their $z$-components similarly, but with $k_1, k_2, k_3$ replaced by constants $l_1, l_2, l_3$ respectively. One has, of course, two focal lengths $f_1$ and $f_2$, and in (88.1) the function $g(\zeta_1, \zeta_2)$ has the linear terms $-\zeta_1/(2m_1 f_1) - \zeta_2/(2m_2 f_2)$.

The number of aberration coefficients which govern the displacement in $\mathscr{I}'$ may be counted easily as follows. We temporarily imagine $v$ to be written in the generic form

$$v = A + \eta_1 B + \eta_2 C + \eta_1\eta_2 D, \quad (88.2)$$

where $A$, $B$, $C$, $D$ are functions of $\xi_1$, $\zeta_1$, $\xi_2$, $\zeta_2$. That there are no coefficients of even order goes without saying. With regard to those of order $2n-1$, one has to add the number of coefficients in one polynomial of degree $n$, two of degree $n-1$, and one of degree $n-2$, each in four variables. Further, the sum of these has to be reduced by the number of coefficients multiplying products of powers of $\zeta_1$

and $\zeta_2$ alone, i.e. by $n+1$. Accordingly, recalling a result stated shortly after equation (11.2), the total number of coefficients is

$$\binom{n+3}{3}+2\binom{n+2}{3}+\binom{n+1}{3}-(n+1)=\tfrac{2}{3}n(n+1)(n+2). \quad (88.3)$$

This will be seen to be in harmony with (11.5), if, in the latter, one replaces $n$ by $2n-1$ and then substracts $n+1$, to allow for the irrelevant coefficients.

We may mention in passing a possible additional symmetry of the doubly plane-symmetric system which has not been hitherto referred to, since it is of a rather academic kind. The symmetry in question is invariance under rotations through 90° about $\mathscr{A}$. When this obtains one has to have

$$V(\xi_1,\eta_1,\zeta_1,\xi_2,\eta_2,\zeta_2)\equiv V(\xi_2,\eta_2,\zeta_2,\xi_1,\eta_1,\zeta_1), \quad (88.4)$$

where, just for the moment, the use of variables $y$, $z$ in place of $y_1$, $z_1$ is understood. It follows immediately that $m_1=m_2$, and $f_1=f_2$; so that *paraxially* $K$ behaves as if it were an $r$-symmetric system. Outside the paraxial region the state of affairs is, of course, more general. Thus, in the third-order region one is still left with *nine* aberration coefficients, as compared with five in the symmetric case.

Returning to the general doubly plane-symmetric system, the third-order displacement is given as usual by

$$\boldsymbol{\epsilon}_3'=\partial v^{(2)}/\partial \mathbf{y}', \quad (88.5)$$

and we write uniformly

$$\begin{aligned}v^{(2)}=p_1\xi_1^2+p_2\xi_1\xi_2+p_3\xi_2^2+p_4\xi_1\eta_1+p_5\xi_1\eta_2+p_6\eta_1\xi_2\\+p_7\xi_2\eta_2+p_8\xi_1\zeta_1+p_9\xi_1\zeta_2+p_{10}\eta_1\eta_2+p_{11}\zeta_1\xi_2+p_{12}\xi_2\zeta_2\\+p_{13}\eta_1\zeta_1+p_{14}\eta_1\zeta_2+p_{15}\zeta_1\eta_2+p_{16}\eta_2\zeta_2.\end{aligned} \quad (88.6)$$

Then, making the substitutions (87.4), one gets the following partial displacements:

(i) *Spherical aberration*

$$\left.\begin{aligned}\hat{y}&=\rho^3(4p_1\cos^2\theta+2p_2\sin^2\theta)\cos\theta,\\ \hat{z}&=\rho^3(2p_2\cos^2\theta+4p_3\sin^2\theta)\sin\theta;\end{aligned}\right\} \quad (88.7)$$

(ii) *Coma*

$$\left.\begin{aligned}\hat{y} &= \rho^2 h'[\tfrac{1}{2}(3p_4+p_6)\cos\psi+\tfrac{1}{2}(3p_4-p_6)\cos\psi\cos 2\theta \\ &\qquad +p_5\sin\psi\sin 2\theta],\\ \hat{z} &= \rho^2 h'[\tfrac{1}{2}(p_5+3p_7)\sin\psi+\tfrac{1}{2}(p_5-3p_7)\sin\psi\cos 2\theta \\ &\qquad +p_6\cos\psi\sin 2\theta];\end{aligned}\right\} \tag{88.8}$$

(iii) *Curvature of field*

$$\left.\begin{aligned}\hat{y} &= \rho h'^2\{[2(p_8\cos^2\psi+p_9\sin^2\psi)-k]\cos\theta+p_{10}\sin\psi\cos\psi\sin\theta\},\\ \hat{z} &= \rho h'^2\{[2(p_{11}\cos^2\psi+p_{12}\sin^2\psi)-k]\sin\theta+p_{10}\sin\psi\cos\psi\cos\theta\},\end{aligned}\right\} \tag{88.9}$$

where $k$ is the usual constant relating to the position of the out-of-focus image plane; and

(iv) *Distortion*

$$\left.\begin{aligned}\hat{y} &= h'^3(p_{13}\cos^2\psi+p_{14}\sin^2\psi)\cos\psi,\\ \hat{z} &= h'^3(p_{15}\cos^2\psi+p_{16}\sin^2\psi)\sin\psi.\end{aligned}\right\} \tag{88.10}$$

We now have the important conclusion that the displacement is just like that represented by (87.5), except that *the interaction between the various types is entirely removed*; this being made possible by the greater number of independent aberration coefficients which the merely doubly plane-symmetric system possesses. It is therefore possible now to have a Petzval surface of a far less formal character than that encountered in Section 87. In short, equations (88.8–10) enable us to see quite clearly how the introduction of additional symmetries generically affects the partial displacements, and the interactions between them. The displacement (19.5) is of course a very special case in which one again has no interaction between the different types; this feature being the result of having a stop which conforms to the symmetries of the system, as remarked at the end of Section 87.

We could now go on to discuss such problems as finding the general conditions imposed upon $K$ by the condition of reversibility. Indeed we would then find that there are *five* relations between the sixteen primary coefficients, and, in general, they cannot all be homogeneous simultaneously. However, as remarked earlier, we prefer to deal with such questions in the more specialized situations as we come to them.

When a system has only a *single* plane of symmetry, the 'meridional plane' $\bar{z} = \bar{z}' = 0$, say, the state of affairs is a good deal more complicated, mainly because there will be aberrations of even as well as of odd orders. One is thus confronted with a great multiplicity of aberration coefficients. The lowest (non-parabasal) order which necessarily has to be considered is the *second*, and this involves already eight distinct coefficients, in general. Consequently the distinction between the effective and the characteristic *third*-order coefficients is no longer of the trivial kind encountered earlier. The base-ray $\mathscr{R}_0$ will naturally be taken to be in the plane of symmetry, a choice which was already implicit in our writing the equation of the meridional plane $\bar{z} = \bar{z}' = 0$. Beyond this, no natural choice of $\mathscr{R}_0$ suggests itself, since one has here no preferred axis. Then

$$V = V(y', y, z'^2, z'z, z^2), \tag{88.11}$$

there being no linear terms in the power series for $V$.

We could now carry through the usual programme to investigate the general properties of singly plane-symmetric systems. This would be little more than an exercise following rather closely the lines of the work of the first part of this section. A few remarks of a more or less general kind will therefore suffice. To this end we imagine $K$ to have been so designed that it forms a sharp parabasal image, the conditions being those considered in Section 10, shortly after equation (10.13). Then the ideal point characteristic is

$$V_0 = g(y, z^2) - [d'^2 + (y' - m_1 y)^2 + (z' - m_2 z)^2]^{\frac{1}{2}}, \tag{88.12}$$

and the aberration function

$$v = \sum_{n=2}^{\infty} v_n \tag{88.13}$$

contains no terms quadratic in the ray-coordinates. One finds that

$$\begin{matrix}\textit{the number of aberration}\\ \textit{coefficients of order}\\ n\,(>1)\end{matrix} = \begin{cases} \frac{1}{12}(n+1)(n+2)(n+6) & (n \text{ even}) \\ \frac{1}{12}(n+1)(n+3)(n+5) & (n \text{ odd}), \end{cases} \tag{88.14}$$

the irrelevant coefficients, i.e. those multiplying products of powers of $y$ and $z$ only, being not counted, as usual.

For the sake of orientation, consider the *second*-order displacement

$$\boldsymbol{\epsilon}_2' = \partial v_2/\partial \mathbf{y}'. \tag{88.15}$$

We write

$$v_2 = \pi_1 y'^3 + \pi_2 y'^2 y + \pi_3 y' y^2 + \pi_4 y' z'^2 + \pi_5 y' z' z + \pi_6 y' z^2 + \pi_7 y z'^2 + \pi_8 y z' z. \tag{88.16}$$

Proceeding as after equation (88.6), and absorbing powers of $m_1$ and $m_2$ in the $\pi_\alpha$, we get

$$\left.\begin{aligned} \epsilon_{y2}' &= \tfrac{1}{2}\rho^2[(3\pi_1+\pi_4)+(3\pi_1-\pi_4)\cos 2\theta] + \rho h'(2\pi_2\cos\theta \\ &\qquad \times \cos\psi + \pi_5 \sin\theta\sin\psi) + h'^2(\pi_3\cos^2\psi + \pi_6\sin^2\psi), \\ \epsilon_{z2}' &= \pi_4\rho^2\sin 2\theta + \rho h'(\pi_5\cos\theta\sin\psi + \pi_7\sin\theta\cos\psi) \\ &\qquad + \pi_8 h'^2 \sin\psi\cos\psi. \end{aligned}\right\} \tag{88.17}$$

We have encountered no displacement of this kind before. Evidently we have three types of aberrations, jointly governed by eight coefficients. The type independent of $h'$ quite closely resembles elliptical coma, whilst the aberration governed by $\pi_2$, $\pi_5$ and $\pi_7$ is a form of curvature of field, with the usual implications. The terms independent of $\rho$ clearly represent distortion. The upshot of all this is that the classification of aberration types of Section 22 can be generalized by allowing $n$ to be half-odd integral, and interchanging the terms 'coma' and 'astigmatism' when it is so. In particular, we have here second-order constant (i.e. zero degree) coma, linear astigmatism, and quadratic coma respectively, in the order in which these were just discussed.

## 89. The displacement when $k_1 \neq k_4$. Line images

Reverting to the $c$-symmetric system, we know that when $k_1 \neq k_4$ a sharp image can never be obtained. It is instructive, however, to consider the image in the plane $\mathscr{I}_m'$ whose axial point is the meridional focus, i.e. the point in which meridional rays from $O$ intersect the axis. This means that we take $d' = -1/k_1$ $(=1)$. In view of (86.4) the aberration function appropriate to the third-order displacement is

$$v = \tfrac{1}{2}(k_4+1)\tau + v^{(2)}, \tag{89.1}$$

where $v^{(2)}$ is given by (87.2). It is, however, advantageous to contemplate, in place of $v$, a modified aberration function $\tilde{v}$—recall the end of Section 36—defined to all orders by the equation

$$V = g(\zeta) - (1 + \xi - 2\eta + \zeta - k_4\tau - 2\tilde{v})^{\frac{1}{2}}. \tag{89.2}$$

Evidently $\tilde{v} = O(4)$, unlike $v$, since all paraxial terms are correctly accounted for. One confirms easily that

$$\tilde{v}^{(2)} = v^{(2)} + \tfrac{1}{4}(1 + k_4)\tau[\xi - 2\eta + \zeta + \tfrac{1}{2}(1 - k_4)\tau]. \tag{89.3}$$

Then

$$\left.\begin{aligned} \epsilon'_y &= \partial\tilde{v}^{(2)}/\partial y' - \tfrac{1}{2}k_4(k_4 + 1)(y' - y_1)\tau + O(5), \\ \epsilon'_z &= \partial\tilde{v}^{(2)}/\partial z' + (k_4 + 1)(z' - z_1)(1 + \tfrac{1}{2}k_4^2\tau) + O(5). \end{aligned}\right\} \tag{89.4}$$

$\tilde{v}^{(2)}$ may be thought of as written like (87.2), with the $p_\alpha$ replaced by $\tilde{p}_\alpha$. Then the generic third-order displacement is essentially that encountered in Section 87, but with an additional *first*-order displacement in the $z'$-direction, of amount proportional to $(z' - z_1)$. We conclude that if $K$ is to produce a (straight) line image of a point source in $\mathscr{I}$, then this line must lie in $\mathscr{I}'_m$, and, not surprisingly, it must be normal to the meridional plane.

If every point of $\mathscr{I}$ is to have a line image, $\epsilon'_y$ must be a constant, depending on $y_1$, for all values of $\mathbf{y}'$ and $\mathbf{y}_1$. To the present order this requires that

$$\tilde{p}_1 = \tilde{p}_2 = \tilde{p}_3 = 0, \quad 2\tilde{p}_4 = -\tilde{p}_6 = \tfrac{1}{2}k_4(k_4 + 1), \tag{89.5}$$

whilst $\tilde{p}_5$, $\tilde{p}_7$ and $\tilde{p}_8$ are not restricted. It is instructive to consider the conditions under which $K$ will form a line image of any point $O$ of $\mathscr{I}$, without any restrictions upon the orders of aberrations to be included. Then we require

$$\epsilon'_y = y_1[D(\zeta) - 1], \tag{89.6}$$

where $D$ is some function such that $D(0) = 1$. (89.6) implies the equations

$$2V_\xi + \alpha' = 0, \quad V_\eta - \alpha' D = 0, \tag{89.7}$$

whence, in the first place,

$$V = V(\xi - 2D\eta, \zeta, \tau). \tag{89.8}$$

Then, with $\xi - 2D\eta = \sigma$,

$$\beta' = 2(y' - Dy_1)V_\sigma, \quad \gamma' = 2(z' - z_1)V_\tau, \tag{89.9}$$

and using these in the first member of (89.7), there comes

$$4(1+\xi-2D\eta+D^2\zeta)\,V_\sigma^2+4V_\tau^2 = 1. \tag{89.10}$$

This may be further simplified by means of a change of variables according to

$$s = (1+\xi-2D\eta+D^2\zeta)^{\frac{1}{2}}, \quad t = \tau^{\frac{1}{2}},$$

so that finally

$$V_s^2+V_t^2 = 1. \tag{89.11}$$

All solutions of this equation which can be written as power series in $s$ and $t$ with only even powers of $t$ present, represent the point characteristic of $c$-symmetric systems with the required property. The set of solutions in question cannot be exhibited explicitly. Probably the most suitable general way of dealing with the problem is to substitute

$$V = \sum_{n=0}^{\infty} v_n(s)\, t^{2n} \tag{89.12}$$

into (89.11), the functions $v_n(s)$ then being found iteratively as the solutions of ordinary differential equations. However, for our purposes it will suffice to consider merely a specific example.

To this end we observe first that $V = \text{const.}-(s^2+t^2)^{\frac{1}{2}}$ satisfies (89.11), but this solution is useless since it cannot accommodate the condition $k_1 \neq k_4$. However, the formally trivial addition of a 'constant' $b$ to $s$ leaves (89.1) unaffected. We are thus led to the result

$$V = \bar{g}(\zeta)-\{[b+(1+u)^{\frac{1}{2}}]^2+\tau\}^{\frac{1}{2}}, \tag{89.13}$$

where

$$u = \xi-2D\eta+D^2\zeta, \tag{89.14}$$

and the bar over the $g$ serves to distinguish this function from that in (89.2). $b$ may be a function of $\zeta$, i.e.

$$b(\zeta) = -[1+(1/k_4)]+b_1\zeta+\ldots, \tag{89.15}$$

where the constant term is so determined that (89.12) accommodates the paraxial form (86.3) of $V$. The displacement corresponding to this particular choice of the solution of (89.11) is then

$$\epsilon_y' = 0, \quad \epsilon_z' = b(z'-z_1)\,[b+(1+u)^{\frac{1}{2}}]^{-1}. \tag{89.16}$$

Further, comparing (89.2) with (89.13), a somewhat tedious calculation shows that

$$\bar{v}^{(2)} = d_1\eta\zeta+\tfrac{1}{2}k_4b_1\zeta\tau+\tfrac{1}{4}k_4(1+k_4)(\xi-2\eta+\zeta-\tfrac{1}{2}k_4\tau)\,\tau. \tag{89.17}$$

Thus

$$\left.\begin{aligned}\tilde{p}_1 = \tilde{p}_2 = \tilde{p}_3 = 0, \quad 2\tilde{p}_4 = -\tilde{p}_6 = \tfrac{1}{2}k_4(1+k_4),\\ \tilde{p}_5 = d_1, \quad \tilde{p}_7 = \tfrac{1}{4}k_4(2b_1+k_4+1), \quad \tilde{p}_8 = -\tfrac{1}{8}k_4^2(1+k_4);\end{aligned}\right\} \tag{89.18}$$

and these are indeed in agreement with (89.5).

## 90. Remark on the sine- or cosine-relations

Suppose that $\mathscr{I}$ and $\mathscr{I}'$ are perfect in the restricted sense described at the beginning of Section 86. Then $v = 0$ in (86.1), and it follows at once that for all rays through $O_0$

$$\beta' - \beta/m = 0, \quad \gamma' - \nu\gamma = 0, \tag{90.1}$$

the second of which holds in any event, as a consequence of translational symmetry. We have, in fact, the usual cosine-relations, which reduce to the sine-relation (27.3) when $m$ happens to have the value $\nu^{-1}$. The validity of (90.1) only requires that the conjugate planes be perfect in a sufficiently small neighbourhood of $O_0'$.

One can now attempt to undertake an investigation designed to yield 'offences against the cosine-conditions', after the fashion of Sections 31 and 82. It turns out, however, that no useful results emerge. This state of affairs is essentially due to the absence of axial symmetry. Whereas when the latter obtains, there is only one $(2n-1)$th-order coefficient of spherical aberration and one of linear coma, one has, in the case of $c$-symmetry, $n+1$ coefficients of spherical aberration which, together with an additional $n$ coefficients, govern linear coma. Indeed, the two types of aberrations are induced by an aberration function of the generic form

$$v = S(\xi, \tau) + \eta C(\xi, \tau), \tag{90.2}$$

where the fact that $\tau$ involves $z_1$ should be borne in mind. Therefore, even in the total absence of spherical aberration, one is still left with a function of *two* variables. When $S = 0$ one finds that

$$\left.\begin{aligned}\epsilon_y' &= y'w\eta[C+2(1+\xi)\,C_\xi+2\tau C_\tau]+y_1 wC,\\ \epsilon_z' &= z'w\eta[C+2\xi C_\xi+2(1+\tau)\,C_\tau]\end{aligned}\right\} \tag{90.3}$$

for sufficiently small $\mathbf{y}_1$, where $w = (1+\xi+\tau)^{\frac{1}{2}}$; whilst when $y_1 = 0$

$$\beta/m - \beta' = y'C. \tag{90.4}$$

No axial ray trace can tell us anything about the values of the derivatives of the function $C$. It is therefore useless to consider anything but the displacement of 'sagittal rays', i.e. rays having $\theta = \pm\frac{1}{2}\pi$, and for these

$$\epsilon'_y = y_1(1+\rho^2)^{\frac{1}{2}}\,C(0,\rho^2), \quad \epsilon'_z = 0. \tag{90.5}$$

Even the usefulness of this result is illusory, for one cannot obtain the value of $C(0,\rho^2)$ from (90.4), since $\xi = 0$ requires that the factor $y'$ on the right be zero. Without numerical interpolation one thus cannot get anywhere in this fashion, and in practice this would be a senseless procedure.

## 91. Reversibility

Let the $c$-symmetric system $K$ be reversible, i.e. possess a *normal* plane of symmetry $\mathscr{C}$ whose axial point is $C$, as usual. For the sake of a little variety we investigate the situation sometimes encountered in practice, namely when $K$ is entirely telescopic ($f_1 = f_2 = \infty$) and the object is at infinity. Now $\nu = 1$, and, recalling (86.6), $K$ is perfect if

$$\beta' = \mu\beta \tag{91.1}$$

for all rays, the condition $\gamma' = \gamma$ being satisfied in any case, on account of the translational symmetry of $K$.

Let the point characteristic be taken with respect to a convenient pair of finitely situated base-points which are disposed symmetrically about $C$. (91.1) requires that $V_0$ depend on $y'$ and $y$ only through the combination $(\mu y' - y)^2$. We therefore write, in the usual way,

$$V = g(\mu^2\xi - 2\mu\eta + \zeta, \tau) + v. \tag{91.2}$$

Reversibility entails that $V$ be invariant under the mutual interchange of $\xi$ and $\zeta$. On momentarily confining our attention to the paraxial terms of $V$ we see that we must have

$$\mu^2 = 1, \tag{91.3}$$

cf. (56.13). From the point of view of the 'displacement', now defined as

$$\epsilon' = \beta' - \mu\beta, \tag{91.4}$$

the previous choice of reflection invariants is not convenient, and we replace them according to

$$\hat{\xi} = \xi, \quad \hat{\eta} = \mu\xi - \eta, \quad \hat{\zeta} = \xi - 2\eta + \zeta, \quad \hat{\tau} = \tau, \tag{91.5}$$

$v$ now being regarded as a function of these variables. It is obvious that $\epsilon'$ involves the derivatives of $v$ with respect to $\hat{\xi}$ and $\hat{\eta}$ only, whence we again arrive at the conclusion stated at the end of Section 86.

Now, upon mutually interchanging $\xi$ and $\zeta$,

$$\hat{\xi} \to \hat{\xi} - 2\mu\hat{\eta} + \hat{\zeta}, \quad \hat{\eta} \to -\hat{\eta} + \mu\hat{\zeta}, \quad \hat{\zeta} \to \hat{\zeta}, \quad \hat{\tau} \to \hat{\tau}. \qquad (91.6)$$

The reversibility of $K$ then requires that, identically,

$$v(\hat{\xi}, \hat{\eta}, \hat{\zeta}, \hat{\tau}) = v(\hat{\xi} - 2\mu\hat{\eta} + \hat{\zeta}, -\hat{\eta} + \mu\hat{\zeta}, \hat{\zeta}, \hat{\tau}). \qquad (91.7)$$

When $\xi$ and $\zeta$ are mutually interchanged, $\hat{\xi} - 2\mu\hat{\eta} + \hat{\zeta}$ and $\hat{\xi}$ interchange place, whilst $\hat{\xi} - \mu\hat{\eta}$, $\hat{\zeta}$ and $\hat{\tau}$ are unaffected. The primary aberration function must therefore have the form

$$v^{(2)} = \hat{p}_1(\hat{\xi} - 2\mu\hat{\eta} + \hat{\zeta})\hat{\xi} + \hat{p}_2(\hat{\xi} - \mu\hat{\eta})\hat{\zeta} + \hat{p}_3(\hat{\xi} - \mu\hat{\eta})\hat{\tau}, \qquad (91.8)$$

the irrelevant terms being omitted, as usual. We thus recognize that, when the object is at infinity, the telescopic $c$-symmetric reversible system has only *three* independent primary aberration coefficients. Of these, two can be regarded as governing the imagery in the meridional section alone.

## 92. Concentricity

It remains to consider the concentric $c$-symmetric system, as defined in Section 84. A single refracting circular cylinder is of this kind, so that the problem is not merely academic. On the other hand, the theory has its own peculiar difficulties, for the following reasons. One would naturally wish to use the angle characteristic $T$, as we did in Section 64, since invariance under rotations—here about a line—is then so easily dealt with. However, $K$ being always telescopic in the sagittal section, the use of $T$ is forbidden. The point characteristic is not easily handled either, since the derivation of its generic form involves the use of the differential equations (3.6). Essentially the same remark also applies to the mixed characteristics $W_1$ and $W_2$. In short, we are led to contemplate the use of one of the 'strange' mixed characteristics first mentioned at the end of Section 5. A suitable choice is

$$W = T + z'\gamma', \qquad (92.1)$$

regarded as a function of $\beta'$, $z'$, $\beta$, $\gamma$. It is assumed that $f_1 \neq \infty$.

From simple geometrical considerations it is apparent that all characteristic functions, with the sole exception of $V$, will not themselves be invariant under translations. The translational symmetry of $K$ is therefore taken into account by requiring $W$ to be such that the condition (84.4) be satisfied; and this will be the case provided

$$W = \nu z'\gamma + A(\beta', \beta, \gamma), \tag{92.2}$$

where $A$ is some function of its arguments.

We have not yet adopted any particular base-points. From the point of view of the object and image planes there is no particular choice which is outstandingly advantageous. The same is not true, however, in the context of rotations, that is to say, the rotations about the line through the centre $C$, normal to the meridional plane. In fact, recalling the work following equation (65.11) it is natural to take both $B$ and $B'$ to coincide with $C$; for then $W$ can depend on $\beta'$ and $\beta$ solely through the function $\alpha\alpha' + \beta\beta'$. Thus now

$$W = \nu z'\gamma + J(\alpha\alpha' + \beta\beta', \gamma^2), \tag{92.3}$$

which, incidentally, is seen to be in harmony with (92.1). The fact that $\gamma'$ occurs in $\alpha'$ need cause us no worry since we can simply replace it by $\nu\gamma$. If

$$\xi = \beta'^2, \quad \eta = \beta\beta', \quad \zeta = \beta^2, \quad \tau = \gamma^2, \tag{92.4}$$

we therefore have the required generic result

$$W(\beta', z', \beta, \gamma) = \nu z'\gamma + G(\chi, \tau), \tag{92.5}$$

where

$$\chi = 1 - (1 - \xi - \nu^2\tau)^{\frac{1}{2}}(1 - \zeta - \tau)^{\frac{1}{2}} - \eta. \tag{92.6}$$

$W$ still involves a function of *two* variables. We conclude that the concentric system has $n+1$ independent aberration coefficients of order $2n-1$. Here it was assumed that the object is not at infinity. When it is, the part $G(0, \tau)$ of (92.5) is irrelevant, and the number of coefficients is therefore reduced to $n$. At the same time it becomes easier to write down the displacement $\epsilon'$ ($\equiv \epsilon'_y$), which becomes

$$\epsilon' = \left(\beta - \frac{\alpha\beta'}{\alpha'}\right) G_\chi + \frac{x'\beta'}{\alpha'} - \frac{f_1\beta}{\alpha}. \tag{92.7}$$

Since $\epsilon'$ must vanish in the paraxial region when $O'_0$ is taken at the meridional focus ($\bar{x}' = x'$),

$$x' = f_1 = G_\chi(0, 0).$$

Setting $f_1 = 1$, by choice of the unit of length, (92.7) becomes

$$\epsilon' = \left(\frac{\beta'}{\alpha'} - \frac{\beta}{\alpha}\right)(1 - \alpha G_\chi). \tag{92.8}$$

In particular, when spherical aberration of all orders is absent, i.e. $\epsilon' = 0$ when $\beta = \gamma = 0$, we have to have

$$G(\chi, \tau) = \chi, \tag{92.9}$$

rejecting an irrelevant additive function of $\tau$; a result which should be compared with (68.8). Under these conditions we then have

$$\epsilon' = \left(\frac{\beta'}{\alpha'} - \frac{\beta}{\alpha}\right)(1 - \alpha), \tag{92.10}$$

which is identical with the $y$-component of (69.2).

## Problems

**P.8 (i).** Investigate the consequences of the reversibility of doubly plane-symmetric systems, obtaining explicit results at least in the third-order region. (The system forms a sharp paraxial image.)

**P.8 (ii).** Obtain the result (88.14).

**P.8 (iii).** How would you define a 'cylindrical point characteristic' analogous to the spherical point characteristic? Define the aberration function, granted that the ideal behaviour of $K$ is that of Section 86. Write down generic expressions for $\epsilon'_y$ and $\epsilon'_z$ analogous to (37.4).

**P.8 (iv).** A doubly plane-symmetric system forms a sharp plane image, the disposition of the axis of $K$ and of the various normal planes being as usual. Find the generic form of the point characteristic. Correctly to the third order, write down the distortion functions which appear, as power series whose coefficients are to be identified with those which occur in (88.6).

**P.8 (v).** Is the name 'circular coma' appropriate to the (third-order) linear coma of a doubly plane-symmetric system? Justify your answer by a detailed discussion.

CHAPTER 9

# SYSTEMS WITH TOROIDAL SYMMETRY

## 93. Definition of the toroidally symmetric system

A system $K$ is *toroidally semi-symmetric* if it has an axis of symmetry $\mathscr{C}$ which does not have points in common with both the object space and the image space. This last qualification is important, for without specification of the situation of the axis of symmetry relative to object and image one does not know whether one is talking about a toroidally semi-symmetric system, or the kind of system simply called semi-symmetric in Chapter 7. $K$ is *toroidally symmetric* if there exists a plane of symmetry containing $\mathscr{C}$. The imagery of such a system is still pretty complex. A considerably simpler situation obtains if $K$ also has a plane of symmetry normal to $\mathscr{C}$; and then we call it *t-symmetric*. One now has two planes of symmetry whose line of intersection $\mathscr{A}$ is the axis of $K$. The object plane, image plane, and pupil planes (when the latter can be properly defined) are all planes normal to $\mathscr{A}$, as usual; and, in the context of the definitions above, they were naturally not regarded as 'parts of $K$'. Our discussion will be brief, and will revolve for the most part about $t$-symmetric systems.

A simple example of a $t$-symmetric system is provided by (a part of) a circular torus made of some refracting medium. This is a very specialized case of course, not so much because only a single dielectric medium is involved, but rather because the meridional cross-section of the system is circular. (In general the cross-section could be bounded by any curves symmetric about $\mathscr{A}$, and then we speak of a *toroid.*) In terms of a set of Cartesian axes, pairwise parallel to those of the usual coordinate basis, and with origin at the axial point $C$ of $\mathscr{C}$, the equation of the torus is

$$[x^2+y^2+z^2-(R_1^2+R_2^2)]^2-4R_1^2(R_2^2-y^2) = 0, \qquad (93.1)$$

where $|R_1 \pm R_2|$ and $R_2$ are the principal radii of the surface at its point of intersection with the sagittal and meridional planes respectively. Actually (93.1) represents a surface $\mathscr{T}$ a good deal more complicated than any we have contemplated hitherto in as far as, taken as a whole, it is not simply connected. This means that we can draw closed curves on $\mathscr{T}$ such that two points on either side of the curve can be connected by a second curve which does not intersect the first.

The kind of complication just remarked upon is largely irrelevant here since we do not consider toroids as a whole but only the parts lying to one side of the normal plane through $C$, and of any such part we again only take a part, namely either that which is convex or that which is concave towards $\mathscr{C}$ near the sagittal plane. More simply, and more generally, any refracting surface (or any surface of constant refractive index when this varies continuously) is to be written in a form in which $x$ appears as a *convergent* series in ascending powers of $y^2$ and $z^2$. With regard to the example of the torus, this gives rise to four alternatives, namely

$$\pm x = (R_1 + R_2) - \tfrac{1}{2}\left(\pm \frac{y^2}{R_2} + \frac{z^2}{R_1 \pm R_2}\right) + \dots, \qquad (93.2)$$

where the alternative choices of sign on the left and on the right are to be made independently of one another; but when $R_1$ is sufficiently nearly equal to $R_2$ one possible choice on the right is excluded to all intents and purposes, i.e. if regularity is to obtain in not too small a region surrounding the sagittal plane. Clearly we can 'generate' a $t$-symmetric system of refracting surfaces by drawing any set of curves in the meridional plane, each curve being symmetric about $\mathscr{A}$; and then rotating these curves about the line $\mathscr{C}$. We observe straight away that if the toroidal property of $K$ is expressed by the condition that $K$ be invariant under rotations about $\mathscr{C}$, then we must restrict these rotations to be sufficiently small. The situation is entirely analogous to that which we met in the context of the concentric $r$-symmetric system in Chapter 6B. There also we had to restrict the magnitude of the rotations, since otherwise we would always have ended up with complete spherical surfaces, and this was not desirable. We remark that if we intend that together with every surface $x = f(y^2, z^2)$ contained in $K$ the branch $x = -f(y^2, z^2)$ occur in it also, then we need only impose upon $K$ the condition that it be reversible as a whole (see Section 97).

It may be remarked that one can take $\mathscr{C}$ as the *axis* of an $r$-symmetric system. However, with the usual choice of coordinate basis (i.e. with the $\bar{x}$- and $\bar{x}'$-axes along $\mathscr{C}$, etc.) one cannot maintain the ray along $\mathscr{C}$ as base-ray: the system is irregular. Thus, again referring to the example of the torus $\mathscr{T}$, if $R_1 > R_2$ there is no paraxial region in the sense that a ray sufficiently close to $\mathscr{C}$ will never encounter $\mathscr{T}$ at all; whilst if $R_1 \leqslant R_2$ the ray along $\mathscr{C}$ will encounter a singularity.

## 94. The angle characteristic

A $t$-symmetric system is, in general, anamorphotic. Not to be burdened with the consideration of too many special cases we suppose once and for all that $f_1$ and $f_2$ are both finite. Then let $T_c$ be the angle characteristic when both base-points are taken at $C$. For reasons by now familiar, $\gamma'$ and $\gamma$ can occur in $T_c$ only through the function $\alpha\alpha' + \gamma\gamma'$. Symmetry about the meridional plane implies that we need not explicitly consider $\alpha\gamma' - \alpha'\gamma$, since this reverses sign when $\gamma$ and $\gamma'$ do so; whilst even powers of $\alpha\gamma' - \alpha'\gamma$ are redundant in virtue of the identity

$$(\alpha\gamma' - \alpha'\gamma)^2 + (\alpha\alpha' + \gamma\gamma')^2 = (1 - \beta^2)(1 - \beta'^2). \tag{94.1}$$

Again, symmetry about the sagittal plane requires that $T_c$ depend upon $\beta'$, $\beta$ only in the combinations $\xi = \beta'^2$, $\eta = \beta'\beta$, $\zeta = \beta^2$. We therefore write

$$T_c = J(\xi, \eta, \zeta, \chi), \tag{94.2}$$

where

$$\chi = 1 - (\alpha\alpha' + \beta\beta' + \gamma\gamma'), \tag{94.3}$$

as in (65.10). The inclusion of the term $\beta\beta'$ in (94.3) at this stage is merely a matter of convenience. If $O_0$, $O_0'$ be at distances $q, q'$ to the left and right of $C$ respectively and $T$ refers to these as base-points, we therefore have the generic result

$$T = J(\xi, \eta, \zeta, \chi) + q'\alpha' + q\alpha. \tag{94.4}$$

Though comparatively simple in principle, the situation is evidently getting rather involved from a computational point of view.

## 95. The aberration function

The sagittal section of a $t$-symmetric system is like that of a concentric $r$-symmetric system, so that one cannot have sharp imagery associated with any pair of conjugate planes. As remarked already several times, this does not prevent us from defining the ideal angle characteristic to correspond to whatever imagery we desire. For the sake of illustration we shall take the object to be at infinity. Suppose then that a sharp image could be formed in $\mathscr{I}'$ such that

$$y' = f_1\beta/\alpha, \quad z' = f_2\gamma/\alpha. \tag{95.1}$$

$K$ must of course have been so designed that the meridional and sagittal sections have coincident foci. Then, with the anterior base-point chosen arbitrarily,

$$T_0 = g(\beta^2, \gamma^2) - (f_1\beta\beta' + f_2\gamma\gamma')/\alpha. \tag{95.2}$$

This incidentally also shows at once that $K$ cannot have the property represented by (95.1), since $T_0$ obviously cannot be written in the form (94.4). Now, if $q$ is the distance between $B$ and $C$,

$$t = J(\xi, \eta, \zeta, \chi) + q'\alpha' + q\alpha + (f_1\beta\beta' + f_2\gamma\gamma')/\alpha, \tag{95.3}$$

terms depending on $\beta$ and $\gamma$ alone being consistently omitted.

At this stage it becomes convenient to write

$$\xi_2 = \gamma'^2, \quad \eta_2 = \gamma'\gamma, \quad \zeta_2 = \gamma^2$$

for the time being. Then set

$$J = (\tfrac{1}{2}k_1\xi + k_2\eta + k_4\chi) + (p_1\xi^2 + p_2\xi\eta + p_3\xi\zeta + 2p_4\xi\chi + p_5\eta\zeta + 2p_6\eta\chi + 2p_7\zeta\chi + 4p_8\chi^2) + O(6). \tag{95.4}$$

The first-order terms of $t$ are therefore

$$t^{(1)} = \tfrac{1}{2}(k_1 + k_4 - q')\xi + (k_2 - k_4 + f_1)\eta + \tfrac{1}{2}(k_4 - q')\xi_2 + (f_2 - k_4)\eta_2. \tag{95.5}$$

Since the imagery is supposed to be in accordance with (95.1) in the paraxial region, this shows that

$$k_1 = 0, \quad k_2 = f_2 - f_1, \quad k_4 = f_2, \quad q' = f_2. \tag{95.6}$$

Then

$$\begin{aligned} t^{(2)} = {} & (p_1 + p_4 + p_8)\xi^2 + (p_2 - 2p_4 + p_6 - 4p_8)\xi\eta \\ & + (p_3 + p_4 - 2p_6 + p_7 + 6p_8 - \tfrac{1}{4}f_2)\xi\zeta \\ & + (p_4 + 2p_8)\xi\xi_2 - 2(p_4 + 2p_8)\xi\eta_2 + (p_4 + 2p_8 - \tfrac{1}{4}f_2)\xi\zeta_2 \\ & + (p_5 + p_6 - 2p_7 - 4p_8 + \tfrac{1}{2}f_1)\eta\zeta + (p_6 - 4p_8)\eta\xi_2 \\ & - 2(p_6 - 4p_8)\eta\eta_2 + (p_6 - 4p_8 + \tfrac{1}{2}f_1)\eta\zeta_2 + (p_7 + 2p_8 - \tfrac{1}{4}f_2)\zeta\xi_2 \\ & - 2(p_7 + 2p_8 - \tfrac{1}{4}f_2)\zeta\eta_2 + p_8\xi_2^2 - 4p_8\xi_2\eta_2 + (6p_8 - \tfrac{1}{4}f_2)\xi_2\zeta_2 \\ & - (4p_8 - \tfrac{1}{2}f_2)\eta_2\zeta_2. \end{aligned} \tag{95.7}$$

This corresponds almost, but not quite, to (88.6). What remains to be done is to express $\beta'$, $\gamma'$ in terms $y'_e$ and $z'_e$ respectively, according to

$$\beta' = f_1\beta - y'_e, \quad \gamma' = f_2\gamma - z'_e, \tag{95.8}$$

where $y'_e$, $z'_e$ are, to the required order, the coordinates of the point of intersection of a ray with a normal plane which lies at a distance $d'(=1)$ to the left of $\mathscr{I}'$. If one appropriately interprets $\xi_1, \ldots$ in (88.6), the latter will then just represent $t^{(2)}$, the sixteen earlier coefficients $p_1, \ldots, p_{16}$ being linear combinations of the *eight* third-order coefficients which occur in (95.4). Since then $\boldsymbol{\epsilon}'_3 = \partial t^{(2)}/\partial \mathbf{y}'_e$, (88.7)–(88.10) will effectively represent the third-order displacement; and the coefficients governing the various partial displacements mutually 'interact' in a way precisely analogous to that described in Section 87.

As regards the number of independent aberration coefficients of order $2n-1$, the situation is once again essentially similar to to that of Section 75, so that the number in question is $n(n+2)$.

## 96. On the formation of a sharp image when the object is at infinity

In the light of the results of the last section one will naturally ask whether a $t$-symmetric system can form a sharp image of the plane $\mathscr{I}$ at infinity; and if so, one will want to know the shape of the image surface $\mathscr{I}^{*\prime}$. Now, if $\mathscr{I}^{*\prime}$ exists, there must be functions $D_1(\zeta, \zeta_2)$, $D_2(\zeta, \zeta_2)$ and $C(\zeta, \zeta_2)$ such that, in the usual way,

$$\left.\begin{aligned} Y' &= \beta D_1(\zeta, \zeta_2) = y' + C(\zeta, \zeta_2)\,\beta'/\alpha', \\ Z' &= \gamma D_2(\zeta, \zeta_2) = z' + C(\zeta, \zeta_2)\,\gamma'/\alpha', \end{aligned}\right\} \tag{96.1}$$

and $$D_1(0,0) = f_1, \quad D_2(0,0) = f_2, \quad C(0,0) = 0. \tag{96.2}$$

Since $\mathbf{y}' = -\,\partial T/\partial \boldsymbol{\beta}'$, (96.1) implies that

$$T = -\alpha' C - \eta D_1 - \eta_2 D_2 + g, \tag{96.3}$$

where $g$ is some function of $\zeta$ and $\zeta_2$ only. (96.3) must now be compared with the generic form of $T$, i.e.

$$T = J(\xi, \eta, \zeta, 1 - \alpha\alpha' - \eta - \eta_2) + q'\alpha' + q\alpha. \tag{96.4}$$

Here $\eta_2$ occurs only in the fourth argument $\chi$ of $J$. If one imagines

$J$ to be written as a power series in $\chi$, $\chi^n$ will, when $n > 1$, contain $\eta_2$ multiplied by a function depending upon $\xi_2$. Since $D_1$ does not depend upon this variable, it follows that $J$ must depend linearly on $\chi$, say

$$J = a(\xi, \eta, \zeta) + \chi b(\xi, \eta, \zeta). \tag{96.5}$$

Comparison with (96.4) then shows in the first place that

$$b(\xi, \eta, \zeta) = D_2(\zeta, \zeta_2),$$

i.e. $D_2$ and $b$ are both functions of $\zeta$ alone. Because $\eta^2 = \xi\zeta$ we can always write

$$a(\xi, \eta, \zeta) = \eta\bar{a}(\xi, \zeta) + \tilde{a}(\xi, \zeta),$$

and then inspection of all terms involving $\eta$ shows that we must have

$$\bar{a}(\xi, \zeta) - b(\zeta) + D_1(\zeta, \zeta_2) = 0,$$

and therefore $\bar{a}$ and $D_1$ are functions of $\zeta$ only. We are now left with the identity

$$\tilde{a}(\xi, \zeta) + b(\zeta) + [q' + C(\zeta, \zeta_2) - \alpha b(\zeta)]\,\alpha' + q\alpha - g(\zeta, \zeta_2) = 0. \tag{96.6}$$

Here $\xi$ occurs only in $\tilde{a}$ and in $\alpha'$. However, in the latter we always have $\xi$ combined linearly with $\xi_2$, whereas $\tilde{a}$ does not depend upon this variable. Hence $\tilde{a}$ cannot depend upon $\xi$, and the factor multiplying $\alpha'$ must vanish, i.e.

$$C(\zeta, \zeta_2) = -q' + b(\zeta)\,(1 - \zeta - \zeta_2)^{\frac{1}{2}}. \tag{96.7}$$

The remaining terms of (96.6) then give

$$g(\zeta, \zeta_2) = \tilde{a}(\zeta) + b(\zeta) + q(1 - \zeta - \zeta_2)^{\frac{1}{2}}. \tag{96.8}$$

The angle characteristic corresponding to the situation under discussion is therefore finally

$$T = q'\alpha' - (\alpha\alpha' + \gamma\gamma')\,D_2(\zeta) - \beta\beta'\,D_1(\zeta) + g(\zeta, \zeta_2). \tag{96.9}$$

We thus see that a sharp image of an object at infinity is indeed attainable. *The parametric equation* of $\mathscr{I}^{*\prime}$ is

$$(X' + q')^2 + Z'^2 = (1 - \zeta)\,D_2^2(\zeta), \quad Y'^2 = \zeta D_1^2(\zeta). \tag{96.10}$$

Here (96.2) should be recalled, along with the last member of (95.6), which also follows from (96.7).

$\mathscr{I}^{*\prime}$ is a surface of revolution with $\mathscr{C}$ as axis, a result which is perhaps not very surprising. It is possible for $\mathscr{I}^{*\prime}$ to be a circular cylinder, namely when

$$D_2 = (1 - \zeta)^{-\frac{1}{2}} f_2. \tag{96.11}$$

At any rate, we have found all possible functions, as given by (96.9), which could serve as ideal characteristic functions $T_0$ (when the object is at infinity) if one requires $T_0$ to represent a kind of sharp imagery whose realization by $K$ is not precluded *a priori*. In later sections we shall see how the shape of $\mathscr{I}^{*\prime}$ is further restricted by various ancillary conditions imposed upon $K$.

## 97. The system reversible as a whole

The condition of reversibility on $K$ as a whole is that the angle characteristic, taken with respect to base-points disposed symmetrically about $C$, must be invariant under the mutual interchange of $\boldsymbol{\beta}$ and $\boldsymbol{\beta}'$. Since under this operation $\chi$ is invariant in any event, we see from (94.4) that reversibility requires the relation

$$J(\xi, \eta, \zeta, \chi) = J(\zeta, \eta, \xi, \chi) \tag{97.1}$$

to be identically satisfied. The number of independent third-order coefficients is then reduced from eight to six, since (97.1) entails that in (95.4)

$$p_2 = p_5, \quad p_4 = p_7; \tag{97.2}$$

only the coefficients relating to fixed conjugates being taken into account as usual. The coefficients $p_\alpha$ here are of a kind somewhat different from those which appear in (56.4), say, in the sense that their vanishing does not entail the absence of aberrations. For this reason the observation that the conditions of reversibility (97.2) are homogeneous is irrelevant. With regard to higher orders, one has five fifth-order relations of the kind (97.2), nine seventh-order relations, and so on.

If the condition of reversibility be imposed under the circumstances contemplated in Section 96 we have to require the function on the right of (96.9) to be invariant when $\boldsymbol{\beta}$ and $\boldsymbol{\beta}'$ are interchanged, $q$ having been chosen to be equal to $q'$. It follows at once that $D_1$ and $D_2$ must be constants, and that $q'\alpha' - g(\xi, \xi_2)$ is a constant. Thus

$$T = \text{const.} + (\alpha + \alpha' - \alpha\alpha' - \gamma\gamma')f_2 - \beta\beta' f_1. \tag{97.3}$$

*The surface $\mathscr{I}^{*\prime}$ has the equation*

$$\frac{(X'+f_2)^2 + Z'^2}{f_2^2} + \frac{Y'^2}{f_1^2} = 1, \tag{97.4}$$

i.e. it is an ellipsoid of revolution.

## 98. The meridional section reversible

Even when $K$ as a whole is not reversible, the meridional section may be, i.e. there may exist a line $\mathscr{L}$, normal to $\mathscr{A}$, about which it is symmetric. In view of the invariance under rotations about $\mathscr{C}$ this implies reversibility, in an obvious sense, of the section of $K$ by any plane containing $\mathscr{C}$. Now let $L$ be the axial point of $\mathscr{L}$ and take base-points which are situated symmetrically about $L$. Then there exists a constant $p$ such that $J(\xi, \eta, \zeta, \chi)+p\alpha$ is invariant under the mutual interchange of $\beta$ and $\beta'$. It should be borne in mind that we now have to take $\xi_2=\eta_2=\zeta_2=0$, and that one can always add a constant multiple of $\alpha+\alpha'$ to any function whose invariance under the substitution in question is under consideration. By way of example, when the system as such lies entirely to the left of the normal plane through $C$, then $p$ is twice the distance between $L$ and $C$. To proceed from here, let us suppose for the sake of simplicity that the object is at infinity. Then, to require the invariance of $J+p\alpha$ is tantamount to requiring the invariance of $J$ alone, since the presence of the additional term will have consequences only as regards relations involving coefficients multiplying powers of $\zeta$ alone, and these are of no interest. The reversibility condition therefore reduces to

$$J(\xi, \eta, \zeta, \chi) = J(\zeta, \eta, \xi, \chi), \tag{98.1}$$

where

$$\chi = 1-\eta-(1-\xi)^{\frac{1}{2}}(1-\zeta)^{\frac{1}{2}}.$$

This is considerably weaker than (97.1), on account of the absence of the variables $\gamma$ and $\gamma'$. Indeed, writing (98.1) out in full in the third-order region, we are concerned with the invariance of the expression

$$p_1\xi^2+p_2\xi\eta+p_3\xi\zeta+p_5\eta\zeta+(p_4\xi+p_6\eta+p_7\zeta)(\xi-2\eta+\zeta)$$
$$+p_8(\xi-2\eta+\zeta)^2+\tfrac{1}{8}k_4(\xi^2-2\xi\zeta+\zeta^2). \tag{98.2}$$

The required equality of the coefficients of $\xi\eta$ and $\eta\zeta$ gives

$$p_2-p_5-2(p_4-p_7) = 0, \tag{98.3}$$

which should be compared with (97.2). It is worthy of note that when $p = 0$, i.e. $\mathscr{C}$ and $\mathscr{L}$ coincide, $K$ is in fact reversible as a whole, but the two separate conditions (97.2) do not emerge here. We shall meet a somewhat analogous situation in Section 99.

When the present kind of reversibility obtains we may again investigate the possibility of a sharp image being formed of an object at infinity. Writing

$$f_2 d(\zeta) = (1-\zeta)^{\frac{1}{2}} D_2(\zeta), \tag{98.4}$$

and $f_2 g(\zeta)$ for the sum of all functions of $\zeta$ alone which appear when $\xi_2 = \eta_2 = \zeta_2 = 0$ in $J + p\alpha$, one has the condition

$$(1-\xi)^{\frac{1}{2}} d(\zeta) - f_2^{-1}\eta D_1(\zeta) + g(\zeta) \equiv (1-\zeta)^{\frac{1}{2}} d(\xi)$$
$$- f_2^{-1}\eta D_1(\xi) + g(\xi). \tag{98.5}$$

Evidently $D_1(\zeta) = \text{const.} = f_1$. Then, setting $\xi = 0$, and recalling that $d(0) = 1$, there comes

$$d(\zeta) + g(\zeta) = (1-\zeta)^{\frac{1}{2}} + g(0).$$

Resubstituting this in (98.5) one obtains an equation which may be written as

$$[(1-\xi)^{\frac{1}{2}} - 1][1 - d(\zeta)] = [(1-\zeta)^{\frac{1}{2}} - 1][1 - d(\xi)],$$

and this shows that the ratio of $1 - d(\zeta)$ to $(1-\zeta)^{\frac{1}{2}} - 1$ must be a constant, $k$, say. Thus

$$d(\zeta) = 1 + k[1 - (1-\zeta)^{\frac{1}{2}}]. \tag{98.6}$$

In the first of equations (96.10) one has on the right just $f_2^2 d^2(\zeta)$, whilst the second here reads $Y'^2 = f_1^2 \zeta$. Hence, under the present conditions, *$\mathscr{I}^{*\prime}$ has the equation*

$$X'^2 + Z'^2 = f_2^2[(1+k) - k(1 - Y'^2/f_1^2)^{\frac{1}{2}}]^2, \tag{98.7}$$

the origin having been moved to $C$. To appreciate the nature of this surface, it suffices to consider its meridional section, since it is a surface of revolution about $\mathscr{C}$. Setting $Z' = 0$ we then have

$$\frac{[\pm X' - f_2(k+1)]^2}{k^2 f_2^2} + \frac{Y'^2}{f_1^2} = 1. \tag{98.8}$$

This is the equation of an ellipse, so that $\mathscr{I}^{*\prime}$ is that part of an *elliptical toroid* which passes through the axial ideal image point. The special case of a torus arises when $k^2 = f_1^2/f_2^2$. When $k = -1$ one has the ellipsoid (97.4), whilst the circular cylinder, to which (96.11) refers, has $k = 0$.

## 99. The meridional section concentric

It remains to study the consequences of the concentricity of the meridional section. Physically the state of affairs now contemplated is that every surface of constant refractive index (or else every refracting or reflecting surface) is the surface of a torus, the generic equation of which is given by (93.4) with $R_1$ fixed, whilst $R_2$ is to be regarded as a variable parameter. If the 'centre' $C^*$ of the meridional section lies at the point $x = p$ (where $p$ may be negative), the angle characteristic $T^*$, referred to base-points coincident at $C^*$, is

$$T^* = J(\xi, \eta, \zeta, \chi) + p(\alpha' - \alpha). \qquad (99.1)$$

Concentricity requires that this function depend on $\beta$ and $\beta'$ only through the function $\alpha\alpha' + \beta\beta'$ when $\xi_2 = \eta_2 = \zeta_2 = 0$, i.e.

$$J(\xi, \eta, \zeta, 1 - \alpha\alpha' - \beta\beta') + p(\alpha' - \alpha) = \Gamma(\alpha\alpha' + \beta\beta'),$$

say. It follows that, with

$$\tau = 1 - (1 - \beta^2)^{\frac{1}{2}} (1 - \beta'^2)^{\frac{1}{2}} - \beta\beta', \qquad (99.2)$$

$$J(\xi, \eta, \zeta, \chi) = G(\tau, \chi) + p(\alpha - \alpha'), \qquad (99.3)$$

where $G$ is some function of two arguments. The angle characteristic of the $t$-symmetric system with concentric meridional section is therefore

$$T = G(\tau, \chi) + (q' - p)\,\alpha' + (q + p)\,\alpha, \qquad (99.4)$$

and, using this result, the imagery may be discussed in the usual way.

One interesting conclusion which follows from (99.4), with $p \neq 0$, is that a system of the kind under consideration can never form a sharp image of an object at infinity at all. To see this, one need only compare (99.4) with (96.9). The linear occurrence of $\gamma\gamma'$ in the latter, coupled with the fact that the factor multiplying it depends upon $\zeta$ only, requires that $G$ depend linearly upon $\chi$. Analogous reasoning concerning the term in $\beta\beta'$ then shows that $G$ must also depend linearly upon $\tau$; and then there remains no way in which the variable $\xi_2$, which occurs in $p\alpha'$, can be accommodated.

The case with $p = 0$ is exceptional, since $\xi_2$ then no longer enters into the argument, and one finds that a sharp image may be attained if, and only if, $D_1(\zeta) = D_2(\zeta) = \text{const.} = f$, say, and $G(\tau, \chi) = f\chi$. Then, however, one has simply the characteristic function of a concentric ($r$-symmetric) system.

The conclusion at which we have just arrived might be held to be obvious, were one to think that when $p = 0$ the system is necessarily concentric in the sense of Chapter 6B. Yet we are made suspicious by the fact that, in (94.4), $J$ gives no indication of reducing to a function of $\chi$ alone when $p = 0$. The fact of the matter is that when $C$ and $C^*$ coincide, a system which is $t$-symmetric with concentric meridional section need not be invariant under arbitrary rotations. However artificially, one can devise a direction-dependent refractive index, e.g. a function of $x^2+y^2+z^2$ and of $(x\gamma - z\alpha)^2$, such that the system, though internally anisotropic, has all the symmetries here required of it. We have, in fact, thus returned to the point made at the end of Section 84.

## Problems

**P.9 (i).** Find the number of aberration coefficients of order $2n-1$ of the $t$-symmetric system with concentric meridional section.

**P.9 (ii).** The spherical aberration of a $t$-symmetric system vanishes to all orders for an object at infinity. What conditions must the function $J$ of (94.4) satisfy? Show that your result is consistent with equations (95.6) and (95.7). (It is given that the system forms a sharp paraxial image.)

**P.9 (iii).** A parallel bundle of rays incident upon a system like that of P.9 (ii) produces a circular image patch ($\beta = \gamma = 0$). What condition must be satisfied by the function $J$ of (94.4) so that this may be the case?

CHAPTER 10

# CHROMATIC DEFECTS OF THE IMAGE

## 100. Introductory remarks. Chromatic aberration coefficients

Hitherto we have concerned ourselves exclusively with the monochromatic image produced by given systems. This means that the light traversing $K$ was always supposed to have a single fixed wavelength. In practice, however, it is quite usual for the light to be white, or, at any rate, to have some spectral distribution in which various colours are represented. It therefore becomes necessary to take due account of the fact that the distribution of refractive index within $K$ depends upon the wavelength $\lambda$. In a sense therefore, every system $K$, imagined in the first place as relating to some fixed value of $\lambda$, generates a whole collection of systems, one for every value of $\lambda$ to be contemplated; granted, of course, that $K$ is not a purely reflecting system.

It is a matter of great convenience to refer the properties of $K$ at wavelength $\lambda$ to those which obtain for light of some fixed colour, henceforth called the *base-colour*. The wavelength of this will be denoted by $\underset{0}{\lambda}$; and, consistently with this notation, if $Q(\lambda)$ is any quantity depending on $\lambda$, then $\underset{0}{Q}$ shall stand for $Q(\underset{0}{\lambda})$. Though the choice of $\underset{0}{\lambda}$ is to a certain extent arbitrary, we shall see in Section 101 that it is, in practice, approximately determined by the extremes of the range of wavelengths admitted in any particular problem.

Now let $\underset{0}{F}(q_1, q_2, q_3, q_4)$ be the characteristic function of $K$ when $\lambda = \underset{0}{\lambda}$, so that $\underset{0}{F}$ is a certain optical distance measured along the ray $\underset{0}{\mathscr{R}}$ specified by the values of the ray-coordinates in question. For a different wavelength the ray $\mathscr{R}$ specified by the same values of the $q_i$ will, in general, differ from $\underset{0}{\mathscr{R}}$, and in any event the optical length in question has to be calculated with the refractive index appropriate to $\lambda$, not $\underset{0}{\lambda}$. It follows that the value of $F$ will differ from that of $\underset{0}{F}$. In short, the characteristic function is now to be regarded as a function of both the ray-coordinates *and* of $\lambda$.

At this stage we meet for the first time a problem of a peculiar kind. It is this: $F$ will usually be a very complicated function of $\lambda$, just as it is a very complicated function of the $q_i$. Therefore one will naturally be inclined to write $F$ now as a power series in the *five* variables $q_1, q_2, q_3, q_4$ *and* $\delta\lambda (= \lambda - \underset{0}{\lambda})$. For the base-ray $\underset{0}{\mathscr{R}_0}$ all five of them of course take the value zero. The dependence of $F$ upon $\delta\lambda$ reflects the dependence of the refractive index $N$ upon $\delta\lambda$, so that $N$ also will then have to be written as a series of ascending powers of $\delta\lambda$. However, if this procedure is to have any usefulness at all, the series in question must converge reasonably quickly for ranges of $\delta\lambda$ encountered in practice. It is unfortunately true that this condition is certainly *not* satisfied by $N$: the convergence is exceedingly slow. This, then, is the difficulty to be overcome; that is to say, we have to find some parameter, $\omega$ say, which is a function of $\delta\lambda$ and vanishes together with it, such that the series for $N$ in ascending powers of $\omega$ will converge rapidly, and that for $F$ along with it. Such a parameter can, indeed, be found in any given situation, and we call it the *chromatic coordinate*; see Section 101.

We now write

$$F = F(q_1, q_2, q_3, q_4, \omega) = \sum_{m=0}^{\infty} \underset{m}{F}(q_1, q_2, q_3, q_4)\, \omega^m. \qquad (100.1)$$

The ideal characteristic function $F_0$ is defined, as always, in terms of the *desired* 'ideal' behaviour of $K$; and one will usually require this *behaviour*, i.e. not $F_0$ itself, to be independent of $\omega$. The aberration function is then

$$f = F - F_0 = \sum_{n=0}^{\infty} f_n = \sum_{n=0}^{\infty} \sum_{m=0}^{\infty} \underset{m}{f_n}\, \omega^m. \qquad (100.2)$$

It contains all the information concerning the extent to which a ray through $O$ will in fact fail to pass through the required point $I'$, whatever its colour may be.

To every aberration function of order $n$ encountered hitherto there now corresponds an infinite sequence $\underset{m}{f_n}$ of aberration functions of *coordinate order n and chromatic order m*. Note that $f_n$ now stands for $\sum_m \underset{m}{f_n}\, \omega^m$, and it must not be confused with $\underset{0}{f_n}$. At this point we need to remark explicitly on the occurrence of the terms $f_0$ and $f_1$ in (100.2). Taking $f_1$ first, this will always be present when, in accordance with the demands of actual practice, displacements are referred to a *fixed* reference point $I'$ lying in a fixed image surface, i.e. that corresponding to the base-colour. Therefore, even when the parabasal rays from $O$ unite in a common point in the image space, this point will, in general, not coincide with $I'$. On the other hand we shall regard quantities such as the magnification and the focal length as fixed, i.e. independent of $\lambda$; all effects of their actual non-constancy already being taken into account by the aberration function.

The presence of $f_0$ has a different origin. It will be recalled that under the most general circumstances the coordinate basis was suitably arranged relative to the base-ray $\mathscr{R}_0$. If we agree that the coordinate basis is to be fixed, it follows that if $\mathscr{R}_0$ refers to $\underset{0}{\lambda}$, then at wavelength $\lambda$ the ray specified by $q_i = 0$ $(i = 1, \ldots, 4)$ will not, in general, coincide with $\mathscr{R}_0$. Thus when $\lambda \neq \underset{0}{\lambda}$ one has the unavoidable complication that the displacement already contains terms of order zero. So much for matters of principle. In order not to overburden ourselves with an excess of detail we henceforth consider only systems which are at least doubly plane-symmetric, in the sense of Section 88. Then the base-ray can be taken along $\mathscr{A}$ and will therefore be independent of $\lambda$, so that $f_0 = 0$. In future all sums over $n$ will thus start with $n = 1$, rather than $n = 0$.

To return to matters of terminology, $\underset{0}{f_n}$ will naturally be called the $n$th-order 'monochromatic aberration function'. Again, since in practice one rarely needs to consider chromatic orders beyond the second, or at worst beyond the third, we refer to $\underset{1}{f_n}$ $(\underset{2}{f_n}, \underset{3}{f_n}, \ldots)$ as the 'primary (secondary, tertiary, ...) chromatic $n$th-order aberration function', the qualification 'coordinate' being suppressed here, since it is largely redundant. Whatever we lose here through long-windedness we more than regain in precision. With regard to the displacement, we have, analogously to (100.2),

$$\boldsymbol{\epsilon}' = \sum_{m=0}^{\infty} \underset{m}{\boldsymbol{\epsilon}'}\, \omega^m = \sum_{n=1}^{\infty} \sum_{m=0}^{\infty} \underset{m}{\boldsymbol{\epsilon}'_n}\, \omega^m, \tag{100.3}$$

so that, for example, $\underset{0}{\epsilon}'$ is the total monochromatic displacement, $\underset{1}{\epsilon}'_n\,\omega$ the primary chromatic $n$th-order displacement, and so on.

We can extend this terminology to *types* of aberrations without difficulty; bearing in mind that the chromatic displacement is essentially similar to the monochromatic displacement from a geometrical point of view. Since we are supposing $K$ to be at least doubly plane-symmetric we may use an extension of the notation of Section 16. Thus

$$f^{(n)} = \sum_{m=0}^{\infty} \sum_{\mu=0}^{n} \sum_{\nu=0}^{\mu} \underset{m}{f}{}^{(n)}_{\mu\nu}\, \xi^{n-\mu}\eta^{\mu-\nu}\zeta^{\nu}\omega^{m}. \tag{100.4}$$

To avoid irrelevant verbal difficulties we proceed in terms of a specific case, namely, when $K$ is symmetric, with $F = V$. Then $\underset{0}{v}{}^{(n)}_{00}$ is the (characteristic) coefficient of monochromatic $(2n-1)$th-order spherical aberration, $\underset{1}{v}{}^{(n)}_{00}$ is that of primary chromatic $(2n-1)$th-order spherical aberration, and so on. Similarly, $\underset{2}{v}{}^{(3)}_{10}$, say, will be the (characteristic) coefficient of secondary chromatic fifth-order circular coma. In short, the way in which the terminology extends itself to the various coefficients, whether characteristic or effective, governing the aberrations of the various types, virtually explains itself.

We may, of course, also continue to speak of 'spherical aberration', 'circular coma', etc., as a whole. Thus we previously had in $v$ functions $S(\xi)$, $C(\xi)$ corresponding to these types, according to equation (26.1). Under the present circumstances we shall have to write

$$v = v^{(1)} + S(\xi, \omega) + \eta C(\xi, \omega), \tag{100.5}$$

but the general situation is essentially the same as before; see Section 107.

With regard to the state of chromatic correction, various terms are in common use, such as achromatism, apochromatism, and the like. It will be more convenient to return to these at a later stage; see Section 104. We may, however, raise at this point the question how remainder terms are now to be denoted. Previously, when truncating a series (say one representing a displacement) after its $n$th-order term, i.e. after the terms of degree $n$ in the ray-coordinates, all rejected terms were collectively denoted by the symbol $O(n+1)$, for $O(n)$ stood for any sequence not containing terms of degree less than $n$. This notation is now amplified by understanding $O(n_m)$ to stand for any collection of terms none of which is of degree less than $n$ in the ray-coordinates *and* less than $m$ in the chromatic coordinate. However, a further slight extension is required in view of the fact that $m$ will generally depend upon $n$, in a sense adequately illustrated by the example of the displacement of axial rays in a symmetric system. Let it be approximated by the expression

$$\epsilon' = (\underset{1}{c_1^*}\,\omega + \underset{2}{c_1^*}\,\omega^2)\,\rho + (\underset{0}{p_1^*} + \underset{1}{p_1^*}\,\omega)\,\rho^3 + \underset{0}{s_1^*}\,\rho^5; \qquad (100.6)$$

that is to say, in some particular case it is assumed that knowledge of the five coefficients which occur on the right implies a sufficiently accurate description of axial chromatism and spherical aberration for the values of $\rho$ and $\omega$ of interest. Evidently the remainder term, if it is to be sufficiently explicit, must be written as $O(1_3, 3_2, 5_1, 7_0)$. Then one knows exactly that the terms not included explicitly consist of the paraxial terms which are at least cubic in $\omega$, of the secondary and higher-order chromatic primary spherical aberration, and so on.

## 101. The chromatic coordinate

As has been explained, to achieve rapid convergence of the various chromatic power series, we have to choose a function $\omega(\delta\lambda)$ such that $N$ (or, for that matter, $1/N$) when written as a series in ascending powers of $\omega$, in some sense converges sufficiently rapidly. We write the series as

$$\delta N = N - \underset{0}{N} = \underset{0}{N} \sum_{m=1}^{\infty} \nu_m \omega^m. \qquad (101.1)$$

$K$ is here supposed to be made up, at least in part, of various homogeneous media, so that each such medium is characterized by the values of $\underset{0}{N}$ and the constants $\nu_m$ which occur on the right of (101.1). $\omega$, on the other hand, must be the *same* function for every medium within $K$. This means incidentally that the results of investigations which attempt to represent $N$ as the linear superposition of several universal functions of $\lambda$ are irrelevant to our task, however interesting they may be in other respects.

Now, within the visual spectrum the refractive index of ordinary crown glasses is quite well represented by Hartmann's equation

$$N = N_1 + \frac{c}{\lambda - \lambda_1}, \tag{101.2}$$

where $N_1$, $c$, $\lambda_1$ are constants of the particular glass. For any such glass, therefore, there are constants $\nu_1$ and $\alpha$ such that

$$\delta N = \frac{\underset{0}{N}\nu_1 \delta\lambda}{1 + \alpha\delta\lambda}, \tag{101.3}$$

for any choice of $\underset{0}{\lambda}$. Taking the micron ($= 10^{-3}$ mm) as a convenient unit of length and $\underset{0}{\lambda}$ to be $\lambda_D$, i.e. the wavelength of the sodium-$D$ line, $\nu_1$ and $\alpha$ may be approximately fitted to a wide range of glasses. It turns out that $\alpha$ never differs very substantially from $\frac{5}{2}$. Accordingly we now define, with $\underset{0}{\lambda} = \lambda_D$,

$$\omega = \frac{\delta\lambda}{1 + \frac{5}{2}\delta\lambda}. \tag{101.4}$$

If for any reason one adopts some other value of $\underset{0}{\lambda}$, e.g. if the range of wavelengths of interest is other than the visible, then the number multiplying $\delta\lambda$ in the denominator should of course be chosen appropriately.

With (101.4) the series (101.1) does indeed converge quite rapidly and with, say, four terms the representation of the refractive index of any glass should be adequate. It must be emphasized that it would be quite wrong to object to (101.4) on the grounds that many glasses do not obey Hartmann's formula well: all that concerns us is that the function $\nu_1\omega$ represents $\delta N/\underset{0}{N}$ vastly better than does $\nu_1\delta\lambda$. In practice one should proceed as follows. Taking the demands of the particular design problem into account, one first decides to what chromatic order one intends to proceed. Let it be $p\,(> 1)$. The series (101.1) is then truncated after the $p$th term, and the resulting polynomial of degree $p$ fitted to $p+1$ known values $\underset{0}{N}, N_1, N_2, \ldots, N_p$ of the refractive index. This at first sight perhaps somewhat strange procedure has the great virtue that the value of

$f_{\mu\nu}^{(n)}$, as given by the truncated series

$$\sum_{m=1}^{p} f_{\mu\nu}^{(n)} \omega^m,$$
$$\quad\;\; m$$

differs from its actual value when $\omega$ takes the values corresponding to $N_1, N_2, \ldots, N_p$ solely on account of this truncation, i.e. it is not due to the inaccurate representation of $N$ by a polynomial of degree $p$. It is, of course, understood that where $1/N$ is required in the course of calculation, it is to be written as a power series whose product with

$$\underset{0}{N}\left(1 + \sum_{m=1}^{p} \nu_m \omega^m\right)$$

is exactly unity.

When the constitution of $K$ differs from that contemplated above we shall take it for granted that an appropriate chromatic coordinate can always be defined. For example, in an electron-optical instrument $\omega$ will be taken proportional to $E - \underset{0}{E}$, where $E$ is the actual energy of electrons, whilst $\underset{0}{E}$ is a 'base-energy', e.g. the mean energy, all these energies being measured in the field-free regions of the system.

## 102. On the reduction of distances

With our original convention concerning the constancy of the refractive indices in the object and image spaces it became possible to eliminate the explicit appearance of $N$ and $N'$ from our equations by introducing reduced distances (Section 6). We can, alas, no longer usefully adhere to this convention when $N$ and $N'$ have to be considered as functions of $\omega$. If this symbol were to stand for the reduced object coordinates we would have the unhappy situation that a given object point would no longer be characterized by the constancy of $\mathbf{y}$. This feature would be likely to lead to endless confusion, and we have no option but to abandon the reduction of distances as previously understood.

However, in order to have equations which become formally identical with those of the purely monochromatic theory when $\omega = 0$, we shall agree to understand all distances to be *quasi-reduced*, meaning that if $X$ is any distance in a region in which the base-value of the refractive index is $\underset{0}{N}$ (independently of the ray-coordinates) then $X$ shall in fact stand for $\underset{0}{N}$ times the actual distance in question. Then $m$, for instance, has exactly the meaning attached to it hitherto, bearing in mind that quantities such as $d'$, $m$, $\hat{f}$, etc., are in any event to be taken as referring to the base-colour only.

Let us see, by way of example, what the ideal point characteristic of a symmetric system looks like, under the circumstances contemplated in Section 15. Beginning with unreduced quantities

$$V_0 = \tilde{V}(O,O') - N'[d'^2 + (y'-my)^2 + (z'-mz)^2]^{\frac{1}{2}}. \quad (102.1)$$

Quasi-reduction implies the transcription

$$d' \to d'/\underset{0}{N'}, \quad \mathbf{y}' \to \mathbf{y}'/\underset{0}{N'}, \quad \mathbf{y} \to \mathbf{y}/\underset{0}{N}, \quad m \to \underset{0}{N}m/\underset{0}{N'},$$

and then (102.1) becomes, on setting the fixed constant $d'$ equal to unity,

$$V_0 = g(\zeta) - \nu'(1+u)^{\frac{1}{2}}, \quad (102.2)$$

where

$$\nu' = N'/\underset{0}{N'} = 1 + \sum_{m=1}^{\infty} \nu'_m \omega^m, \quad (102.3)$$

in view of (101.1); and the other symbols have their familiar meanings, though $g$ may now depend also upon $\omega$. In particular $\mathbf{y}_1 = m\mathbf{y}$ is the quasi-reduced ideal image height, which is as it should be, since the ideal image point for all colours is the same as that for the base-colour under the procedure here adopted.

It will be seen that $V_0$ does depend upon $\omega$; but, as regards ideal imagery, this dependence is compensated by the fact that in place of (6.1) we must now use the equations

$$\partial V/\partial \mathbf{y}' = \nu' \boldsymbol{\beta}', \quad \partial V/\partial \mathbf{y} = -\nu \boldsymbol{\beta}. \quad (102.4)$$

## 103. Paraxial aberrations

The relatively detailed discussion of chromatic defects will be entirely confined to symmetric systems, not least because it will turn out to be an adequate prototype of what has to be done when one is confronted with other symmetries.

The occurrence of the factor $\nu'$ on the right of (102.2) suggests that we write the first-order aberration function in the form

$$v^{(1)} = \nu'(\tfrac{1}{2}c_1\xi + c_2\eta), \quad (103.1)$$

where $c_1$ and $c_2$ are 'constants' depending upon $\omega$, and $\underset{0}{c_1} = \underset{0}{c_2} = 0$. The displacement is then easily found to be

$$\boldsymbol{\epsilon}_1' = c_1\mathbf{y}' + c_2\mathbf{y}_1. \tag{103.2}$$

We note in passing that even paraxially rays grazing the rim of the stop will not, in general, intersect the plane $\mathscr{E}'$ in a circle concentric with $E'$ when $\omega \neq 0$, since for arbitrary values of $\omega$ $\mathscr{E}'$ is not the image of the axial point of the plane of the stop. However, we are once again disregarding aberrations associated with the pupil planes, since to do so in the present context is surely justified in all but the most extreme cases.

According to (103.2) there are *two* chromatic paraxial defects known traditionally as *longitudinal* and *transverse* chromatic aberration respectively, the qualification 'paraxial' usually being omitted. The reasons for this terminology are pretty obvious. In an out-of-focus image plane which has $\chi = c_1/(1-c_1)$ the displacement is independent of $\mathbf{y}'$, so that axial rays come to a focus in the axial point of this plane. Moreover, one then has $\tilde{\mathbf{Y}}' = [(1+c_2)/(1-c_1)]\,\mathbf{y}_1$, so that the effective magnification exceeds $m$ by an amount

$$m_{\text{eff}} - m = (c_1 + c_2)/(1-c_1),$$

which reduces to $c_2$ when $c_1 = 0$. For this reason one also speaks of $c_2$ as governing the 'chromatic difference of magnification'.

These results are valid for all values of $\omega$, i.e.

$$\boldsymbol{\epsilon}_1' = \sum_{m=1}^{\infty} (\underset{m}{c_1}\mathbf{y}' + \underset{m}{c_2}\mathbf{y}_1)\,\omega^m. \tag{103.3}$$

Now take $\mathbf{y}_1 = 0$ for the moment. Then, when $\underset{1}{c_1}$ is arranged to be zero, $\boldsymbol{\epsilon}_1'$ will depend quadratically upon $\omega$ when $\omega$ is sufficiently small. This means that, near $\omega = 0$, rays whose colours are given, respectively, by the values $\omega$ and $-\omega$ of the chromatic coordinate are brought to a common focus, $\boldsymbol{\epsilon}_1'$ being stationary at $\omega = 0$. If practical demands require $\boldsymbol{\epsilon}_1'$ to be stationary at the particular wavelength $\lambda$, then one may either take this as the base-colour, or else design $K$ in such a way that $\underset{1}{c_1}$ takes a certain very small value, depending upon that of $\underset{2}{c_1}$. Under these circumstances one then says that $K$ is *achromatic*. In more general contexts the traditional terminology is rather vague, a question to which we shall return in the next section. At any rate, when this kind of achromatism obtains, the residual chromatic displacement is often referred to as the 'secondary spectrum'. (This is an ugly terminology, and we shall avoid it.) A much better state of (axial, paraxial) correction is obtained by designing $K$ in such a way that $\underset{1}{c_1}$ and $\underset{2}{c_1}$ have values relative to each other and to $\underset{3}{c_1}$ such that the displacement is stationary at two points within the range of wavelengths of interest.

With regard to the transverse aberration coefficients $\underset{m}{c_2}$ one has, in principle, the same situation, but in practice one is usually content with reducing $\underset{1}{c_2}$ to a sufficiently small value.

## 104. The meaning of achromatism and apochromatism

We have already remarked on the very restricted connotation of 'achromatism', i.e. a system is achromatic if the longitudinal paraxial displacement is stationary at one point of the range of wavelengths of interest. In other words, the condition of achromatism in effect places a limitation on only *one* of the aberration coefficients, namely $\underset{1}{c_1}$. This may be good enough in some cases, but in the case of systems of large aperture, for example, this implicit over-emphasis on paraxial chromatism is out of place. It is useless to devote endless effort into arranging the displacement of axial rays to be stationary at two, or even three, points of the range (i.e. 'corrected for three, or four, colours') if one ends up with a system in which the spherical aberration varies appreciably with $\omega$, or as one sometimes says, a large amount of *sphero-chromatism* remains. It is, indeed, pointless not to consider these types of aberrations jointly in the course of design. Furthermore, systems corrected for three colours in the restricted sense above, have sometimes mistakenly been called apochromatic, whilst those corrected for four colours are referred to as superachromatic. This state of affairs makes it desirable to develop a more detailed terminology.

Accordingly we call $K$ *simply achromatic with respect to a given type of aberration* if the latter is stationary at an appropriate point of the range of wavelengths being contemplated. Similarly, if it is stationary at two, three, ..., points we call it doubly, triply, ..., achromatic (with respect to the type in question). In the case of simple achromatism one may leave the qualification 'simple' understood whenever there is no risk of confusion. Further, one may speak of a given property of $K$ as being achromatic (or achromatized) rather than of $K$ as being achromatic with respect to this property. Thus we may say either that $K$ is achromatic with respect to spherical aberration, or else that spherical aberration is achromatic.

In the example just cited we have spoken simply of 'spherical aberration.' Now it is clear that, save in exceptional cases, if spherical aberration is achromatic it cannot be so for all zones simultaneously. The statement above is therefore ambiguous unless a particular zone be specified. Explicitly, recall that the displacement due to spherical aberration has the generic form

$$\begin{aligned}\epsilon' &= (\underset{0}{p_1^*} + \underset{1}{p_1^*}\omega + \underset{2}{p_1^*}\omega^2 + \dots)\rho^3 + (\underset{0}{s_1^*} + \underset{1}{s_1^*}\omega + \dots)\rho^5 \\ &\qquad + (\underset{0}{t_1^*} + \underset{1}{t_1^*}\omega + \dots)\rho^7 + \dots \\ &\equiv \sum_{n=2}^{\infty}\sum_{m=0}^{\infty} \underset{m}{u_{00}^{(n)}}\rho^{2n-1}\omega^m. \qquad (104.1)\end{aligned}$$

If $\epsilon'$ is to be stationary at $\omega = 0$ (say) for all values of $\rho$ one has to have

$$\sum_{n=2}^{\infty} \underset{1}{u_{00}^{(n)}}\rho^{2n-1} = 0, \qquad (104.2)$$

which requires that $\underset{1}{u_{00}^{(n)}} = 0$ (all $n \geqslant 2$); and this is asking too much in practice. However, the situation is not as serious as it looks, since, on account of the way in which third-order spherical aberration is usually balanced against the higher orders, it will often suffice to require that $\underset{1}{u_{00}^{(n)}} = 0$ only for $n = 2$ and $3$.

The classical definition of *apochromatism* amounts to the condition that the (paraxial) longitudinal aberration be doubly achromatic *and* that both spherical aberration and circular coma be simply achromatic. Nothing is said about the (paraxial) transverse aberration. Moreover, with regard to the non-paraxial aberrations, one still has to say something about what particular zone one has in mind, or else interpret their 'absence' as meaning for instance $\underset{1}{p_1^*} = \underset{1}{p_2^*} = 0$ and $\underset{1}{s_1^*} = \underset{1}{s_2^*} = 0$, as already discussed; the inclusion of chromatic conditions of coordinate order greater than three depending upon the monochromatic state of correction. Terms such as superachromatism, and superapochromatism are also used occasionally, the former meaning that $c_1$ is triply achromatic, the latter apochromatism but with $c_1$ triply, instead of doubly, achromatic.

## 105. The third- and higher-order aberrations

A factor $\nu'$ was included in the right-hand member of (101.3) merely as a matter of convenience. The extension of this formal device to all orders clearly suggests itself. Accordingly we define a chromatically modified aberration function—also denoted by $v$—by the relation

$$V = V_0 + \nu' v. \tag{105.1}$$

The displacement is then given as usual by (17.1). Owing to the presence of first-order aberrations, the third-order coefficients are no longer given by (23.2), but contain the coefficients $c_1$ and $c_2$ as well. One proceeds essentially as in Section 23, and, since the work is rather tedious but straightforward, it suffices to quote the result:

$$\left.\begin{aligned} p_1^* &= 4p_1 + \tfrac{1}{2}(3c_1 - 3c_1^2 + c_1^3), \\ p_2^* &= 2\bar{p}_1^* = 2p_2 + (-2c_1 + c_2 + c_1^2 - 2c_1c_2 + c_1^2c_2), \\ p_3^* &= 2p_3 + \tfrac{1}{2}(c_1 - 2c_2 + 2c_1c_2 - c_2^2 + c_1c_2^2), \\ \bar{p}_2^* &= 2p_4 + (c_1 - 2c_2 + 2c_1c_2 - c_2^2 + c_1c_2^2), \\ \bar{p}_3^* &= p_5 + \tfrac{1}{2}(3c_2 + 3c_2^2 + c_2^3). \end{aligned}\right\} \tag{105.2}$$

These hold for all values of $\omega$, so that the corresponding equations for the coefficients of the various orders may be read off from them. For example,

$$\underset{1}{p_1^*} = 4\underset{1}{p_1} + \tfrac{3}{2}\underset{1}{c_1}, \quad \underset{2}{p_1^*} = 4\underset{2}{p_1} + \tfrac{3}{2}(\underset{2}{c_1} - \underset{1}{c_1^2}); \tag{105.3}$$

and so on.

Under conditions in which one has to worry about chromatic defects outside the paraxial region, one will surely have achromatized the paraxial coefficients, i.e. $\underset{1}{c_1} \approx 0$, $\underset{1}{c_2} \approx 0$. Then (105.2) become equations for the primary chromatic third-order coefficients which are effectively identical with (23.2). A similar remark applies to the fifth-order coefficients: one will not need to know them unless $\underset{1}{c_1}$ and $\underset{1}{c_2}$ are (approximately) zero. It follows that the *primary* chromatic fifth-order coefficients are given by equations (23.3), if one imagines a subscript 1 to be placed under every coefficient which occurs in them; and one will hardly ever need to consider higher chromatic orders.

In practice one must not forget that the coefficients which occur in this section relate to the modified aberration function. If, in the absence of this modification, one distinguishes the corresponding coefficients by placing dots over them, the only change required in the equations for the primary chromatic third-order coefficients, for instance, is the replacement

$$\underset{1}{p_\alpha} \to \underset{1}{\dot{p}_\alpha} - \nu'_1 \underset{0}{p_\alpha}. \tag{105.4}$$

In any event, $\nu'$ is very often unity in practice.

## 106. Stop shifts

To investigate the effects of stop shifts we go over to the angle characteristic. Our first task to this end is to inquire into the generic form of $T$, and of that of $T_0$ in particular. Since the aberrations are to vanish independently of the value of $\omega$ for the pair of conjugate planes characterized by the magnification $m$ (which relates to the base-colour) we must have

$$\kappa \frac{\partial T_0}{\partial \boldsymbol{\beta}'} + m \frac{\partial T_0}{\partial \boldsymbol{\beta}} = 0 \tag{106.1}$$

for all rays, where $\kappa = \nu/\nu'$. (106.1) implies that $T_0$ must be a function of $\zeta = (\overline{m}\beta' - \beta)^2 + (\overline{m}\gamma' - \gamma)^2$ alone. Here $\overline{m} = m/\kappa$, and though this is formally a (properly) reduced magnification it is defined in terms of $m$, that is to say, we are not concerned with planes which are conjugate for the colour $\omega$, rather than the base-colour. We naturally adopt the rotational invariants $\xi$, $\eta$, $\zeta$ as given by (15.11), but with $m$, $s$ replaced by $\overline{m}$ and $\bar{s}(= s/\kappa)$ respectively, $s$ itself also retaining its previous significance.

Paraxially $T_0$ is proportional to $\zeta$, and we take the factor of proportionality to be $\nu/2m$, for this entails that

$$\mathbf{y}' = \boldsymbol{\beta} - \overline{m}\boldsymbol{\beta}'. \qquad (106.2)$$

An infinite object distance can therefore be accommodated by letting $m$ tend to zero. Any other choice of the factor of proportionality creates difficulties in the context of this limiting process, and these we want to avoid. When aberrations are present, we write

$$T = g(\omega) + \nu(\zeta/2m + t), \qquad (106.3)$$

having absorbed all terms of $T_0$ which depend non-linearly upon $\zeta$ into the aberration function $t$. Note that the latter is defined by

$$T - T_0 = \nu t, \qquad (106.4)$$

i.e. a factor $\nu$, rather than $\nu'$, is included on the right. This has the consequence that the displacement

$$\boldsymbol{\epsilon}' = (s-m)(2\boldsymbol{\sigma} t_\xi + \boldsymbol{\mu} t_\eta) \qquad (106.5)$$

has a factor $s-m$ which is independent of $\omega$. Here $\boldsymbol{\sigma}$ and $\boldsymbol{\mu}$ are defined as in (24.2), but with $\bar{s}$, $\overline{m}$ replacing $s$ and $m$ respectively. Like (24.4), equation (106.5) is only superficially simple, for we have all the complications which we met in Section 24. Indeed, if higher-order chromatic aberrations are to be taken into account, the situation is worse, for one has to go through the kind of work following equation (24.5) even in the paraxial region. (See, however, the alternative procedure outlined at the end of this section.)

To see what is involved let us therefore consider only paraxial imagery for the time being. We write

$$t = \tfrac{1}{2}c_1\xi + c_2\eta + \tfrac{1}{2}c_3\zeta, \qquad (106.6)$$

where, we note, the term $\frac{1}{2}c_3\zeta$, which depends upon $\zeta$ alone, has to be included, just as the term $p_6\zeta^2$ had to be retained in equation (24.1). Then

$$\left.\begin{aligned} \mathbf{y}'_e &= (\kappa + sc_1 + mc_2)\,\boldsymbol{\sigma} + (1 - \kappa + sc_2 + mc_3)\,\boldsymbol{\mu}, \\ \mathbf{y}_1 &= m(c_1 + c_2)\,\boldsymbol{\sigma} + [1 + m(c_2 + c_3)]\,\boldsymbol{\mu}. \end{aligned}\right\} \qquad (106.7)$$

When $m = 0$ these reduce to

$$\mathbf{y}'_e = (\kappa + sc_1)\,\boldsymbol{\sigma} + (1 - \kappa + sc_2)\,\boldsymbol{\beta}, \quad \mathbf{y}_1 = \boldsymbol{\beta}. \qquad (106.8)$$

The paraxial displacement is, in view of (106.5), given by

$$\boldsymbol{\epsilon}_1' = (s-m)(c_1\boldsymbol{\sigma}+c_2\boldsymbol{\mu}), \tag{106.9}$$

the primary chromatic part of which is

$$\underset{1}{\boldsymbol{\epsilon}_1'} = (s-m)(\underset{1}{c_1}\boldsymbol{\sigma}+\underset{1}{c_2}\boldsymbol{\mu}) = (s-m)(\underset{1}{c_1}\mathbf{y}_e'+\underset{1}{c_2}\mathbf{y}_1), \tag{106.10}$$

since, according to (106.7), $\mathbf{y}_e' = \boldsymbol{\sigma}+O(1_1, 3_0)$, $\mathbf{y}_1 = \boldsymbol{\mu}+O(1_1, 3_0)$. These equations may also be used in the primary chromatic parts of the right-hand members of (106.7); cf. the step following equation (24.8). This suffices to obtain the secondary chromatic first-order displacement

$$\begin{aligned}\underset{2}{\boldsymbol{\epsilon}_1'} = (s-m)\{&[\underset{2}{c_1}-(s\underset{1}{c_1^2}+2m\underset{1}{c_1}\underset{1}{c_2}+m\underset{1}{c_2^2}+\kappa\underset{11}{c_1})]\,\mathbf{y}_e' \\ &+[\underset{2}{c_2}-(s\underset{1}{c_1}\underset{1}{c_2}+m\underset{1}{c_1}\underset{1}{c_3}+m\underset{1}{c_2^2}+m\underset{1}{c_2}\underset{1}{c_3}-\kappa\underset{11}{c_1})]\,\mathbf{y}_1\}.\end{aligned} \tag{106.11}$$

These results should be sufficient for most purposes. However, in principle one can include the terms of all chromatic orders by simply solving the pair of linear equations (106.8) for $\boldsymbol{\sigma}$ and $\boldsymbol{\mu}$, and inserting the solution in (106.9).

When we go on to the non-paraxial aberrations the complications are less tractable, though one can still cope with those of the first chromatic order without excessive labour, mainly because the $\underset{0}{c_\alpha}$ are absent. The kind of result one gets may be illustrated by the coefficient of primary chromatic third-order spherical aberration:

$$\underset{1}{p_1^*} = (s-m)\{4\underset{1}{p_1}-3[4(s\underset{1}{c_1}+m\underset{1}{c_2}+\underset{1}{\kappa})\underset{0}{p_1}+m(\underset{1}{c_1}+\underset{1}{c_2})\underset{0}{p_2}]\}. \tag{106.12}$$

It should always be borne in mind that in the various equations above we have set $f = 1$ throughout.

There is not much point in continuing to write down equations relating to the most general circumstances, for in practice one is not likely to have to deal with chromatic coefficients of higher order unless the system is at least paraxially highly achromatized ($c_1 \approx c_2 \approx 0$). At any rate, the relations between the characteristic coefficients $\underset{m}{t_{\mu\nu}^{(n)}}$ and the effective coefficients have been dealt with at sufficient length to show what is involved.

We can now use exactly the method of Section 44 to study the effects of stop shifts. We only need to observe that the position of the stop enters into $T$ only implicitly through the presence of $s$ in the relations between $\xi$, $\eta$, $\zeta$ on the one hand and $\bar{\xi}, \bar{\eta}, \bar{\zeta}$ on the other. That $\bar{s}, \overline{m}$ now replaces $s, m$ has no relevant consequences whatsoever, bearing in mind that $\kappa$ will not occur in $q$, which continues to be given by (44.4). It follows that all equations such as (35.13), (44.8) and (44.10) may be taken over as they stand. Since all of these are linear in the various coefficients, any such equation decomposes into a set of similar equations relating to chromatic orders 1, 2, .... It only remains to add the equations for coordinate order 1, and they are

$$\hat{c}_1 = c^2 c_1, \quad \hat{c}_2 = c(c_2 + qc_1), \quad \hat{c}_3 = c_3 + 2qc_2 + q^2 c_1. \qquad (106.13)$$

Moreover, the investigations of Chapter 5D concerning invariant and semi-invariant aberrations evidently extend to the chromatic coefficients without effective modification.

In view of the complexity of the relations between the characteristic and effective coefficients it is well worth while to spend a little time on considering the effects of stop shifts on the displacement by a direct method, i.e. one which does not explicitly involve the characteristic coefficients at all. Taking (14.27) and (35.10) into account, a quite elementary geometrical argument yields the relation

$$\mathbf{y}' = c\hat{\mathbf{y}}' + q(\mathbf{y}_1 + \hat{\boldsymbol{\epsilon}}'); \qquad (106.14)$$

where $\mathbf{y}'$ now stands for the quantity hitherto denoted by $\mathbf{y}'_e$.

Recall that we can here write $\boldsymbol{\epsilon}'$ or $\hat{\boldsymbol{\epsilon}}'$ for the displacement, whichever is the more convenient, and in (106.14) we made the second choice; cf. the discussion following equation (35.11), as well as that leading to equation (41.3). Now, paraxially, writing $c_1^*, \bar{c}_1^*$ for the effective coefficients,

$$\hat{\boldsymbol{\epsilon}}' = \hat{c}_1^* \hat{\mathbf{y}}' + \hat{\bar{c}}_1^* \mathbf{y}_1 \equiv \boldsymbol{\epsilon}' = c_1^*[c\hat{\mathbf{y}}' + q(\mathbf{y}_1 + \hat{c}_1^* \hat{\mathbf{y}}' + \hat{\bar{c}}_1^* \mathbf{y}_1)] + \bar{c}_1^* \mathbf{y}_1, \qquad (106.15)$$

where (106.14) has been used on the right. It follows at once that

$$\hat{c}_1^* = \frac{cc_1^*}{1 - qc_1^*}, \quad \hat{\bar{c}}_1^* = \frac{\bar{c}_1^* + qc_1^*}{1 - qc_1^*}. \qquad (106.16)$$

The primary and secondary chromatic relations therefore are

$$\underset{1}{\hat{c}_1^*} = c\underset{1}{c_1^*}, \quad \underset{1}{\hat{\bar{c}}_1^*} = \underset{1}{\bar{c}_1^*} + q\underset{1}{c_1^*}, \tag{106.17}$$

$$\underset{2}{\hat{c}_1^*} = c(\underset{2}{c_1^*} + q\underset{1}{c_1^{*2}}), \quad \underset{2}{\hat{\bar{c}}_1^*} = \underset{2}{\bar{c}_1^*} + q(\underset{2}{c_1^*} + \underset{1}{c_1^*}\underset{1}{\bar{c}_1^*}) + q^2\underset{1}{c_1^{*2}}. \tag{106.18}$$

As a matter of fact, to obtain even the expression for $\underset{2}{\hat{c}_1^*}$ just given, starting from $\underset{2}{c_1^*}$ as it appears in (106.11) as the factor multiplying $\mathbf{y}_e'$ there, is quite a laborious task; compared with which the work involved in the method just used is almost negligible.

As already remarked, it will generally be good enough to consider the non-paraxial coefficients only when $\underset{1}{c_1}$ and $\underset{1}{c_2}$ can be taken as effectively zero; and then the effects of stop shifts on the primary chromatic third-order coefficients are just as simple as those on the monochromatic third-order coefficients.

## 107. Offence against the sine-condition

It is natural to inquire how one can incorporate chromatic effects into our previous investigations of total circular coma in Chapter 5A. Fortunately it turns out that the 'monochromatic equations' are of a kind almost ideally suited to the present context. To begin with, we take the aberration function $v$ to be modified in the sense implied by equation (105.1). Further, equation (100.5) is unnecessarily clumsy: it is much better here to incorporate $v^{(1)}$ in the remaining terms on the right, so that we again have (26.1), but now

$$S(\xi) = \sum_{n=0}^{\infty} s_m \xi^{n+1}, \quad C(\xi) = \sum_{n=0}^{\infty} \bar{c}_n \xi^n, \tag{107.1}$$

i.e. the summations extend from $n = 0$ rather than $n = 1$, as they did in (26.2). Also, we have written $\bar{c}_n$ in place of $c_n$ to avoid confusion with the constants which occur in (103.1). Thus

$$s_0 = \tfrac{1}{2}c_1, \quad \bar{c}_0 = c_2, \tag{107.2}$$

and the $s_n$, $\bar{c}_n$ of course all depend upon $\omega$. This incorporation of the paraxial terms in the spherical and circular comatic aberration functions is quite consistent with our general classification of Section 22, for the types of aberrations governed by $c_1$ and $c_2$ clearly have the required character. In particular, $c_2$ is a comatic coefficient, for the zonal curve, which is in fact a single point, lies asymmetrically with respect to the ideal image point.

Granted, then, that every displacement under discussion is understood to include the part induced by $v^{(1)}$, everything remains essentially as before. The only formal change is the replacement of $\boldsymbol{\beta}$ throughout by $\kappa\boldsymbol{\beta}$, which entails that in relations such as (31.5) $m$ must be replaced by $m/\kappa$. The central result of Section 31 then retains its previous significance, i.e. $Kh'$ is the exact total sagittal circular comatic displacement for whatever colour $\omega$ to which the various quantities in (31.5) relate. This conclusion is not entirely trivial. It is true that we could use (31.5) as it stands for any colour $\omega$ we please, but in so doing we would be considering the displacement in the variable ideal image plane corresponding to $\omega$ rather than in a fixed image plane.

For the sake of illustration suppose that, with an appropriate choice of $\underset{0}{\lambda}$, $K$ is so designed that paraxial rays of the colour $\lambda_1$ unite in $I'$. This means that one has complete paraxial simple achromatism: $c_1(\omega_1) = c_2(\omega_1) = 0$. At these colours $\underset{0}{\lambda}$ and $\lambda_1$ the values of the comatic displacements then differ from each solely on account of the variation with $\omega$ of the third-, fifth-, ... order terms of $S$ and $C$; but at intermediate colours the effects of residual (paraxial) chromatism are correctly taken into account. Again, from a slightly different point of view, one has quite generally

$$K = \underset{0}{K} + w\underset{0}{J}(\underset{1}{C} - 2\xi \underset{0}{R}\underset{0}{J}^2\underset{0}{P}\underset{1}{\dot{S}})\,\omega + \dots, \tag{107.3}$$

and one will achieve a high state of chromatic correction of the circular comatic asymmetry to the extent that one can cause the factor multiplying $\omega$ to vanish.

## 108. Reversibility and concentricity

A moment's reflection shows that the angle characteristic of a reversible symmetric system $K$ must satisfy the identity (54.2) for all colours, bearing in mind that $\kappa = 1$ here. Paraxially one has

$$T - d^*\alpha = \tfrac{1}{2}c_1\xi + c_2\eta + \tfrac{1}{2}c_3\zeta + \tfrac{1}{2}d^*\bar{\zeta}.$$

Using (15.11) the equality of the coefficients of $\bar{\xi}$ and $\bar{\zeta}$ immediately gives the relation

$$(1-s^2)\,c_1+2(1-sm)\,c_2+(1-m^2)\,c_3 = 0. \qquad (108.1)$$

For general values of $s$ and $m$ this is not very interesting on account of the presence of the coefficient $c_3$ which does not enter into the paraxial pseudo-displacement. Only when $m^2 = 1$ does one get a useful result, namely when $m = \pm 1$:

$$(1 \pm s)\,c_1+2c_2 = 0. \qquad (108.2)$$

In particular, when the system is completely reversible the (characteristic) coefficient of transverse chromatic aberration vanishes for all values of $\omega$:

$$\underset{m}{c_2} = 0 \quad (m = 1, 2, \ldots). \qquad (108.3)$$

The exact paraxial displacement is then

$$\boldsymbol{\epsilon}_1' = 2c_1(1+c_1-c_3-2c_1c_3)^{-1}[(1-c_3)\,\mathbf{y}_e'+c_3\mathbf{y}_1]. \qquad (108.4)$$

With regard to the non-paraxial coefficients, reversibility imposes precisely all the limitations described in Chapter 6A. (56.4) and (59.4), in particular, are the relations between the coefficients of coordinate order three, and these split up into the corresponding relations for every chromatic order, taken separately. The precise limitations imposed upon, say, the primary chromatic third-order *effective* coefficients are, of course, of a rather complicated kind. However, this is a somewhat esoteric problem, since in practice one will be confronted with more specialized situations in which certain coefficients, such as $\underset{1}{c_1}$ or $\underset{1}{c_2}$, are already known to be of negligible magnitude; and then the work becomes much more tractable.

We go on to consider the concentric system. It is obvious that the argument which led to the appropriate generic form (65.11) of $T$ can be taken over into the present context effectively unaltered. Of course, the undetermined function of $\chi$ will in general depend now also upon $\omega$, and we write it as $\nu G(\chi, \omega)$, the factor $\nu$ being inserted once again merely for convenience. Then

$$T = \nu[G(\chi, \omega)+q'\kappa^{-1}(1-\bar{\xi})^{\frac{1}{2}}+q(1-\bar{\zeta})^{\frac{1}{2}}]. \qquad (108.5)$$

One evidently has the usual paucity of independent aberration coefficients in every (coordinate and chromatic) order: there is only one, namely $\underset{m}{G_n}$. As an elementary consequence of this (taking the case $\nu = \nu'$ only), if $K$ produces an ideal image of an axial object point irrespectively of colour, then all chromatic defects of whatever kind will be absent.

From (106.4), (106.6), (65.12) and (65.14) we have, paraxially,

$$c_1\xi + 2c_2\eta + (c_3 + m^{-1})\zeta = G_1(\omega)(\bar{\xi} - 2\bar{\eta} + \bar{\zeta}) - q'\kappa^{-1}\bar{\xi} - q\bar{\zeta}. \quad (108.6)$$

We can now express $c_1, c_2, c_3$ in terms of $G_1$. Thus, concisely,

$$\begin{aligned} c_1 &= (\kappa - m)^2\,\Gamma, \\ c_2 &= -(\kappa - m)(\kappa - s)\,\Gamma, \\ c_3 &= (\kappa - s)^2\,\Gamma - (\kappa - 1)/(\kappa - m), \end{aligned} \quad (108.7)$$

where $$(s - m)^2\,\Gamma = G_1 - (1 - m)/(\kappa - m). \quad (108.8)$$

From these we can determine the exact effective paraxial coefficients. In particular, if $D$ is the discriminant of the pair of equations (106.7), it turns out that

$$\bar{c}_1^* = D^{-1}\kappa(\kappa - m)(s - 1)\,\Gamma. \quad (108.9)$$

Unlike the characteristic coefficient $c_2$, the effective coefficient $\bar{c}_1^*$ vanishes when $s = 1$. This is as it should be, for the general result stated just after equation (67.5), namely, that when the stop is central all barred effective coefficients vanish, applies also to the chromatic coefficients of all types.

## 109. Remark on the so-called $D - d$ method

We saw in Section 37 that the spherical point characteristic $V^\dagger$ had one great advantage over $V$, namely, that the displacement, given by

$$\boldsymbol{\epsilon}' = X'\frac{\partial V^\dagger}{\partial \mathbf{y}'} = X'\frac{\partial v^\dagger}{\partial \mathbf{y}'}, \quad (109.1)$$

is simply the derivative of $v^\dagger$ to within a factor which, in all but the most extreme cases, is virtually constant for a given object point; reflecting the near-constancy of the optical distance between $O$ and points of $\mathscr{W}_0$. It is therefore of advantage to use $V^\dagger$ now in the context of chromatic aberrations, though the following discussion would not alter in essence if it proceeded in terms of $V$. (109.1) continues to apply, but $v^\dagger$ will now contain paraxial terms of chromatic origin. (In order not to encumber the discussion unnecessarily we are supposing that $N = N' = 1$.)

We may write

$$V^\dagger(\omega) = \underset{0}{V^\dagger} + \omega \underset{1}{V^\dagger} + \omega^2 \underset{2}{V^\dagger} + \dots. \qquad (109.2)$$

For colours sufficiently close to $\underset{0}{\lambda}$ chromatic defects are contained in the term linear in $\omega$; so that in a small range of wavelengths about $\omega = 0$ we merely require to know the derivative

$$\underset{1}{V^\dagger} = (\partial V^\dagger / \partial \omega)_{\omega=0}. \qquad (109.3)$$

Should this happen to vanish for a certain range of values of the ray-coordinates then $K$ is simultaneously achromatized with respect to *all* types of aberrations which sensibly contribute to the monochromatic defects within the coordinate range in question.

Now let $O$, $D'$ be points in $\mathscr{I}$ and $\mathscr{W}_0$ respectively. (The base-surface $\mathscr{W}_0$ is exactly that of Section 37, and it is the same for all colours.) Further, let $\underset{0}{\mathscr{R}}$ and $\mathscr{R}_\omega$ be two rays through $O$ and $D'$, the chromatic coordinates having the values 0 and $\omega$ respectively. Then

$$V^\dagger(\omega) = \int N(\omega)\, ds(\omega), \quad V^\dagger(0) = \int N(0)\, ds(0), \qquad (109.4)$$

where $ds(\omega)$ and $ds(0)$ are elements of arc of $R_\omega$ and $\underset{0}{\mathscr{R}}$ respectively, and both integrals are extended from $O$ to $D'$. Hence

$$V^\dagger(\omega) - V^\dagger(0) = \int [N(\omega) - N(0)]\, ds(0) + \int N(\omega)\, ds(\omega) - \int N(\omega)\, ds(0). \qquad (109.5)$$

The last two terms on the right represent the difference between the integral of $N(\omega)$ taken along the actual ray $\mathscr{R}_\omega$ joining $O$ and $D'$ and the integral of the same function $N(\omega)$ along another curve (i.e. $\underset{0}{\mathscr{R}}$) joining the same points. When $\omega$ is sufficiently small the *curve* $\underset{0}{\mathscr{R}}$ is neighbouring to the *ray* $\mathscr{R}_\omega$, and, in view of Fermat's Principle (2.8), the two terms in question, taken together, must vanish. At the same time $\delta N = N(\omega) - N(0)$ becomes $\underset{0}{N}\nu_1\omega$, because of (101.1), so that finally

$$\underset{1}{V}^{\dagger} = \int \underset{0}{N}\nu_1 ds(0). \tag{109.6}$$

In practice the most common situation is that the ray in the course of its passage through $K$ encounters a number of refracting and reflecting surfaces. Some of its segments will be in air (to which we ascribe the refractive index $N = 1$) and some within homogeneous media (e.g. glasses). The length of the segment of $\underset{0}{\mathscr{R}}$ in a particular such medium is traditionally denoted by $D$, and then (109.6) may be written

$$\omega \underset{1}{V}^{\dagger} = \sum D\delta N, \tag{109.7}$$

the sum being taken over the various 'elements' of $K$. Since it is $v^{\dagger}$ rather than $V^{\dagger}$ which is relevant to the displacement, we may subtract out the irrelevant part $\underset{1}{g}^{\dagger}(\zeta)$ of $\underset{1}{V}^{\dagger}$. With the convention that $v^{\dagger}(0, 0, \zeta)$ is to be understood as having been absorbed in $g^{\dagger}(\zeta)$, one has

$$\omega \underset{1}{g}^{\dagger}(\zeta) = \sum d\delta N, \tag{109.8}$$

where $d$ denotes the length of a segment of the ray through $E'$, $\omega$ of course being zero for this ray. Then

$$\omega \underset{1}{v}^{\dagger} = \sum (D-d)\,\delta N. \tag{109.9}$$

We denote the sum on the right by $\Omega$. Its use is very popular on account of the ease with which it may be calculated; and its form explains the reason for the name '$(D-d)$-methods' for procedures of chromatically correcting systems on the basis of information gained from evaluating $\Omega$ for a number of rays.

In practice it has been customary to take for $\delta N$ the actual value of $N - \underset{0}{N}$ as given by glass tables, the fact that $\omega$ may not be 'small' (less than 0.1, say) being disregarded. In so doing one calculates a quantity which differs from

$$\Omega = \omega \sum (D-d) \underset{0}{N}\nu_1 \tag{109.10}$$

by an amount whose dominant term is (at least) quadratic in $\omega$. This difference will, however, be very small, just because $\omega$ was so defined as to lead to the rapid convergence of the series (101.1), which certainly ensures that $|\nu_1\omega| \gg |\nu_2\omega^2|$. Granted this, the use of the exact values of $\delta N$ implies the abandonment of consistent definitions in the context of aberration coefficients, without any corresponding gain with respect to *exact* inferences which may be drawn from the calculated values of $\Omega$. The reason for this state of affairs is that in any event we can only assert that

$$\Omega = \omega \underset{1}{v}^{\dagger} + O(2_2), \tag{109.11}$$

on account of the last two terms on the right of (109.5) which were rejected. These contribute an amount which itself varies with the second and higher powers of $\omega$: and that is all we know of it. In short, we can, in general, draw no quantitative inferences about $\underset{2}{v}^{\dagger}$ from a knowledge of $\Omega$ alone.

It may be apposite to mention at this point that one occasionally comes across claims which seem to contradict what has just been said. These are usually supported by evidence which revolves about the results of computations relating to systems actually encountered in practice. The reason why the latter fail to expose the fallacies involved is that, as the result of much empirical experience, systems are deliberately designed from the outset in such a way that rays of different colours, connecting the same pairs of points, will not differ widely from each other anywhere in the system. In other words, one aims, *by prescription*, at keeping down the value of the sum of the terms rejected in (109.5).

A much-used application of the $(D-d)$-sum may be illustrated by considering a single axial ray, which we may take to be meridional. When $\omega = 0$ the total displacement of this ray is $\underset{0}{\epsilon}'(y')$. Another ray through $O$ with the same value of the coordinate $y'$, but this time having the colour $\omega$, will have the same displacement as the first ray if

$$\dot{V}^{\dagger}(y', \omega) - \dot{V}^{\dagger}(y', 0) = 0,$$

where a dot denotes differentiation with respect to $y'$. Retaining only the chromatically dominant term, i.e. that linear in $\omega$, achromatism therefore obtains at $y'_b$, say, when

$$\dot{\Omega}(y'_b) = 0. \tag{109.12}$$

Suppose now that actual calculation shows $\Omega$ to be zero for the ray having $y' = y'_a$, say. Then, since $\Omega$ vanishes at $y' = 0$ and at $y' = y'_a$, its derivative must vanish somewhere between $y'_a$ and 0, i.e. there must exist a constant $\theta$ between 0 and 1 such that

$$\dot{\Omega}(\theta y'_a) = 0. \qquad (109.13)$$

The vanishing of $\Omega$ at $y'_a$ therefore ensures that $K$ *is axially achromatized for the intermediate zone* $y'_b = \theta y'_a$; but, in general, as regards the actual value of $\theta$, nothing more can be said.

The specialization to axial rays is unnecessary; everything goes through equally well when $y_1 \neq 0$. The same is, however, not true for a completely general ray, which is, perhaps, not very surprising. The vanishing of the primary chromatic displacement now requires

$$\Omega_{y'}(y'_b, z'_b) = 0, \quad \Omega_{z'}(y'_b, z'_b) = 0, \qquad (109.14)$$

and these replace the single condition (109.12). (The fixed coordinates $\mathbf{y}_1$ have been left understood.) On the other hand, if $\Omega(y'_a, z'_a) = 0$ we can only infer that there exists a constant $\theta$ between 0 and 1 such that

$$y'_a \Omega_{y'}(\theta y'_a, \theta z'_a) + z'_a \Omega_{z'}(\theta y'_a, \theta z'_a) = 0. \qquad (109.15)$$

From this (109.14) does not follow, so that in principle no property of achromatism can be inferred from the vanishing of $\Omega$, calculated for a *single* skew ray.

## Problems

**P.10 (i).** Derive an expression for the exact paraxial displacement $\boldsymbol{\epsilon}'_1(\mathbf{y}'_e, |\mathbf{y}_1)$ when the object is at infinity, given equations (106.7–9).

**P.10 (ii).** Show that the primary chromatic central angle characteristic of a single spherical refracting surface of unit radius can be written in the form

$$\underset{1}{T} = \underset{0}{N'}\nu'_1[(\underset{0}{N'} - 1) + \chi]\underset{0}{T^{-1}}.$$

($N$ has been taken to be unity, and the qualification 'central' means that both base-points are taken at the centre of the sphere.)

**P.10 (iii).** Given the definition (101.4) of the chromatic coordinate, show that one can choose a new base-colour $\underset{0}{\lambda^*}$ and at the same time a new constant $\alpha^*$ in place of $\frac{5}{2}$ in the denominator in such a way that the new coordinate $\omega^*$ is simply a linear function of the old.

**P.10 (iv).** The monochromatic offence against the sine-condition of a certain system has been removed. Show that to a sufficient degree of approximation it will have been removed also for neighbouring colours if the condition

$$\underset{1}{C} = 2\xi \underset{1}{\dot{S}}$$

is satisfied.

CHAPTER II

# ANISOTROPY

## 110. Fermat's Principle and the propagation of light in anisotropic media

Hitherto we have always supposed the media in the object and image spaces to be isotropic, i.e. the refractive indices $N$ and $N'$ at points $A$ and $A'$ of these spaces were taken to be independent of the directions $\mathbf{e}$ and $\mathbf{e}'$ of the ray through $A$ and $A'$. On a number of occasions it was pointed out, however, that when one discusses the imagery of a given system $K$ merely upon the basis of a characteristic function whose generic form reflects the over-all symmetries of $K$, the possibility of $K$ being internally anisotropic is by no means excluded. In other words, at points within $K$, $N$ may very well be a function of the directions of the rays through them; so that it is then a function of the *six* variables $x, y, z, \alpha, \beta, \gamma$, the last three of which are connected by the identity $\alpha^2+\beta^2+\gamma^2 = 1$.

What has just been said can be meaningful only if a characteristic function can still be defined when light is propagated through anisotropic media. This being so it is incumbent upon us to inquire into the actual state of affairs. This chapter is, on the whole, directed mainly towards this end; for which reason it will deal with little more than general principles. It continues for the time being the simple, phenomenological kind of argument which was pursued in the early sections of this book; which is, after all, in keeping with the simplified picture of the propagation of light which geometrical optics presents.

The essentially new ingredient which has to be added to the basic picture presented in Chapter 1 is this: exceptional circumstances apart, the light—now supposed monochromatic—traversing a given system consists of two distinct parts, *which propagate independently of each other*. We arbitrarily distinguish the two kinds of rays and all quantities associated with them by the subscripts $A$ and $B$. In particular, we have two kinds of rays $\mathscr{R}_A$ and $\mathscr{R}_B$, which have their usual significance. For instance, the tangent to $\mathscr{R}_A$ at $P$ gives the direction in which $A$-light (i.e. radiant energy of the $A$-kind) is being transported at $P$; with an entirely analogous meaning of $\mathscr{R}_B$.

Now, as we know, if the propagation of light is governed by a variational principle, the existence of a point characteristic is immediately guaranteed. We are therefore led to ask whether, or in what form, Fermat's Principle survives in anisotropic media. The answer to this is quite simple: *each kind of ray is governed by an extremal principle exactly of the form* (2.8), granted that the optical medium as such is, in general, now defined by two distinct refractive index functions $N_A(x, \ldots, \gamma)$ and $N_B(x, \ldots, \gamma)$. (See also the end of Section 113.) Then the ray $\mathscr{R}_A$ joining two points $A$ and $A'$ is that curve for which

$$V_A^* = \int_A^{A'} N_A \, ds_A$$

is stationary in the sense explained at length in Section 2, whilst for $\mathscr{R}_B$

$$V_B^* = \int_A^{A'} N_B \, ds_B$$

is likewise stationary. (In special cases, e.g. in an electron-optical instrument with magnetic focusing, there may be only one kind of ray.) The optical distance between two points is not a function of these points alone, but depends also upon the kind of light with respect to which it is defined; but this is nothing new, for we encountered an analogous situation already in the context of chromatic effects.

Since the two optical distances $\tilde{V}_A(A, A')$ and $\tilde{V}_B(A, A')$ which are now in hand depend only on the coordinates of $A$ and $A'$, we recognize that there are now *two* point characteristics,

$$V_A(x', y', z', x, y, z) \quad \text{and} \quad V_B(x', y', z', x, y, z)$$

associated with $K$, one for each kind of light; and they jointly characterize completely the purely geometrical-optical behaviour of $K$. Evidently when we come to derive the fundamental equations corresponding to (3.5) we can simplify the notation by suppressing the indices $A$ and $B$ on most occasions. They are easily restored by inspection, should this be required. At any rate, we now see that all our previous work remains relevant even when $K$ is internally anisotropic, a state of affairs we had hitherto taken for granted: we merely need to understand the characteristic function in every case to stand alternatively for that appropriate to $A$-light and $B$-light. The basic equations (3.5), it should be recalled, have been used only in isotropic regions.

It will be noted that we have here adopted a course different from that followed in Section 2, in as far as we have now simply postulated the validity of Fermat's Principle, without first contemplating the refraction of light at the boundary between two homogeneous media. The reason for this is that the detailed, explicit rules which describe such refraction are very cumbersome: it seems better to consider the problem after accepting Fermat's Principle, and this we do very briefly in the next section.

## 111. Refraction at a boundary between homogeneous media

To deal with the principles of the refraction of light at the boundary of two homogeneous media one or other or both of which may be anisotropic, we need to consider the state of affairs to which Fig. 2.1 relates. The optical length $V^*(APA')$ is still given by (2.4), but upon varying $P$ we must now write

$$\delta V^* = \delta(N's') + \delta(Ns) = N'\delta s' + N\delta s + s'\delta N' + s\delta N, \quad (111.1)$$

since, in general, $\delta N$ and $\delta N'$ will differ from zero on account of the fact that the directions of the lines $AP$, $PA'$ differ from those of $AP_1$, $P_1A'$ respectively, where $P_1$ is the displaced point of incidence. If we define a vector $\mathsf{n}$ as

$$\mathsf{n} = \left(\frac{\partial N}{\partial \alpha}, \ \frac{\partial N}{\partial \beta}, \ \frac{\partial N}{\partial \gamma}\right), \quad (111.2)$$

(111.1) becomes

$$\delta V^* = N'\mathsf{e}'.\delta\mathsf{s}' + N\mathsf{e}.\delta\mathsf{s} + s'\mathsf{n}'.\delta\mathsf{e}' + s\mathsf{n}.\delta\mathsf{e}. \quad (111.3)$$

Consider merely the second and fourth terms on the right. We have $s\delta\mathsf{e} = \delta\mathsf{s} - \mathsf{e}\delta s = \delta\mathsf{s} - \mathsf{e}(\mathsf{e}.\delta\mathsf{s})$. The two terms in question become

$$N\mathsf{e}.\delta\mathsf{s} + s\mathsf{n}.\delta\mathsf{e} = [(N - \mathsf{n}.\mathsf{e})\mathsf{e} + \mathsf{n}].\delta\mathsf{s}. \quad (111.4)$$

At this stage we recall that the variables $\alpha$, $\beta$, $\gamma$ are not mutually independent. Indeed, with the aid of the identity $\alpha^2 + \beta^2 + \gamma^2 = 1$

which holds between them, we can always arrange $N$ to become a homogeneous function of any desired degree $q$ in $\alpha$, $\beta$, $\gamma$. The choice $q = 1$ entails just that

$$\mathbf{e}.\mathbf{n} = N, \tag{111.5}$$

which is a convenient result. In future we therefore understand that *all refractive indices are to be written as homogeneous functions of degree* 1 *in* $\alpha, \beta, \gamma$. (111.3) then becomes

$$\delta V^* = \mathbf{n}'.\delta\mathbf{s}' - \mathbf{n}.\delta\mathbf{s}. \tag{111.6}$$

Proceeding now as in Section 2 we obtain the laws of refraction in the form

$$\mathbf{n}' - \mathbf{n} = \sigma\mathbf{p}, \tag{111.7}$$

where $\sigma$ is some scalar factor, since, in view of Fermat's Principle, $\delta V^*$ must vanish for all small displacements of $P$, which leave this point in the boundary.

For an isotropic medium of constant refractive index $N_i$, say, we have to take

$$N = N_i(\alpha^2 + \beta^2 + \gamma^2)^{\frac{1}{2}}, \tag{111.8}$$

and then

$$\mathbf{n} = N_i\mathbf{e}(\alpha^2 + \beta^2 + \gamma^2)^{-\frac{1}{2}} = N_i\mathbf{e}, \tag{111.9}$$

and with this (111.7) reduces to (2.6). The simplest formal anisotropic counterpart of (111.8) is the case in which $N^2$ is a quadratic form in the components of $\mathbf{e}$. (By 'formal' we mean that we need not necessarily be able to find a physical realization of such a refractive index.) Writing $e_1, e_2, e_3$ in place of $\alpha, \beta, \gamma$ whenever convenient, we have

$$N = \left(\sum_{k,\,l=1}^{3} a_{kl}e_k e_l\right)^{\frac{1}{2}}, \tag{111.10}$$

where the $a_{kl}(\equiv a_{lk})$ are a set of six constants such that $N^2 > 0$ for all values of the $e_k$. Now the laws of refraction are

$$N'^{-1}\sum_l a'_{kl}e'_l - N^{-1}\sum_l a_{kl}e_l = \sigma p_k, \tag{111.11}$$

where $\mathbf{p} = (p_1, p_2, p_3)$. It is always possible to make a rotation of coordinates so that after the rotation the constants in (111.10) have the values

$$a_{kl} = \begin{cases} a_k & (k = l) \\ 0 & (k \neq l), \end{cases} \tag{111.12}$$

where the $a_k$ are non-negative; but this can in general be done for only one of the media to which (111.11) refers. As a specially simple case, let $N$ be given by (111.10), whilst the second medium shall be a vacuum, so that $N' = (\alpha'^2+\beta'^2+\gamma'^2)^{\frac{1}{2}}$. With the coordinates so chosen that (111.12) obtains, we therefore have

$$e'_k - N^{-1}a_k e_k = \sigma p_k, \qquad (111.13)$$

which expresses $\mathbf{e}'$ directly as a function of the components of $\mathbf{e}$ and $\mathbf{p}$. Moreover, it shows particularly clearly that $\mathbf{e}$, $\mathbf{e}'$ and $\mathbf{p}$ will in general not be coplanar, for otherwise there would have to be constants $q$ and $q'$ such that the left-hand member of (111.13)—or, more generally, of (111.11)—takes the form $q'\mathbf{e}' - q\mathbf{e}$.

As a matter of fact (111.10) is far from academic. The so-called *uniaxial crystal* has the refractive indices

$$N_A = b^{\frac{1}{2}}(\alpha^2+\beta^2+\gamma^2)^{\frac{1}{2}}, \quad N_B = [b\alpha^2+a(\beta^2+\gamma^2)]^{\frac{1}{2}}, \qquad (111.14)$$

where $a$ and $b$ are positive constants, and the axes have been appropriately chosen. Evidently the crystal behaves as an isotropic medium for $A$-rays. It will be noted that $N_B$ is invariant under rotations about the $x$-axis, and also under reflections in any plane containing this axis: so that we have a kind of counterpart to the refractive index function of an isotropic inhomogeneous symmetric system. A particularly simple situation exists when the boundary at which the refraction described by (111.13) occurs is a plane normal to the $x$-axis. Then (111.13) becomes, for the $B$-ray,

$$\alpha' - (b/N)\alpha = \sigma, \quad \boldsymbol{\beta}' = (a/N)\boldsymbol{\beta}. \qquad (111.15)$$

If the incident and refracted rays make angles $\phi$, $\phi'$ with the $x$-axis respectively one finds very easily that

$$\sin\phi' = a[b+(a-b)\sin^2\phi]^{-\frac{1}{2}}\sin\phi. \qquad (111.16)$$

Further, in this special case $\mathbf{e}$, $\mathbf{e}'$ and $\mathbf{p}$ are coplanar. Thus, the angle between the normals of the planes of incidence and refraction vanishes if and only if $\mathbf{e}\times\mathbf{e}'.\mathbf{p} = 0$, which implies here merely the condition that $\beta\gamma' - \beta'\gamma = 0$, and this is satisfied.

These simple remarks will suffice to show how one can deal with the passage of rays through homogeneous anisotropic media by means which do not differ in principle from those to be used when isotropy obtains.

## 112. The point characteristics

It has already been remarked that when the optical medium is anisotropic one is, in general, confronted with two point characteristics $V_A$ and $V_B$. The time has come to derive equations corresponding to (3.4) and (3.5) for them. However, according to our previous agreement, we can deal with both functions at the same time by just omitting the subscripts $A$ and $B$, and understanding every equation in which $V$ occurs really to stand for two such similar equations with the indices $A$ and $B$ supplied in the appropriate places.

An elementary argument entirely similar to that pursued in Section 3 can be used to obtain the required result. The construction to which Fig. 2.1 relates can be taken over unchanged, and everything remains the same up to and including equation (3.2). The pair of equations following this is, however, no longer valid. The reason is this: the optical distances between $B$ and $A$ on the one hand and between $B$ and $C$ on the other ($C$ being the foot of the normal from $A$ on to $\mathscr{S}_1$) were previously taken to be equal. This is no longer the case since the lines $BA$ and $BC$ have different directions. In fact, in the notation of equation (111.2), if $BA = l$,

$$\tilde{V}(B, C) - \tilde{V}(B, A) = l\mathsf{n}\,.\,\delta\mathsf{e} = \mathsf{n}\,.\,(\delta\mathsf{s} - \mathsf{e}\delta l),$$

$$= \mathsf{n}\,.\,[\delta\mathsf{s} - (\mathsf{e}\,.\,\delta\mathsf{s})\,\mathsf{e}]. \qquad (112.1)$$

There comes

$$\tilde{V}(B, A_1) - \tilde{V}(B, A) = [\mathsf{n} + [N - (\mathsf{n}\,.\,\mathsf{e})]\,\mathsf{e}]\,.\,\delta\mathsf{s}. \qquad (112.2)$$

Evidently we have done little more than to re-derive equation (111.4). We now recall our convention that *all refractive indices are to be written as homogeneous functions of degree 1 in the direction cosines.* The right-hand member of (112.2) then reduces to $\mathsf{n}\,.\,\delta\mathsf{s}$. Accordingly, dealing in an entirely similiar way with the difference $\tilde{V}(C', B') - \tilde{V}(A_1', B')$, we obtain in place of (3.4)

$$\delta V = \mathsf{n}'\,.\,\delta\mathsf{s}' - \mathsf{n}\,.\,\delta\mathsf{s}. \qquad (112.3)$$

The *basic equations of Hamiltonian optics*, including now propagation in anisotropic media, are therefore

$$\left.\begin{aligned} \frac{\partial N'}{\partial \alpha'} = \frac{\partial V}{\partial x'}, \quad \frac{\partial N'}{\partial \beta'} = \frac{\partial V}{\partial y'}, \quad \frac{\partial N'}{\partial \gamma'} = \frac{\partial V}{\partial z'}, \\ \frac{\partial N}{\partial \alpha} = -\frac{\partial V}{\partial x}, \quad \frac{\partial N}{\partial \beta} = -\frac{\partial V}{\partial y}, \quad \frac{\partial N}{\partial \gamma} = -\frac{\partial V}{\partial z}. \end{aligned}\right\} \tag{112.4}$$

These equations may be resolved algebraically for the components of $\mathbf{e}'$ and $\mathbf{e}$. Then the equations $\mathbf{e}'.\mathbf{e}' = 1$ and $\mathbf{e}.\mathbf{e} = 1$ yield a pair of non-linear first-order partial differential equations which $V$ must satisfy. For example, if a homogeneous medium has a refractive index given by the second member of (111.14),

$$b^{-\frac{1}{2}}V_x = -b^{\frac{1}{2}}\alpha/N, \quad a^{-\frac{1}{2}}V_y = -a^{\frac{1}{2}}\beta/N, \quad a^{-\frac{1}{2}}V_z = -a^{\frac{1}{2}}\gamma/N. \tag{112.5}$$

Squaring and adding these, one obtains the equation

$$\frac{1}{b}\left(\frac{\partial V}{\partial x}\right)^2 + \frac{1}{a}\left[\left(\frac{\partial V}{\partial y}\right)^2 + \left(\frac{\partial V}{\partial z}\right)^2\right] = 1; \tag{112.6}$$

and there is a similar equation in which the coordinates are primed. The two, taken together, are satisfied by the function

$$V = \{b(x'-x)^2 + a[(y'-y)^2 + (z'-z)^2]\}^{\frac{1}{2}}. \tag{112.7}$$

Given $x, y, z$, the surface $V = \text{const.}$ is evidently an ellipsoid of revolution.

It is possible to define other characteristic functions in much the same way as was done in the context of isotropy, though the details are more complicated. We do not consider them, however, since we shall have no occasion to use them.

## 113. Rays and normals

The consideration of surfaces of constant $V$ is of course in no way restricted to the special case (112.7). Given any characteristic function we may always consider the family of surfaces $V = \text{const.}$ regarding either $x, y, z$, or $x', y', z'$ as fixed. Merely from a notational point of view we contemplate the latter alternative, thereby avoiding the appearance of a large number of primes. At the same time we shall with advantage take primed and unprimed variables to refer to object and image space respectively. (This notation would have been better all along in any event—it was the one originally adopted by Hamilton—but tradition dictated otherwise.) If $P(x, y, z)$ is any point, *two* directions of interest are associated with it, namely (i) the unit normal $\mathsf{u}$ at $P$ to the surface $\mathscr{W}$ defined by

$$V(\bar{x}, \bar{y}, \bar{z}) = V(x, y, z); \tag{113.1}$$

and (ii) the direction $\mathsf{e}$ of the ray through $P$ and the fixed point $O'(x', y', z')$. What has to be constantly borne in mind is that $\mathsf{e}$ *and* $\mathsf{u}$ *are, in general, distinct*. The only exception occurs when $N$ has the form appropriate to an isotropic medium (recall equations (111.14)):

$$N = N_i(\bar{x}, \bar{y}, \bar{z})(\alpha^2 + \beta^2 + \gamma^2)^{\frac{1}{2}}. \tag{113.2}$$

We are of course talking about full (i.e. two-parameter) pencils of rays through $O'$; at certain points of $\mathscr{W}$ $\mathsf{e}$ and $\mathsf{u}$ may happen to coincide. More formally, since, according to (112.4),

$$\mathsf{n} = n\mathsf{u}, \tag{113.3}$$

where $n$ is the magnitude of $\mathsf{n}$, $\mathsf{u}$ and $\mathsf{e}$ coincide if and only if

$$\mathsf{e} \times \mathsf{n} = 0, \tag{113.4}$$

and if no restriction be placed upon the values of $\beta$ and $\gamma$ this condition is satisfied only when (113.2) obtains. As a somewhat trivial example of the coincidence of $\mathsf{u}$ and $\mathsf{e}$ for isolated rays we infer from (112.5) that in the case of the uniaxial crystal ($N_B$ given by (111.14)), $\mathsf{e} \times \mathsf{u} = 0$ only if either $\alpha = 0$ or $\alpha = 1$.

We return briefly to the law of refraction (111.7). We may write it as

$$n'\mathsf{u}' - n\mathsf{u} = \sigma\mathsf{p}. \tag{113.5}$$

If desired this may be regarded as governing in the first instance the refraction of the normal $\mathsf{u}$, rather than of the ray $\mathsf{e}$. In that case the angles $I_n$, $I'_n$ of incidence and refraction of the normal *formally* obey Snell's Law:

$$n' \sin I'_n = n \sin I_n, \tag{113.6}$$

whilst $\mathsf{u}$, $\mathsf{u}'$ and $\mathsf{p}$ are coplanar. However, the apparent refractive index $n$ (sometimes called the *wave-index*) depends upon the direction of the ray, and the simplicity of (113.6) is deceptive. The quantity $N$, here simply called 'refractive index', is more often referred to as the '*ray-index*'. Our terminology is appropriate to geometrical optics, bearing in mind, on the one hand, the central position which $N$ occupies as the function to be varied in the basic integral $\int N ds$, and, on the other, that one is ultimately interested in the direction of transport of light rather than in formal constructions.

## 114. Remark on the occurrence of anisotropy

Although it is not our purpose to study the details of the propagation of light in anisotropic media, it is convenient, for the sake of orientation, to remark upon situations in which anisotropy is encountered in practice, and to comment upon the form of the functions $N_A$ and $N_B$. We have already had one special example, i.e. that of the homogeneous medium to which (111.14) relates. This may be expected to be a degenerate case of a more general medium, on account of the fact that only two constants occur in $N_A$ and $N_B$ together. This is indeed the case. Thus, the so-called *biaxial crystal* is a homogeneous medium for which

$$\left.\begin{matrix} N_A^2 \\ N_B^2 \end{matrix}\right\} = \tfrac{1}{2}\{(b+c)\alpha^2+(c+a)\beta^2+(a+b)\gamma^2 \pm [(b-c)^2\alpha^4+(c-a)^2\beta^4 + (a-b)^2\gamma^4 - 2(c-a)(a-b)\beta^2\gamma^2 - 2(a-b)(b-c)\gamma^2\alpha^2 - 2(b-c)(c-a)\alpha^2\beta^2]^{\frac{1}{2}}\}. \quad (114.1)$$

Here $a$, $b$, $c$ are positive constants, and we may suppose that

$$a \geqslant b \geqslant c. \quad (114.2)$$

This can always be achieved by choosing an appropriate orientation of the coordinate axes. In any event, it is important to realize that the comparative simplicity of (114.1) may be destroyed by allowing the coordinates to undergo some general rotation. In that case the expression in square brackets for instance becomes a more general quartic form; see equation (115.33).

When $b = c$ (114.1) is easily seen to reduce exactly to (111.14). Also, we note that $N_A = N_B$ for those rays for which the square root in (114.1) vanishes. One most conveniently replaces $\alpha^2$ by $1-\beta^2-\gamma^2$ in the latter, and it then turns out that the rays in question have

$$\mathsf{e} = \left([(a-b)/(a-c)]^{\frac{1}{2}},\quad 0,\quad \pm[(b-c)/(a-c)]^{\frac{1}{2}}\right). \quad (114.3)$$

These directions, of which there are two, define the so-called *optical* (*ray-*) *axes* of the crystal; and this explains the qualification 'biaxial'.

So much for homogeneous media. A prominent example of an inhomogeneous anisotropic 'medium' is that of the electron-optical system, as has been mentioned before. As with light optics, the geometrical theory can be used in circumstances such that diffraction effects are negligible. For scalar electrons there will certainly be only one refractive index; and when axial symmetry of the system obtains, it has the generic form

$$N = (\alpha^2+\beta^2+\gamma^2)^{\frac{1}{2}} N^{\#}(\bar{x}, \bar{y}^2+\bar{z}^2)+(\gamma\bar{y}-\beta\bar{z})\tilde{N}^{\#}(\bar{x}, \bar{y}^2+\bar{z}^2), \tag{114.4}$$

in a notation closely adapted to that of equation (72.1). The 'skew refractive index' $\tilde{N}^{\#}$ vanishes in the absence of a magnetic field, whereas the absence of the electric field entails the constancy of the 'normal refractive index' $N^{\#}$. The equations for $V$ are

$$V_x^2+V_y^2+V_z^2+2(zV_y-yV_z)\tilde{N}^{\#}+(y^2+z^2)\tilde{N}^{\#2} = N^{\#2}, \tag{114.5}$$

together with that in which all variables are primed. The presence of the factor $zV_y-yV_z$ reflects the necessity for the explicit occurrence of the skew-symmetric invariant $\tau$ in (72.1).

More complex instances of anisotropy arise for instance when optical media are placed into electrostatic fields, or are subjected to mechanical stresses. When an initially optically isotropic medium —optical isotropy does not necessarily require that the medium be amorphous, i.e. non-crystalline—is elastically deformed, it will in general become anisotropic. If the deformation is non-uniform the medium will then no longer be optically homogeneous or isotropic. Still, in principle one need only know the functions $N_A$ and $N_B$: once they are available one can hope to gain insight into the general optical behaviour of the system in question without having to bother with cumbersome and in the first place irrelevant detail. The material of Chapter 7 is a good illustration of this remark. There the generic form of (72.1), which could also be taken to be a direct consequence of (114.4), told us a good deal in a very simple way about the imagery associated with axially symmetric electron-optical systems; these being merely special semi-symmetric systems. At the same time we were not overwhelmed by having to consider the very complex relationship which exists between the explicit form of $V$ of any particular system and the disposition of electric and magnetic fields within it.

## 115. Considerations relating to Maxwell's equations

With the preceding section we have essentially completed our treatment of the generalities of Hamiltonian optics. It may, however, not come amiss to deal very briefly with the relation between geometrical optics and the electromagnetic equations, if for no other reason than to amplify a little the remarks of Section 1. The material to be presented is to some extent out of keeping with the more phenomenological character of the rest of this work. For this reason it will be kept as brief as possible, partly by remaining on a mathematically superficial level, and partly by allowing ourselves to be satisfied with considerations of plausibility rather than with strict and detailed proofs. In this way we may hope to gain insight into the problem as easily as possible.

The propagation of electromagnetic fields through a non-absorbing, source-free medium is fully described by Maxwell's equations

$$\left.\begin{aligned} c\,\mathrm{curl}\,\mathsf{E} &= -\partial\mathsf{H}/\partial t, & c\,\mathrm{curl}\,\mathsf{H} &= \partial\mathsf{D}/\partial t,\\ \mathrm{div}\,\mathsf{H} &= 0, & \mathrm{div}\,\mathsf{D} &= 0. \end{aligned}\right\} \qquad (115.1)$$

It has been assumed that the magnetic permeability is unity, so that no distinction need be drawn between $\mathsf{B}$ and $\mathsf{H}$. To make (115.1) determinate one still has to add an equation giving the relation between $\mathsf{D}$ and $\mathsf{E}$. In general this relation may be non-linear, but we shall suppose this not to be the case. Then one has

$$D_k = \epsilon_{kl}(x, y, z)\, E_l, \qquad (115.2)$$

where the $\epsilon_{kl}$ are *six* given functions, for one can show that $\epsilon_{kl}$ must be identically equal to $\epsilon_{lk}$. Note that we are using the summation convention: if an index is repeated in any expression, summation over that index over the range 1, 2, 3 is understood.

Let the radiation be monochromatic, wavelength $\lambda$. The $\epsilon_{kl}$, which may be functions of the wavelength, are to have the values appropriate to $\lambda$. Then write

$$\mathsf{D} = \mathsf{D}_0(x, y, z)\,\Phi, \quad \mathsf{E} = \mathsf{E}_0(x, y, z)\,\Phi, \quad \mathsf{H} = \mathsf{H}_0(x, y, z)\,\Phi, \qquad (115.3)$$

with

$$\Phi = \exp\left[(2\pi i/\lambda)(V - ct)\right], \qquad (115.4)$$

where $V$ is some function of $x$, $y$, $z$ yet to be determined. For the purpose of orientation, consider the last of equations (115.1). With (115.3, 4) it becomes

$$\lambda \operatorname{div} \mathsf{D}_0 + 2\pi i \mathsf{D}_0 . \operatorname{grad} V = 0. \tag{115.5}$$

The transition to the geometrical-optical limit corresponds to allowing $\lambda$ to tend to zero. In so doing it is assumed that $\lambda \operatorname{div} \mathsf{D}_0$ tends to zero; or, more generally, that the derivatives of the components of the amplitude vectors do not tend to infinity as fast as or faster than $\lambda^{-1}$. In a more realistic picture the assumption is that, for values of $\lambda$ to be contemplated, terms (in equations such as (115.5)) which have $\lambda$, or some higher power of $\lambda$, as factors can be neglected when compared with the terms independent of $\lambda$. Physically this assumption is not justified, for instance, at the rim of a stop, to give but one example; a state of affairs which, in a sense, ultimately reveals itself in the fact that the shadow of a sharp edge does not itself have a sharp edge.

Dealing with the remaining equations of (115.1) in the manner just discussed, the vector grad $V$ turns up again and again, and we therefore write

$$\operatorname{grad} V = \mathsf{n} \tag{115.6}$$

for convenience. Then we have

$$\mathsf{n} \times \mathsf{E}_0 = \mathsf{H}_0, \quad \mathsf{n} \times \mathsf{H}_0 = -\mathsf{D}_0, \tag{115.7}$$

$$\mathsf{n} . \mathsf{H}_0 = 0, \qquad \mathsf{n} . \mathsf{D}_0 = 0, \tag{115.8}$$

in the formal limit $\lambda \to 0$. Note that equations (115.8) are already contained in (115.7). The local energy density $W$ in the medium is given by

$$W = \tfrac{1}{2}(\mathsf{E} . \mathsf{D} + \mathsf{H} . \mathsf{H}), \tag{115.9}$$

whilst the energy flux (Poynting vector) is

$$\mathsf{S} = \mathsf{E} \times \mathsf{H}. \tag{115.10}$$

(The usual factors $1/4\pi$ and $c/4\pi$ have been omitted from the right-hand members of the last two equations. These factors can always be removed by a suitable choice of units.) Then

$$\mathsf{E} . \mathsf{D} = \mathsf{E}_0 . \mathsf{D}_0 \Phi^2 = -\mathsf{n} \times \mathsf{H}_0 . \mathsf{E}_0 \Phi^2 = \mathsf{n} . \mathsf{E}_0 \times \mathsf{H}_0 \Phi^2,$$

and

$$\mathsf{H} . \mathsf{H} = (\mathsf{n} \times \mathsf{E}_0) . \mathsf{H}_0 \Phi^2,$$

so that

$$W = \mathsf{n} . \mathsf{S} = S \mathsf{n} . \mathsf{e}, \tag{115.11}$$

where $S$ is the magnitude of $\mathsf{S}$ and $\mathsf{e}$ is the usual tangent vector along the ray at the point considered, since, by definition, $\mathsf{e}$ is in the direction of the energy flow. Writing

$$N = W/S, \tag{115.12}$$

(115.11) becomes

$$\mathsf{n}\,.\,\mathsf{e} = N. \tag{115.13}$$

The differential equation for $V$ follows directly from (115.7). Inserting the first of these equations into the second, there comes

$$(\mathsf{n}\,.\,\mathsf{E}_0)\,\mathsf{n} - n^2\mathsf{E}_0 + \mathsf{D}_0 = 0, \tag{115.14}$$

where $n$ is the magnitude of $\mathsf{n}$. Writing this in component form and using (115.2) one gets the set of equations

$$(n_k n_l + \epsilon_{kl} - n^2\delta_{kl})\,E_{0l} = 0, \tag{115.15}$$

where the Kronecker delta $\delta_{kl}$ has the value 1 or 0 according as $k$ is or is not equal to $l$. These equations are homogeneous, and so have a non-trivial solution only if their discriminant vanishes:

$$\det(n_k n_l + \epsilon_{kl} - n^2\delta_{kl}) = 0. \tag{115.16}$$

This equation is quadratic in $n^2$. Upon solving it for $n^2$ one has, in effect, two distinct equations (115.14), and therefore two functions $V$ each of which satisfies (115.6).

It is possible to express $N$ explicitly as a function of $\mathsf{e}$ (and, of course, of $x$, $y$, $z$) in the following manner. If one multiplies the first of equations (115.7) vectorially by $\mathsf{E}_0$, one obtains a linear relation between $\mathsf{S}$, $\mathsf{E}_0$ and $\mathsf{n}$, which means that $\mathsf{e}$, $\mathsf{n}$ and $\mathsf{E}_0$ are coplanar. So, however, are $\mathsf{D}_0$, $\mathsf{n}$ and $\mathsf{E}_0$ also, according to (115.14). It follows that $\mathsf{e}$, $\mathsf{D}_0$ and $\mathsf{E}_0$ are coplanar, i.e. there exist constants $a$ and $b$ such that

$$\mathsf{e} = a\mathsf{D}_0 + b\mathsf{E}_0.$$

Scalar multiplication alternatively by $\mathsf{e}$ and $\mathsf{n}$ allows us to determine $a$ and $b$, and then

$$\mathsf{e} = \frac{\mathsf{D}_0}{\mathsf{e}\,.\,\mathsf{D}_0} + \frac{N\mathsf{E}_0}{\mathsf{n}\,.\,\mathsf{E}_0}, \tag{115.17}$$

where (115.8) and (115.13) have been used. Scalar multiplication of (115.14) by $\mathsf{e}$ yields the relation

$$\mathsf{e}\,.\,\mathsf{D}_0 = -N\mathsf{n}\,.\,\mathsf{E}_0,$$

and on combining this with (115.17) there follows the result

$$N^2\mathsf{E}_0 = \mathsf{D}_0 - \mathsf{e}(\mathsf{e}\,.\,\mathsf{D}_0). \tag{115.18}$$

Here it is convenient to invert the relation (115.2), i.e. to take the latter in the equivalent form

$$E_k = \eta_{kl}(x, y, z)\, D_l, \tag{115.19}$$

where

$$\epsilon_{kl}\eta_{lm} = \delta_{km}. \tag{115.20}$$

We note in passing that if $\epsilon$ and $\eta$ are the determinants of $\epsilon_{kl}$ and $\eta_{kl}$ repectively, then

$$\epsilon\eta = 1. \tag{115.21}$$

Equation (115.18) now becomes

$$(N^2\eta_{kl} + e_k e_l - \delta_{kl})\, D_{0l} = 0, \tag{115.22}$$

which should be compared with (115.15). Here again the discriminant must be zero, i.e.

$$\det(N^2\eta_{kl} + e_k e_l - \delta_{kl}) = 0, \tag{115.23}$$

which is an equation for $N^2$. It is merely a question of algebraic detail (involving the use of (115.21)) to show that (115.23) may be written as

$$N^4 + (\epsilon_{kl}e_k e_l - \operatorname{tr}\epsilon)\, N^2 + \epsilon\eta_{kl}e_k e_l = 0, \tag{115.24}$$

where $\operatorname{tr}\epsilon = \epsilon_{11} + \epsilon_{22} + \epsilon_{33}$. (115.24) is the final form of the equation which allows us to write down $N^2$ as a function of the $\epsilon_{kl}$; and here also we have two possibilities. (It is left understood that the last term on the right and $\operatorname{tr}\epsilon$ must each be supplied with an additional factor $e_m e_m (\equiv 1)$ to make $N$ homogeneous of degree 1.)

Return now to equation (115.13). If $\operatorname{Grad} N$ is the vector whose components are $N_\alpha$, $N_\beta$, $N_\gamma$, just as $\operatorname{grad} N$ denotes the vector whose components are $N_x$, $N_y$, $N_z$, we obtain, by differentiation,

$$\mathsf{n}\,.\,d\mathsf{e} + \mathsf{e}\,.\,d\mathsf{n} = \operatorname{Grad} N\,.\,d\mathsf{e} + \operatorname{grad} N\,.\,d\mathsf{x}, \tag{115.25}$$

$d\mathsf{x}$ being the vector joining arbitrary neighbouring points. On the other hand, using (115.6),

$$\mathsf{e}\,.\,d\mathsf{n} = e_k\,dn_k = e_k\frac{\partial n_k}{\partial x_l}dx_l = e_k\frac{\partial n_l}{\partial x_k}dx_l = d\mathsf{x}\,.\frac{d\mathsf{n}}{ds}, \tag{115.26}$$

where $d/ds$ stands for the derivative along the ray. (115.25) now becomes

$$(d\mathsf{n}/ds - \operatorname{grad} N)\,.\,d\mathsf{x} = (\operatorname{Grad} N - \mathsf{n})\,.\,d\mathsf{e}. \tag{115.27}$$

Observe now that in *this* section the quantities $N$, $V$ and $\mathbf{e}$ were introduced *ab initio*, that is to say, as they arose in the context of Maxwell's equations. Complete correspondence with the equations of the previous sections of this chapter is now achieved by identifying $N$ with the refractive index as it occurs in Fermat's Principle, and $V$ with the point characteristic as it occurs in equation (113.1). If this be done the first triplet of the equations (112.4)—with primes omitted—entails that the right-hand member of (115.27) vanishes, and then, since $d\mathbf{x}$ is arbitrary, we must have

$$d(\operatorname{Grad} N)/ds = \operatorname{grad} N, \tag{115.28}$$

which is just the equation of the ray in the form which emerges directly from Fermat's Principle.

Having established the interpretation in terms of Maxwell's theory of various quantities which first arose in the context of geometrical optics as such, we can go on to attach a more detailed physical meaning to the notion of the 'two kinds of light' introduced in Section 110; to relate the refractive indices $N_A$ and $N_B$ to the components of the dielectric tensor $\epsilon_{kl}$, defined by (115.2); and so on. Thus, very briefly, one has just *two* refractive indices because (115.24) is quadratic in $N^2$; and that this is so is in turn related to the fact that any electromagnetic wave can be regarded, at any point, as the superposition of just two independent linearly polarized components. As a matter of fact, given the unit vector $\mathbf{u}$ $(= \mathbf{n}/n)$ at a point, there will be two vectors $\mathbf{E}_0$ (determined to within a scalar factor) which satisfy (115.15), namely one for each of the two values of $n^2$, $n_A^2$ and $n_B^2$, say, which satisfy (115.16). The corresponding vectors $\mathbf{D}_{0A}$ and $\mathbf{D}_{0B}$ then turn out to be mutually perpendicular. To show that this is so, introduce $\mathbf{D}_0$ in place of $\mathbf{E}_0$ in (115.15), which becomes, in view of (115.20),

$$(\eta_{km} - u_k u_l \eta_{lm} - n^{-2}\delta_{km})\, D_{0m} = 0. \tag{115.29}$$

Now rotate the Cartesian axes so that at the point under consideration the transformed components of $\mathbf{u}$ are $(1, 0, 0)$, and then, since $\mathbf{u}\,.\,\mathbf{D}_0 = 0$, one also has $D_{01} = 0$. (115.29) then reduces to

$$(\eta_{km} - n^{-2}\delta_{km})\, D_{0m} = 0, \tag{115.30}$$

all quantities referring to the new axes, of course. We therefore have the pair of equations

$$\eta_{km} D_{0mA} = n_A^{-2} D_{0kA}, \quad \eta_{km} D_{0mB} = n_B^{-2} D_{0kB}.$$

Multiply the first of these by $D_{0kB}$ and the second by $D_{0kA}$ (summation over $k$ being implied as usual). Then, upon subtracting the resulting equations from one another, and bearing in mind that $\eta_{km} = \eta_{mk}$, it follows that

$$(n_A^{-2} - n_B^{-2})\, \mathsf{D}_{0A} \cdot \mathsf{D}_{0B} = 0. \quad (115.31)$$

Granted that $n_A \neq n_B$, this constitutes the required result, for it is invariant under rotations of the coordinates, so that the special choice of axes which was made in the course of the proof is irrelevant.

The usual interpretation of $N$ follows from (115.12), i.e. the ratio of the local speed with which energy is transported to the speed of light is $1/N$. It remains to consider for a moment the relationship of $N$ to $\epsilon_{kl}$. Writing

$$\bar{\epsilon}_{kl} = \delta_{kl} \operatorname{tr} \epsilon - \epsilon_{kl}, \quad (115.32)$$

(115.24) gives in the most general case

$$\left.\begin{matrix} 2N_A^2 \\ 2N_B^2 \end{matrix}\right\} = \bar{\epsilon}_{kl} e_k e_l \pm [(\bar{\epsilon}_{kl} \bar{\epsilon}_{mn} - 4\epsilon \eta_{kl} \delta_{mn}) e_k e_l e_m e_n]^{\frac{1}{2}}. \quad (115.33)$$

Purely formally, this equation is more general than that which appears in the customary accounts of crystal optics. The reason for this is as follows. When the medium is homogeneous it is possible to arrange one's coordinate axes in such a way that, relative to them, $\epsilon_{kl} = 0$ when $k \neq l$; and when this has been done we write $\epsilon_{11} = \epsilon_1$, $\epsilon_{22} = \epsilon_2$, $\epsilon_{33} = \epsilon_3$ ($\epsilon_k > 0$). In other words, one can always make a principal axis transformation, meaning a rotation of coordinates such that, after it has been carried out, the axes are along the principal axes of the ellipsoid whose equation, before the rotation, was $\epsilon_{kl} x_k x_l = 1$. However, we have throughout considered inhomogeneous media, i.e. the $\epsilon_{kl}$ were given functions of $x$, $y$, $z$, rather than mere given constants. This means that although one can make a principal axis transformation at any selected point, the refractive indices will retain the general form (115.33) elsewhere; cf. the remark following equation (111.12). If we do make a principal axis transformation at some point, then (115.33) reduces there exactly to (114.1), if we make the identification

$$(a, b, c) = (\epsilon_1, \epsilon_2, \epsilon_3). \quad (115.34)$$

A general medium therefore behaves *locally* either as an isotropic medium, a uniaxial crystal, or a biaxial crystal, as the case may be; but, whichever alternative applies at any selected point, the same alternative need not apply at neighbouring points.

## Problems

**P.11 (i).** Show that Fermat's Principle yields the equation

$$d(\operatorname{Grad} N)/ds = \operatorname{grad} N$$

for the equation of the ray, $d/ds$ denoting the derivative in the direction of the ray.

**P.11 (ii).** Two uniaxial crystals have a common plane boundary, and both optical ray-axes are along the normal to the boundary. If $\phi$ and $\phi'$ are the angles of incidence and refraction of a $B$-ray (for which $N_B$ is given by (111.14)), show that

$$\sin\phi' = ab'^{\frac{1}{2}}\sin\phi\{a'^2 b + [a'^2(a-b) + a^2(b'-a')]\sin^2\phi\}^{-\frac{1}{2}}.$$

CHAPTER 12

# THE COMPUTATION OF ABERRATION COEFFICIENTS

## 116. Preliminary remarks. Lagrangian and Hamiltonian methods

To solve a particular design problem in practice one requires the actual numerical values of the aberration coefficients of systems whose physical constitution is given in detail. We therefore need to inquire in the first place how such numerical values might be calculated. Now, we have before us a task so extensive that we can do little more than to present, a mere intimation as it were, of how we might go about carrying it through; and this is quite appropriate to the context of this work, which was from the outset intended to concern itself with general principles rather than with the solution of particular problems. In any event, there arises a general issue of policy which may be described as follows.

The passage of families of rays through a given optical system may be analyzed in two ways which differ from each other in the manner in which particular rays are thought of as selected. Thus one may *either* specify two points $A$ and $A'$, one in the object space and one in the image space, and then seek to discover the directions at $A$ and $A'$ of the ray $\mathscr{R}$ through these points; *or* one may select a ray $\mathscr{R}$ by prescribing, in the object space, a point $A$ on it and its direction there, and then seek to find, in the image space, a point on $\mathscr{R}$ and its direction at this point. A precisely analogous situation exists in dynamics. In describing the motion of a particle $P$ we can ask the question: if $P$ was at the point $A(x, y, z)$ at time $t$, and it was at the point $A'(x', y', z')$ at time $t'$, what were the components of momentum at $A$ and at $A'$ respectively? Alternatively we may ask: if $P$ was at $A$ at time $t$ and its momentum was then $\mathsf{p}$, where will $P$ be at time $t'$ and what will its momentum $\mathsf{p}'$ be at that time?

In the optical context we know, of course, that the first of the alternatives described above amounts to requiring the determination of the point characteristic $V$; and in the dynamical context one likewise has to find a function called Hamilton's *Principal Function* (usually also denoted by $V$) which is the precise dynamical analogue of the point characteristic. In either case a method of solving a given problem, optical or dynamical, which rests upon the calculation of $V$ is called a *Hamiltonian method* of computation. On the other hand, if a dynamical problem is to be solved in which the initial configuration *and* state of motion are prescribed, then this is achieved essentially by solving Lagrange's equations of motion; and one thus provides the answer to the second of the alternative questions above. Correspondingly, one has to solve, in the isotropic optical case, the differential equations $d(N\mathbf{e})/ds = \text{grad}\, N$, or what amounts to a set of finite difference equations, based on (2.6), when the optical system consists of a set of homogeneous media; the given *initial* conditions leading to a unique solution. It is therefore natural to speak of this type of procedure as a *Lagrangian method* of computation in the optical context. It is at this point that we have to ask ourselves whether it is the Hamiltonian or the Lagrangian method which is to be preferred from a purely computational point of view: one suspects it is the latter, simply on the grounds of analogy with the dynamical case. There can be little doubt that 'everyday' dynamical problems are most easily dealt with by solving Lagrange's equations rather than by using the Hamilton–Jacobi theory, which essentially revolves about the Principal Function. Moreover, the specification of rays by coordinates all of which lie in the object space is undoubtedly convenient in practice.

Some years ago (1954), in a monograph entitled *Optical Aberration Coefficients*, and in a sequence of thirteen papers under the same general title, which appeared during the years 1956–67 in the *Journal of the Optical Society of America*, I gave a fairly detailed account of the Lagrangian method as applied to the symmetric system. (All this material has recently been reissued in one volume by Dover Publications, Inc.) It is based on the use of so-called *quasi-invariants* (the paraxial limit of which is essentially given by the joint paraxial invariants (10.12), see Section 119), and it leads to a systematic iterative procedure which allows one to compute with comparative ease the aberration coefficients of the various orders. Specifically, it was shown in explicit detail—which includes numerical examples—how to compute (i) all (effective) monochromatic aberration coefficients of orders 3, 5 and 7; (ii) the (monochromatic) coefficients of spherical aberration and circular coma of order 9 and of spherical aberration of order 11; (iii) the paraxial chromatic coefficients; (iv) all primary chromatic third-order coefficients; (v) the coefficients of secondary chromatic third-order and primary chromatic fifth-order spherical aberration. It must be emphasized that it is quite feasible to compute, for instance, *all* the monochromatic ninth-order coefficients: the equations relating to these turn out to be still quite manageable, and hold no terrors for a high-speed digital computer at all. I further showed how one may compute the (exact) first and higher *derivatives* of the coefficients of the various types and orders with respect to the parameters which define the constitution of the system, i.e. the axial and extra-axial curvatures, the separations, and the refractive indices. The availability of these derivatives is of great value in the adjustment of aberrations, particularly if this is to be achieved by varying several parameters simultaneously and not necessarily by very small amounts: whereas the experienced designer will often know what to do in the context of the monochromatic and the paraxial chromatic aberrations of systems of spherical surfaces, this is probably not the case to the same extent when it comes to aspherical surfaces and non-paraxial chromatic aberrations. These observations do not necessarily lose their validity in the context of automatic design programmes, especially if economy of machine time, as well as the need to achieve an absolute, rather than a relative, optimization are taken into account.

The great simplicity of the Lagrangian method of computation leads me to believe that it, rather than the Hamiltonian method, is best adapted to the problem of practical calculations: though this is admittedly a mere expression of opinion. Furthermore, one can quite easily use Lagrangian methods to calculate characteristic aberration coefficients. To be a little more specific, it is convenient to refer to the monograph and the various papers already quoted as OAC, OAC I, ..., OAC XIII respectively; and to numbered equations of these merely by the additional prefix O, I, II, ..., XIII. The coordinates of the Lagrangian theory may be chosen in such a way that, $\epsilon'$ having been computed to the required order, a rudimentary integration gives the aberration function $t$ to within what is essentially the ubiquitous function $g(\zeta)$ we have so often encountered before; cf. Sections 49, 50 and 205 of OAC, and also OAC VI. Various alternative procedures are possible such as those which avoid the appearance of $g(\zeta)$ (see Section 205*b* of OAC), or which concern themselves directly with the deformation or retardation of the wavefront. With regard to the latter OAC VII may be consulted, but it should be noted that the retardation $R$ considered there is not identical with the retardation $\Delta$ of Section 38. In fact, in terms of Fig. 5.1, $R = Q'G_0'$, so that, recalling (38.3), $R$ is identical with $v$, i.e. the aberration function associated with the function $V(\hat{x}', \hat{y}', \hat{z}', y, z)$ of equation (38.1). Further, on account of (39.3), $\Delta - R = O(10)$. Attention must also be drawn at this point to the fact that throughout the Lagrangian developments just described the sign conventions differ from those used here. To achieve consistency one has to reverse, in the former, the signs of $\beta$, $\gamma$ and every object or image height, and therefore of every (actual) displacement.

Even though it be granted that Lagrangian methods of computation are to be preferred in practice it will not be out of place to present here a brief treatment of one possible procedure of calculating characteristic functions, that is to say, the coefficients of their power series, with exclusive concentration on the monochromatic angle characteristic of symmetric systems, save for a brief remark relating to the point characteristic at the end of Section 123. To do more would surely lead us altogether too far afield and would perhaps be unjustified in view of the opinions already expressed. At any rate, there will be strong echoes of OAC; and we shall have the opportunity to become aware explicitly of some of the sources of irrelevant complexity of Hamiltonian methods, such as the fact that one is precluded from proceeding step by step from object space to image space, or the occurrence of the variables $\boldsymbol{\beta}/\alpha$ where one would like $\boldsymbol{\beta}$ to appear; and so on.

## 117. The angle characteristic of a surface of revolution

Considered geometrically, every characteristic function $F$ represents an optical distance. Let there be given a symmetric system $K$ consisting of homogeneous media whose mutual boundaries are surfaces of revolution $\mathscr{S}_j\,(j = 1, \ldots, k)$, and let $B_j, B'_j$ ($B'_j \equiv B_{j+1}$; $j = 1, \ldots, k$) be a suitable set of points upon the axis $\mathscr{A}$ of $K$. Then we may regard the characteristic function $F(B, B')$ referred to the base-points $B\,(\equiv B_1)$ and $B'\,(\equiv B'_k)$ as the sum of the appropriate optical distances referred to the base-points $B_j, B'_j$:

$$F(B, B') = \sum_{j=1}^{k} F_j(B_j, B'_j), \tag{117.1}$$

and we speak of $F_j$ as the 'characteristic function of $\mathscr{S}_j$'. The sum on the right of (117.1) is in the first place a function of all the variables $q_{ij}$ ($i = 1, \ldots, 4$; $j = 1, \ldots, k$), using the notation introduced at the end of Section 6. Eventually, therefore, all the variables other than $q_{31}$, $q_{41}$, $q_{1k}$ and $q_{2k}$ will have to be eliminated; and when that has been done, $F$ will be the required characteristic function of $K$.

Evidently the first task to be accomplished is to obtain an expression for $F_j$; but before this can be done a definite choice of the various base-points must be made. Amongst the possibilities which naturally commend themselves are the points $O_{0j}$, $O'_{0j}$ and the points $E_j$, $E'_j$ in which paraxial rays through $O_0$ and through the axial point of the plane of the stop respectively intersect $\mathscr{A}$; where, of course, the points $O_{01}$, $O'_{0k}$, $E_1$, $E'_k$ are those hitherto denoted by $O_0$, $O'_0$, $E$, $E'$ respectively. One will naturally want to take $O_{01}$ as the first anterior base-point, but when $F$ is $V$, the choice of $O'_{01}$ as first posterior base-point is precluded. One might take $E'_1$, but then the corresponding choice of $O_{02}$ and $E'_2$ as the next pair leaves that part of the ray which lies between the normal planes through $E'_1$ and $O'_{01}$ unaccounted for. In short, the missing segment must be restored in some way, and here we already have one complication, albeit not a serious one. If $F$ is $T$ we have no such difficulty: we can take $B_j$, $B'_j$ to be $O_{0j}$, $O'_{0j}$. Only when $\mathscr{S}_j$ is plane would there appear to be a difficulty. This is, however, easily overcome by taking $\mathscr{S}_j$ to be spherical, and letting its curvature $c_j$ tend to zero only at the end of one's formal work. There is, furthermore, another, much greater, advantage which $T_j$ has over $V_j$: whereas $T_j$ can easily be obtained in closed form for various surfaces of practical importance, for spherical surfaces in particular, one must necessarily be content with $V_j$ in the form of a power series. To obtain the latter is, moreover, no easy task.

With the preceding remarks in mind, we go on to consider the problem of obtaining the form of $T_j$ for a surface of revolution. Since we are only concerned with a fixed value of $j$, this index may be omitted in the remainder of this section. We take local axes $x, y, z$ which have the usual orientation, their origin being at the axial point (the pole) $A$ of $\mathscr{S}$. The coordinates of the point of incidence $P$ of the ray $\mathscr{R}$ shall be $x_P, y_P, z_P$, whilst the equation of $\mathscr{S}$ will be written as

$$x = \tfrac{1}{2}s(y^2+z^2), \tag{117.2}$$

with $s(0) = 0$.

The following notation is very useful. Let some quantity, whatever it may be, which is associated with $\mathscr{R}$ have the values $X$, $X'$ respectively before and after refraction of $\mathscr{R}$ by $\mathscr{S}$. Then we write

$$\Delta X = X'-X, \quad \nabla X = X'+X. \tag{117.3}$$

(2.1) and (2.6), for instance, now read

$$\Delta N\mathsf{e} = \sigma\mathsf{p} = \mathsf{p}\Delta N\cos I. \tag{117.4}$$

If the base-points have the ($x$-) coordinates $q$ and $q'$, an elementary geometrical argument shows that

$$T = \Delta\{N[(q-x_P)\alpha-(y_P\beta+z_P\gamma)]\}. \tag{117.5}$$

Next, we have from (117.2)

$$\mathsf{p} = R^{-1}(1, -\dot{s}y_P, -\dot{s}z_P), \tag{117.6}$$

where $$\dot{s} = ds(u)/du, \quad u = y_P^2+z_P^2, \tag{117.7}$$

and $$R = 1+u\dot{s}^2. \tag{117.8}$$

Upon scalar multiplication of (117.4) throughout by $\mathsf{p}$, using (117.6) on the left, one gets

$$\sigma = \Delta(N\mathsf{e}.\mathsf{p}) = R^{-1}\Delta\{N[\alpha-\dot{s}(y_P\beta+z_P\gamma)]\},$$

whence $$-\Delta[N(y_P\beta+z_P\gamma)] = (R\sigma-\Delta N\alpha)/\dot{s}. \tag{117.9}$$

On the other hand, the $x$-component of (117.4) gives the relation

$$\sigma = R\Delta N\alpha, \tag{117.10}$$

and with this (117.9) becomes

$$-\Delta[N(y_P\beta+z_P\gamma)] = (R^2-1)(\Delta N\alpha)/\dot{s} = u\dot{s}\Delta N\alpha, \tag{117.11}$$

in view of (117.6). Inserting this in (117.5) there comes

$$T = \Delta(Nq\alpha) + (u\dot{s} - \tfrac{1}{2}s)\,\Delta N\alpha. \qquad (117.12)$$

It remains to express the function of $u$ which appears on the right as a function of the components of e and e′.

To begin with, taking the scalar product of each member of (117.4) with itself, we have

$$\sigma^2 = N'^2 + N^2 - 2NN'(\alpha\alpha' + \beta\beta' + \gamma\gamma'),$$

whence
$$\sigma = (N' - N)(1 + 2\kappa\chi)^{\frac{1}{2}}, \qquad (117.13)$$

where
$$\kappa = NN'/(N' - N)^2, \qquad (117.14)$$

and $\chi$ is defined by (65.10). Then (117.10) becomes

$$u\dot{s}^2 = [(\Delta N)/(\Delta N\alpha)]^2(1 + 2\kappa\chi) - 1 = \psi, \qquad (117.15)$$

say. The left- and right-hand members of this equation are respectively functions of $u$ alone and of the direction cosines alone, so that $u$ may be directly obtained as a function of the direction cosines. Upon inserting this function in (117.12) one finally has the explicit form of the angle characteristic.

The most important specific example is that of the *spherical surface*, of radius $r$, say. Here
$$s = 2[r - (r^2 - u)^{\frac{1}{2}}], \qquad (117.16)$$

whence
$$\psi = u/(r^2 - u),$$

and
$$u\dot{s} - \tfrac{1}{2}s = r[r(r^2 - u)^{-\frac{1}{2}} - 1] = r[(1 + \psi)^{\frac{1}{2}} - 1].$$

Using this result in (117.12), there comes

$$T = r(\Delta N)(1 + 2\kappa\chi)^{\frac{1}{2}} + \Delta[N(q - r)\,\alpha], \qquad (117.17)$$

and this is, of course, consistent with (65.11).

Another very simple example is provided by the *paraboloid*

$$s = u/2a, \qquad (117.18)$$

where $a$ is a constant. Then

$$u = 4a^2\psi, \quad u\dot{s} - \tfrac{1}{2}s = u/4a = a\psi,$$

whence

$$T = \Delta[N(q - a)\,\alpha] + a(\Delta N)^2(1 + 2\kappa\chi)/(\Delta N\alpha). \qquad (117.19)$$

Finally, consider the *conicoid*, i.e. a surface of revolution whose meridional section is a conic section. We take its equation in the form

$$(a-x)^2+bu = a^2, \tag{117.20}$$

so that

$$s = 2[a-(a^2-bu)^{\frac{1}{2}}]. \tag{117.21}$$

It then follows without difficulty that

$$u\dot{s}-\tfrac{1}{2}s = a[(1+\psi/b)^{\frac{1}{2}}-1], \tag{117.22}$$

which is to be inserted into (117.12).

Quite generally one may write $s$ as a power series:

$$s(u) = s_1u+s_2u^2+\dots. \tag{117.23}$$

The resulting series on the left of (117.15) may be inverted iteratively, and the result substituted in the power series for $u\dot{s}-\frac{1}{2}s$. One thus obtains

$$u\dot{s}-\tfrac{1}{2}s = \tfrac{1}{2}[s_1^{-1}\psi-s_1^{-4}s_2\psi^2-(s_1^{-6}s_3-4s_1^{-7}s_2^2)\psi^3]+O(8); \tag{117.24}$$

and this is all one requires as long as one does not intend to proceed beyond the fifth order.

## 118. Paraxial precalculation

Paraxial rays may be traced through $K$ by an elementary application of the laws of refraction (2.1). At the $j$th surface we take local axes $x_j, y_j, z_j$ as before, origin at $A_j$, and it will do no harm here to denote the coordinates of the point of incidence $P_j$ of the ray by these symbols, i.e. to omit the subscript $P$. To the first order, (2.1) reduces to

$$\Delta N_j \mathsf{e}_j = \mathsf{p}_j \Delta N_j, \tag{118.1}$$

which gives $\mathsf{e}'_j$ in terms of $\mathsf{e}_j$, granted that $\mathsf{p}_j$ is known:

$$\mathsf{p}_j = (1, -c_jy_j, -c_jz_j), \tag{118.2}$$

where $c_j(=1/r_j)$ is the curvature of $\mathscr{S}_j$. Henceforth (except at the end of Section 120) we consider only spherical surfaces, but even when $\mathscr{S}_j$ is not spherical, (118.2) remains valid if $c_j$ be understood to be the curvature of $\mathscr{S}_j$ at $A_j$. To proceed to the next surface we only require the relations.

$$\mathsf{e}_{j+1} = \mathsf{e}'_j, \quad \mathbf{y}_{j+1} = \mathbf{y}'_j + d'_j \boldsymbol{\beta}'_j, \tag{118.3}$$

where $d'_j$ is the axial separation between $\mathscr{S}_{j+1}$ and $\mathscr{S}_j$. (The symbols

$d_1$ and $d'_k$ will evidently not occur, so that confusion with the symbols $d$ and $d'$ in equations (14.27–30) is not likely.)

Recall the variables $\boldsymbol{\sigma}$ and $\boldsymbol{\mu}$ defined, for arbitrary rays, by (24.2):

$$\boldsymbol{\sigma} = \boldsymbol{\beta}_1 - s\boldsymbol{\beta}'_k, \quad \boldsymbol{\mu} = \boldsymbol{\beta}_1 - m\boldsymbol{\beta}'_k. \tag{118.4}$$

$s$ and $m$ are, as always, the reduced magnifications associated with the pupil planes, and the object and image planes, respectively. The values of these constants are supposed known. If meridional paraxial rays through $O_0$ and $E$ be traced through $K$, then $\beta_1/\beta'_k$ has the value $m$ in the first case, and $s$ in the second.

Now, on account of the linearity of the relations (118.1) and (118.3), there must exist *paraxial constants* $v_{aj}, v_{bj}, v'_{aj}, v'_{bj}$ such that paraxially

$$\boldsymbol{\beta}_j = v_{aj}\boldsymbol{\sigma} + v_{bj}\boldsymbol{\mu}, \quad \boldsymbol{\beta}'_j = v'_{aj}\boldsymbol{\sigma} + v'_{bj}\boldsymbol{\mu}. \tag{118.5}$$

These constants are easily calculated as follows. Assigning to $j$ alternatively the values 1 and $k$, one obtains two identities involving only $\boldsymbol{\beta}_1$ and $\boldsymbol{\beta}'_k$. Inspection shows that

$$v_{a1} = -m/(s-m), \quad v'_{ak} = -1/(s-m),$$
$$v_{b1} = s/(s-m), \qquad v'_{bk} = 1/(s-m). \tag{118.6}$$

The ratios $v_{a1}/v'_{ak}$ and $v_{b1}/v'_{bk}$ have just the values which $\beta_1/\beta'_k$ takes for rays through $O_0$ on the one hand and through $E$ on the other. Therefore, if these points are located at $x_1 = l_0$ and $x_1 = p$ respectively, trace two meridional (paraxial) rays, the starting data of which are

$$\text{(i) } a\text{-}ray\text{: } y_{a1} = ml_0/(s-m), \qquad v_{a1} = -m/(s-m), \tag{118.7}$$

$$\text{(ii) } b\text{-}ray\text{: } y_{b1} = -sp/(s-m), \quad v_{b1} = s/(s-m). \tag{118.8}$$

The value of $\beta_j$ in the trace of the $a$-ray is then just $v_{aj}$, the value of $\beta'_j$ is $v'_{aj}$, whilst the trace of the $b$-ray similarly allows one simply to read off the values of $v_{bj}$ and $v'_{bj}$. Again, the values of $y_j$ given by the $a$-ray and $b$-ray define paraxial constants $y_{aj}$ and $y_{bj}$, such that for *any* paraxial ray

$$\mathbf{y}_j = y_{aj}\boldsymbol{\sigma} + y_{bj}\boldsymbol{\mu}. \tag{118.9}$$

The notation quite naturally extends itself to any other variables which are linearly related to $\mathbf{y}$ and $\boldsymbol{\beta}$. One such case is represented by the quantity

$$\mathbf{i} = c\mathbf{y} + \boldsymbol{\beta}, \tag{118.10}$$

which is of frequent occurrence. If the $a$-trace and $b$-trace are so set out that $\mathbf{i}$, regarded as an auxiliary variable, appears explicitly in them, then the constants $i_{aj}$, $i_{bj}$ may simply be read off, and then

$$\mathbf{i}_j = i_{aj}\boldsymbol{\sigma} + i_{bj}\boldsymbol{\mu}. \tag{118.11}$$

We remark in passing that upon using (118.10) in (118.1) the latter gives

$$\Delta N\mathbf{i} = 0, \tag{118.12}$$

which is a paraxial expression of the invariance of the vector $N\mathsf{e} \times \mathsf{p}$ under refraction.

It may be useful to comment upon an apparent inconsistency in the notation introduced above, represented by the use of the symbols $v_{aj}, \ldots$ rather than $\beta_{aj}, \ldots$ on the right of (118.5). The point is that the right-hand member of the first of these, for instance, obviously can be thought of as giving at the same time the 'paraxial' value of any variable which reduces to $\boldsymbol{\beta}_j$ in the paraxial limit. We shall indeed have occasion later to introduce variables $\mathbf{V}(\equiv V, W)$ defined as

$$\mathbf{V} = \boldsymbol{\beta}/\alpha, \tag{118.13}$$

and then we shall have

$$\mathbf{V}_j = v_{aj}\boldsymbol{\sigma} + v_{bj}\boldsymbol{\mu} \tag{118.14}$$

in the paraxial limit. If we define four variables $\mathbf{S}(\equiv S_y, S_z)$ and $\mathbf{M}(\equiv M_y, M_z)$ through

$$\mathbf{S} = \mathbf{V}_1 - s\mathbf{V}'_k, \quad \mathbf{M} = \mathbf{V}_1 - m\mathbf{V}'_k, \tag{118.15}$$

we shall still have

$$\mathbf{V}_j = v_{aj}\mathbf{S} + v_{bj}\mathbf{M} \tag{118.16}$$

in the paraxial limit, since $\mathbf{S}$ then coincides with $\boldsymbol{\sigma}$, and $\mathbf{M}$ with $\boldsymbol{\mu}$.

Under some circumstances it is desirable to relate the paraxial $b$-constants to the corresponding $a$-constants in such a way that the $b$-ray enters into the various relations solely through the ratio

$$\theta = i_b/i_a. \tag{118.17}$$

To this end recall that, omitting the surface index $j$,

$$N(y_a v_b - y_b v_a) = N_1(y_{a1}v_{b1} - y_{b1}v_{a1}), \tag{118.18}$$

with a similar relation in which on the left all quantities are primed. Because of (118.7–8) the right-hand member becomes

$$-N_1 sm(s-m)^{-2}(p-l_0) = -dsm(s-m)^{-2} = (s-m)^{-1}f_K, \tag{118.19}$$

where (14.28) has been used, $f_K$ denoting the focal length of $K$ as a whole. With (118.10) the left-hand member of (118.18) may be written as

$$Nr(i_a v_b - i_b v_a) = Nri_a(v_b - \theta v_a).$$

If we put
$$\phi = f_K/[(s-m)\,Nri_a], \tag{118.20}$$

we finally arrive at the relation

$$v_b = \theta v_a + \phi. \tag{118.21}$$

(118.21) yields a valid result if one simply attaches primes to both $v_a$ and $v_b$, since $\Delta\theta = 0$ and $\Delta\phi = 0$. One also has

$$y_b = \theta y_a - r\phi. \tag{118.22}$$

Once all the $b$-constants have been formally eliminated in this way one can simplify the notation further by omitting the subscript $a$ from the $a$-constants, without risk of confusion.

## 119. The quasi-invariants $\boldsymbol{\Lambda}$ and $\boldsymbol{\Lambda}^*$

Let $T$ be the angle characteristic of the $j$th surface $\mathscr{S}_j$. The index $j$ is again omitted for the time being, since only a particular, though arbitrarily selected, surface is in question. (This convention will generally be adhered to, except where special emphasis requires the restoration of this index.) The base-points $B$ and $B'$ shall be mutually conjugate, but otherwise unrestricted. Further, let $v$ and $v'$ be the paraxial constants—exactly analogous to $v_a$, $v'_a$ or to $v_b$, $v'_b$— defined by a paraxial ray which passes through $B$, and therefore also through $B'$. Then, contemplate the quantity $\boldsymbol{\Lambda}$ defined by

$$\boldsymbol{\Lambda} = -Nv\mathbf{H}, \tag{119.1}$$

where $H_y, H_z$ are the coordinates of the point of intersection of a (finite) ray $\mathscr{R}$ with the anterior base-plane $\mathscr{B}$. The change in $\boldsymbol{\Lambda}$ consequent upon refraction of $\mathscr{R}$ by $\mathscr{S}$ will be written as

$$\mathbf{g} = \Delta\boldsymbol{\Lambda} = -\Delta(Nv\mathbf{H}). \tag{119.2}$$

Observe now that in the limit when $\mathscr{R}$ itself becomes paraxial, $\boldsymbol{\Lambda}$ reduces to a joint paraxial invariant $\lambda$ of the kind defined by (10.12). Under these circumstances $\mathbf{g}$ must vanish, according to (10.13). We therefore call $\boldsymbol{\Lambda}$ a *quasi-invariant*, this nomenclature being intended to apply to any quantity, associated with a ray, which vanishes in the paraxial limit. In short, $\mathbf{\Lambda}$ has the valuable property that

$$\Delta\mathbf{\Lambda} = O(3), \tag{119.3}$$

i.e. the dominant terms of $\mathbf{g}$ are not paraxial but of order three.

In view of (6.2) we have quite generally

$$\mathbf{g} = \nabla\left(v\frac{\partial T}{\partial \boldsymbol{\beta}}\right). \tag{119.4}$$

The expression on the right may be evaluated in general terms, by using the explicit expression for $T$ which, taking now the case of a spherical surface, is given by (117.17). Differentiating it, one finds with little labour that

$$\mathbf{g} = \kappa r(\Delta N)(1+2\kappa\chi)^{-\frac{1}{2}}[v(\alpha'\boldsymbol{\beta}/\alpha-\boldsymbol{\beta}')+v'(\alpha\boldsymbol{\beta}'/\alpha'-\boldsymbol{\beta})]$$
$$+[N(q-r)v\boldsymbol{\beta}/\alpha-N'(q'-r)v'\boldsymbol{\beta}'/\alpha']. \tag{119.5}$$

Now, since the paraxial ray through which $v$ is defined passes through the point $x = q$, we have $qv = -y$. In view of (118.10) we therefore have

$$N(q-r)v = -Nri, \tag{119.6}$$

and in exactly the same way,

$$N'(q'-r)v' = -N'ri' = -Nri, \tag{119.7}$$

where (118.12) has been used on the right. Also

$$\kappa r\Delta N = \frac{rNN'}{N'-N} = f, \tag{119.8}$$

where $f$ stands for the mean focal length (in the sense of equation (14.13)) of $\mathscr{S}$, regarded as an optical system in its own right. If we now go over at the same time to the variables defined by (118.13), (119.5) becomes

$$\mathbf{g} = [f(1+2\kappa\chi)^{-\frac{1}{2}}(v'\alpha-v\alpha')+Nri]\,\Delta\mathbf{V}. \tag{119.9}$$

With regard to the first-order term of $\mathbf{g}$, this is $(f\Delta v+Nri)\Delta\mathbf{V}$, since, as regards dominant terms, we need not distinguish between $\Delta\mathbf{V}$ and $\Delta\boldsymbol{\beta}$. However, $f\Delta v = f\Delta i = -Nri$, on account of (118.12) and (119.8), and so the term in question vanishes, confirming what we already know.

It will be seen that $\mathbf{g}$ factorizes, a result which appears in an equivalent form in Section 14 of OAC. This factorization leads to considerable simplifications in the computation of higher-order coefficients (see OAC, Section 84), but unfortunately no such decomposition of $\mathbf{g}$ is possible when the surface is aspherical (cf. O(60.3), and equations (120.10), (120.22) below).

It is of some importance to investigate also the properties of the quasi-invariant

$$\mathbf{\Lambda}^* = \alpha\mathbf{\Lambda}. \tag{119.10}$$

The equations corresponding to (119.2) and (119.4) are

$$\mathbf{g}^* = \Delta\mathbf{\Lambda}^* = -\Delta(Nv\mathbf{H}^*) = \nabla\left(v\alpha\frac{\partial T}{\partial \boldsymbol{\beta}}\right), \tag{119.11}$$

where $\mathbf{H}^* = \alpha\mathbf{H}$. Proceeding as before it turns out that

$$\mathbf{g}^* = e[-(1+2\kappa\chi)^{-\frac{1}{2}}(\alpha\boldsymbol{\beta}'-\alpha'\boldsymbol{\beta})+(\boldsymbol{\beta}'-\boldsymbol{\beta})], \tag{119.12}$$

where

$$e = Nri. \tag{119.13}$$

Here $\boldsymbol{\beta}$ and $\boldsymbol{\beta}'$ now appear as the 'natural' variables. $\mathbf{g}^*$ moreover has again decomposed into two factors, of which one relates only to the paraxial ray, whereas the second does not contain it at all; cf. O(50.21). Accordingly we write

$$\mathbf{g}^* = e\mathbf{w}. \tag{119.14}$$

However, as before, this decomposition is peculiar to spherical surfaces alone.

## 120. The third-order aberrations

Before we go on to questions relating to higher-order aberrations it may be as well to derive expressions for the five third-order coefficients from which their numerical values may be computed; bearing in mind their outstanding importance in practice. Two possibilities arise, namely, we focus our attention either on the third-order aberration function $t^{(2)}$ as such, or else directly on the third-order displacement $\boldsymbol{\epsilon}_3'$. In view of the trivial nature of the relations (23.2) between the effective and the characteristic coefficients it is largely a matter of indifference which approach we choose. We shall, indeed, adopt the second of them for reasons to be explained in a moment; and it will be more convenient to postpone certain remarks concerning the first method until such time as we have arrived at the desired results.

From (119.2) we have at once

$$\mathbf{\Lambda}'_k = \mathbf{\Lambda}_1 + \sum_{j=1}^{k} \mathbf{g}_j, \tag{120.1}$$

since $\mathbf{\Lambda}'_j \equiv \mathbf{\Lambda}_{j+1}$; granted, of course, that the same paraxial ray serves to define $q_j, q'_j, v_j, v'_j$ for all values of $j$. Let this ray now be taken specifically to be the $a$-ray of (118.7), the corresponding angle characteristic being $T_a$. Then (120.1) becomes

$$\mathbf{H}'_k = \frac{N_1 v_{a1}}{N'_k v'_{ak}} \mathbf{H}_1 - \frac{1}{N'_k v'_{ak}} \sum_{j=1}^{k} \mathbf{g}_j. \tag{120.2}$$

Now let $\mathbf{H}'_k$ and $\mathbf{H}_1$ stand for the *reduced* coordinates of $O'$ and $O$ respectively—all other lengths remaining unreduced. Then

$$\boldsymbol{\epsilon}' = \mathbf{H}'_k - m\mathbf{H}_1$$

is exactly the displacement hitherto denoted by this symbol, and, bearing in mind the second member of (118.6), (120.2) at once reduces to

$$\boldsymbol{\epsilon}' = (s-m) \sum_{j=1}^{k} \mathbf{g}_j; \tag{120.3}$$

an equation which, of course, involves no approximations.

On the right we must now use (119.9) for the individual *contributions* $\mathbf{g}_j$. The latter are still functions of the *intermediate* variables $\mathbf{V}_j$, $\mathbf{V}'_j$, and these must be eliminated in favour of $\mathbf{S}$ and $\mathbf{M}$. As far as the third-order displacement is concerned, we evidently only require the dominant terms of (119.9) for we already know these to be of the third order. Consequently, if in them the intermediate variables be expressed in terms of the *paraxial* relations, the error so committed is at least of the fifth order, and is therefore irrelevant in the present context. The task before us is therefore a very simple one.

Since $\chi = O(2)$, we have from (119.1)

$$\mathbf{g} = f[-\kappa\chi\Delta v_a + \tfrac{1}{2}(v_a\bar{\xi} - v'_a\bar{\zeta})]\,\Delta\mathbf{V} + O(5), \tag{120.4}$$

where, of course, $\bar{\xi}, \bar{\eta}, \bar{\zeta}$ refer to $\mathscr{S}_j$, e.g. $\bar{\xi} = \beta_j'^2 + \gamma_j'^2$. Using (65.12) this becomes

$$\mathbf{g} = \tfrac{1}{2}f[-\kappa(k-1)\,i_a(\bar{\xi} - 2\bar{\eta} + \bar{\zeta}) + (v_a\bar{\xi} - v'_a\bar{\zeta})]\,\Delta\boldsymbol{\beta} + O(5), \tag{120.5}$$

$\Delta\mathbf{V}$ having been replaced by $\Delta\boldsymbol{\beta}$, which is permissible because the difference between these expressions is $O(3)$. The constant $k$ is defined as

$$k = N/N'. \tag{120.6}$$

Let us take the various terms in turn. We have

$$\Delta\boldsymbol{\beta} = (\Delta v_a)\,\boldsymbol{\sigma}+(\Delta v_b)\,\boldsymbol{\mu} = (\Delta i_a)\,\boldsymbol{\sigma}+(\Delta i_b)\,\boldsymbol{\mu}$$
$$= (k-1)\,i(\boldsymbol{\sigma}+\theta\boldsymbol{\mu}), \qquad (120.7)$$

the convention introduced at the end of Section 118 being understood. In just the same way

$$\bar{\xi}-2\bar{\eta}+\bar{\zeta} = (\Delta\beta)^2+(\Delta\gamma)^2 = (k-1)^2 i^2(\xi+2\theta\eta+\theta^2\zeta). \quad (120.8)$$

Again,

$$v_a\bar{\xi}-v_a'\bar{\zeta} = v_a v_a'(\Delta v_a)\,\xi+2v_a v_a'(\Delta v_b)\,\eta+(v_a v_b'^2-v_a' v_b^2)\,\zeta.$$

Upon using (118.21), this becomes quite easily

$$v_a\bar{\xi}-v_a'\bar{\zeta} = (k-1)\,i[vv'(\xi+2\theta\eta+\theta^2\zeta)-\phi^2\zeta]. \qquad (120.9)$$

(120.7–9) are now to be inserted into (120.5). When we remember that $\kappa = k(k-1)^{-2}$ and $f = -Nr(k-1)^{-1}$, we thus find that

$$\mathbf{g} = \tfrac{1}{2}Nr(k-1)\,i^2[(ii'-vv')(\xi+2\theta\eta+\theta^2\zeta)+\phi^2\zeta](\boldsymbol{\sigma}+\theta\boldsymbol{\mu}). \quad (120.10)$$

Next we note that $\qquad r(ii'-vv') = y(i'+v), \qquad (120.11)$

whilst $\qquad \tfrac{1}{2}N(k-1)\,ri^2\phi^2 = \tfrac{1}{2}(s-m)^{-2}f_K^2\,\tilde{\omega}, \qquad (120.12)$

where $\qquad \tilde{\omega} = c\Delta(1/N). \qquad (120.13)$

At this point we set $f_K = 1$ as we usually did throughout our earlier work. It is quite natural to introduce the abbreviation

$$\mathfrak{p} = \tfrac{1}{2}N(k-1)\,i^2y(i'+v), \qquad (120.14)$$

and then (120.10) reads

$$\mathbf{g} = [\mathfrak{p}(\xi+2\theta\eta+\theta^2\zeta)+\tfrac{1}{2}(s-m)^{-2}\tilde{\omega}\zeta](\boldsymbol{\sigma}+\theta\boldsymbol{\mu}). \qquad (120.15)$$

Recalling the meaning of the effective coefficients, together with the relations $\boldsymbol{\sigma} = \mathbf{y}_e'+O(3)$, $\boldsymbol{\mu} = \mathbf{y}_1+O(3)$ which appear just after equations (24.8), we have finally

$$\left.\begin{aligned}
p_1^* &= (s-m)\,\Sigma\mathfrak{p},\\
p_2^* &= 2(s-m)\,\Sigma\theta\mathfrak{p},\\
p_3^* &= (s-m)\,\Sigma[\theta^2\mathfrak{p}+\tfrac{1}{2}(s-m)^{-2}\tilde{\omega}],\\
\bar{p}_1^* &= (s-m)\,\Sigma\theta\mathfrak{p},\\
\bar{p}_2^* &= 2(s-m)\,\Sigma\theta^2\mathfrak{p},\\
\bar{p}_3^* &= (s-m)\,\Sigma\theta[\theta^2\mathfrak{p}+\tfrac{1}{2}(s-m)^{-2}\tilde{\omega}].
\end{aligned}\right\} \qquad (120.16)$$

The summations of course go over all the surfaces of the system.

We may also remind ourselves that the Seidel coefficients $\sigma_1, \ldots, \sigma_5$ in (19.7) are given by

$$\sigma_1 = p_1^*, \quad \sigma_2 = \tfrac{1}{2}p_2^*, \quad \sigma_3 = \tfrac{1}{2}\bar{p}_2^*, \quad \sigma_4 = p_3^* - \tfrac{1}{2}\bar{p}_2^*, \quad \sigma_5 = \bar{p}_3^*. \tag{120.17}$$

To the present order the complications which arise from the asphericity of refracting surfaces are minor, and we proceed to investigate them. Since the equation of a spherical surface is $s = cu + \frac{1}{4}c^3u^2 + O(6)$, we write that of an aspherical surface as

$$s = cu + \tfrac{1}{4}c^3(1+a)\,u^2 + O(6), \tag{120.18}$$

where the constant $a$ is a measure of the asphericity near the pole, whilst $c$ is now the *axial* curvature of $\mathscr{S}$, i.e. the curvature at its pole. In (117.23) we now have

$$s_1 = c, \quad s_2 = \tfrac{1}{4}c^3(1+a), \tag{120.19}$$

and (117.24) may then be written conveniently as

$$u\dot{s} - \tfrac{1}{2}s = (u\dot{s} - \tfrac{1}{2}s)_s - \tfrac{1}{8}ra\psi^2 + O(6), \tag{120.20}$$

the first term on the right representing that part of the complete expression which remains when $a = 0$. Now $\psi$ is defined by (117.15), and upon expanding it in powers of $\bar{\xi}, \bar{\eta}, \bar{\zeta}$ it turns out that

$$\psi = (\Delta N)^{-2}\,[(\Delta N\beta)^2 + (\Delta N\gamma)^2] + O(4). \tag{120.21}$$

If we denote by $D(\mathbf{g})$ the terms which have to be added to the right-hand member of (120.5), we have, in view of (117.12) and (119.4)

$$\begin{aligned} D(\mathbf{g}) &= -\tfrac{1}{4}(\Delta N)\,ar\,\psi\nabla\left(v\frac{\partial\psi}{\partial\boldsymbol{\beta}}\right) + O(5) \\ &= -\tfrac{1}{2}ra(\Delta N)^{-3}(\Delta Nv)(\Delta N\boldsymbol{\beta})\,[(\Delta N\beta)^2 + (\Delta N\gamma)^2] + O(5). \end{aligned}$$

However, $\quad \Delta Nv = cy\Delta N, \quad \Delta N\boldsymbol{\beta} = c\mathbf{y}_P\Delta N + O(3),$

and therefore

$$D(\mathbf{g}) = -\tfrac{1}{2}c^3a(\Delta N)\,y\mathbf{y}_P(y_P^2 + z_P^2) + O(5). \tag{120.22}$$

On account of the particular form of this result it is pointless to use (118.22) here. It is better to set

$$\tilde{\theta} = y_b/y_a, \tag{120.23}$$

introducing the abbreviation

$$\mathfrak{c} = -\tfrac{1}{2}c^3ay^4\Delta N \tag{120.24}$$

at the same time. Then

$$D(\mathbf{g}) = (\xi + 2\tilde{\theta}\eta + \tilde{\theta}^2\zeta)(\boldsymbol{\sigma} + \tilde{\theta}\boldsymbol{\mu}) + O(5), \tag{120.25}$$

and then the primary effective coefficients are given by (120.16) if one merely makes the formal replacements

$$\theta^s\mathfrak{p} \to \theta^s\mathfrak{p} + \tilde{\theta}^s\mathfrak{c} \quad (s = 0, 1, 2, 3) \tag{120.26}$$

throughout.

## 121. The third-order aberration function

We write

$$t^{(2)} = p_1\xi^2 + p_2\xi\eta + p_3\xi\zeta + p_4\eta^2 + p_5\eta\zeta + p_6\zeta^2, \tag{121.1}$$

as was done after equation (44.7), i.e. the factor $(s-m)^{-1}$ on the right of (19.24) is omitted. Then, referring to (23.2), we know that

$$\left.\begin{aligned} p_1^* &= 4(s-m)p_1, \quad p_2^* = 2(s-m)p_2, \quad p_3^* = 2(s-m)p_3,\\ \bar{p}_1^* &= (s-m)p_2, \quad \bar{p}_2^* = 2(s-m)p_4, \quad \bar{p}_3^* = (s-m)p_5. \end{aligned}\right\} \tag{121.2}$$

By inspection of (120.16) we see that the integrability condition (23.8), which to this order requires $2\bar{p}_1^* - p_2^* = 0$, is satisfied, as it must be. Comparison of (121.2) with (120.16) then shows that

$$\left.\begin{aligned} p_1 &= \tfrac{1}{4}\Sigma\mathfrak{p}, \quad p_2 = \Sigma\theta\mathfrak{p}, \quad p_3 = \tfrac{1}{2}\Sigma[\theta^2\mathfrak{p} + \tfrac{1}{2}(s-m)^{-2}\tilde{\omega}],\\ p_4 &= \Sigma\theta^2\mathfrak{p}, \quad p_5 = \Sigma\theta[\theta^2\mathfrak{p} + \tfrac{1}{2}(s-m)^{-2}\tilde{\omega}]. \end{aligned}\right\} \tag{121.3}$$

These, then, are the expressions required for the calculation of the coefficients of $t^{(2)}$, $p_6$ excepted. Their structure is evidently exceedingly simple; and this is still true when the surface is aspherical, a situation which is allowed for by the formal replacements (120.26). In Section 125 we shall return to the 'missing' coefficient $p_6$.

The route we have pursued to arrive at (121.3) seems somewhat roundabout at first sight. *In principle* we start with the angle characteristic, use its derivatives to compute the displacement, and then return to the angle characteristic by integration (for that is what 'the comparison of (121.2) with (120.16)' amounts to). We therefore naturally ask ourselves whether, to the required order, we might not have eliminated the redundant variables directly from $t_j$. Here an *apparent* complication arises. Thus, write

$$T_j = T_j^{(0)} + T_j^{(1)} + t_j, \qquad (121.4)$$

where $T_j^{(0)}$ is a constant, and $T_j^{(1)}$ stands for the terms of $T_j$ linear in $\bar{\xi}, \bar{\eta}, \bar{\zeta}$. Then, recalling (44.1), we have

$$t = \sum_{j=1}^{k} (T_j^{(1)} + t_j) - \zeta/2m. \qquad (121.5)$$

It is from the sum which appears here that the redundant variables have to be removed; but now, upon using the appropriate relations such as

$$\boldsymbol{\beta}_j = v_{aj}\boldsymbol{\sigma} + v_{bj}\boldsymbol{\mu} + O(3), \qquad (121.6)$$

it seems at first sight that the terms of order three in them will cause $T_j^{(1)}$ to contribute third- (and higher-) order terms to $t_j$. In other words, it appears that one cannot simply take $t^{(2)} = \sum t_j^{(2)}$ and use the paraxial relations to eliminate the intermediate variables. Happily it turns out that, as far as $t^{(2)}$ (but not $t^{(3)}, \ldots$) is concerned, one can do what has just been described after all. That this is so will be shown in Section 124, for by that time we shall have established certain incidental results which will enable us to elaborate upon the point in question in the detail which its importance demands.

## 122. The absolute invariant $\alpha_3$

In Chapter 5D we studied at length the so-called invariant and semi-invariant aberrations of symmetric systems. We found that there was only one such *third*-order invariant, namely $\alpha_3$, and it was absolute, i.e. independent of both $s$ and $m$. According to (50.3)

$$\alpha_3 = 2(s-m)^2(2p_3 - p_4). \qquad (122.1)$$

However, from (121.3) we know that

$$p_3 = \tfrac{1}{2}\sum[\theta^2\mathfrak{p} + \tilde{\theta}^2\mathfrak{c} + \tfrac{1}{2}(s-m)^{-2}\tilde{\omega}], \quad p_4 = \sum(\theta^2\mathfrak{p} + \tilde{\theta}^2\mathfrak{c}),$$

from which it follows that $\quad \alpha_3 = \sum\tilde{\omega}. \qquad (122.2)$

The sum on the right is called the *Petzval Sum*. Since it depends solely on the axial curvatures of the surfaces and the refractive indices of the various media, it is obviously independent of $s$ and $m$, as it must be.

## 123. Intermediate variables as functions of the coordinates

If one wishes to go beyond the third order the paraxial relations between the intermediate variables and the coordinates are no longer adequate. The question therefore arises as to the exact relations which subsist between them, taking, for the time being, the $\mathbf{V}_j$ and $\mathbf{V}'_j$ as intermediate variables, and $\mathbf{S}$ and $\mathbf{M}$ as 'coordinates'. To answer it, we begin by simultaneously contemplating two angle characteristics $T_a$ and $T_b$ which differ from each other with respect to the choice of base-points. $T_a$ is the function considered hitherto, i.e. $T_a$ has base-points $O_0$, $O'_0$, and the individual parts $T_{aj}$ of course relate to the base-points $O_{0j}$, $O'_{0j}$; so that, for instance, $q_j$ in (117.17) is $-y_{aj}/v_{aj}$ as before. $T_b$ on the other hand refers to the base-points $E$, $E'$, and $T_{bj}$ to $E_j$, $E'_j$; and in this case the constant $q_j$ is to be taken as $-y_{bj}/v_{bj}$. The two functions are simply related to each other:

$$T_{bj} - T_{aj} = \Delta[N(q_b - q_a)\,\alpha]_j. \tag{123.1}$$

Using (118.18–19) this becomes

$$T_{bj} - T_{aj} = (s-m)^{-1} f_K \Delta(\alpha/v_a v_b)_j. \tag{123.2}$$

If one sums this over the whole system, and takes the definitions of $s$ and $m$ into account, one gets

$$T_b - T_a = -(s-m) f_K(\alpha'_k - \alpha_1/sm) = -d'\alpha'_k - d\alpha_1, \tag{123.3}$$

a result which can be seen to hold from first principles, bearing in mind the meaning of $d$ and $d'$ in the standard diagram represented by Fig. 4.1.

The various definitions and formulae of Section 119 are now to be thought of as duplicated. In particular, in (119.4) $v_a$ goes with $T_a$ and $v_b$ with $T_b$, i.e.

$$\mathbf{g}_a = \nabla\left(v_a \frac{\partial T_a}{\partial \boldsymbol{\beta}}\right), \quad \mathbf{g}_b = \nabla\left(v_b \frac{\partial T_b}{\partial \boldsymbol{\beta}}\right), \tag{123.4}$$

whilst in place of (119.2),

$$\mathbf{g}_a = \Delta\boldsymbol{\Lambda}_a = -\Delta(Nv_a\mathbf{H}), \quad \mathbf{g}_b = \Delta\boldsymbol{\Lambda}_b = -\Delta(Nv_b\mathbf{H}_E), \tag{123.5}$$

where, in accordance with previous definitions, $\mathbf{H}$ and $\mathbf{H}_E$ are respectively the coordinates of the points of intersection of $\mathscr{R}$ with the normal planes through $O_{0j}$ and $E_j$.

The following notation is well adapted to the problem in hand. We write

$$\mathbf{G}_{a;j} = \sum_{i=1}^{j-1} \mathbf{g}_{ai}, \quad \mathbf{G}_{a:j} = \sum_{i=j}^{k} \mathbf{g}_{ai}, \tag{123.6}$$

with analogous definitions when the subscript $a$ is replaced by $b$. When the symbols on the left are primed, the summations go from 1 to $j$ and from $j+1$ to $k$ respectively, and the sum is to be regarded as zero when the upper limit of the summation is less than the lower. (If one wishes to omit the subscripts $j$, the semicolon and colon must of course be retained.) We also set

$$\mathbf{G}_{a;j} + \mathbf{G}_{a:j} \equiv \sum_{i=1}^{k} \mathbf{g}_{ai} \equiv \mathbf{G}_{a}. \tag{123.7}$$

In view of (119.2) one can now write down the somewhat trivial set of four equations

$$\begin{aligned} N_1 v_{a1} \mathbf{H}_1 - N_j v_{aj} \mathbf{H}_j &= \mathbf{G}_{a;j}, \\ N_1 v_{b1} \mathbf{H}_{E1} - N_j v_{bj} \mathbf{H}_{Ej} &= \mathbf{G}_{b;j}, \\ N_j v_{aj} \mathbf{H}_j - N'_k v'_{ak} \mathbf{H}'_k &= \mathbf{G}_{a:j}, \\ N_j v_{bj} \mathbf{H}_{Ej} - N'_k v'_{bk} \mathbf{H}'_{Ek} &= \mathbf{G}_{b:j}. \end{aligned} \tag{123.8}$$

From the first and second pairs of these one finds without difficulty that

$$\begin{aligned} s\mathbf{G}_{a;j} + m\mathbf{G}_{b;j} &= -N_j(v_{aj} s\mathbf{H}_j + v_{bj} m\mathbf{H}_{Ej}) - f_K \mathbf{V}_1, \\ \mathbf{G}_{a:j} + \mathbf{G}_{b:j} &= N_j(v_{aj} \mathbf{H}_j + v_{bj} \mathbf{H}_{Ej}) + f_K \mathbf{V}'_k. \end{aligned} \tag{123.9}$$

Now multiply the second of these alternatively by $s$ and $m$ and then add it to the first. Setting $f_K = 1$ as often before, one then gets

$$\left.\begin{aligned} N_j v_{bj}(s-m) \mathbf{H}_{Ej} &= \mathbf{S} + (s\mathbf{G}_a + m\mathbf{G}_{b;j} + s\mathbf{G}_{b:j}), \\ N_j v_{aj}(s-m) \mathbf{H}_j &= -\mathbf{M} - (s\mathbf{G}_{a;j} + m\mathbf{G}_{a:j} + m\mathbf{G}_b). \end{aligned}\right\} \tag{123.10}$$

Upon multiplying the first throughout by $v_{aj}$ and the second by $v_{bj}$ and mutually subtracting the resulting equations, there comes finally

$$\mathbf{V}_j = v_{aj}(\mathbf{S} + \boldsymbol{\delta}_{sj}) + v_{bj}(\mathbf{M} + \boldsymbol{\delta}_{mj}), \tag{123.11}$$

where

$$\boldsymbol{\delta}_{sj} = m\mathbf{G}_{b;j} + s\mathbf{G}_{b:j} + s\mathbf{G}_a, \quad \boldsymbol{\delta}_{mj} = s\mathbf{G}_{a;j} + m\mathbf{G}_{a:j} + m\mathbf{G}_b. \tag{123.12}$$

Clearly $\mathbf{V}'_j$ is given by an expression just like (123.11), the only formal change being the appearance of primes attached to the paraxial coefficients and to the functions $\boldsymbol{\delta}_{sj}$, $\boldsymbol{\delta}_{mj}$. The latter are in turn given by expressions like (123.12), but with primes attached to all functions which have an index $j$, new ranges of summation being so implied, in accordance with the remark following equations (123.6).

For quantities linearly related to $\mathbf{V}_j$ and $\mathbf{H}_j$ one can write down convenient equations exactly analogous to (123.11), the only change being in the paraxial coefficients. For example, if $\mathbf{Y}_j$ denotes the coordinates of the point of intersection of $\mathscr{R}$ with the polar tangent plane of $\mathscr{S}_j$, we have

$$\mathbf{Y} = \mathbf{H} - l_0 \mathbf{V}, \tag{123.13}$$

where $l_0 = -y_a/v_a$. The second member of (123.10), together with (123.11), then gives

$$\mathbf{Y}_j = y_{aj}(\mathbf{S} + \boldsymbol{\delta}_{sj}) + y_{bj}(\mathbf{M} + \boldsymbol{\delta}_{mj}), \tag{123.14}$$

(118.18–19) having been used to deal with the various paraxial quantities. Paraxially there is of course no distinction between $\mathbf{Y}$ and $\mathbf{y}(\equiv \mathbf{y}_P)$.

It may be noted that in the course of computation one may obtain the so-called *increments* $\boldsymbol{\delta}_{sj}$ and $\boldsymbol{\delta}_{mj}$ by referring them back to $\boldsymbol{\delta}_{s1}$ and $\boldsymbol{\delta}_{m1}$. Thus, if one simply writes

$$\mathbf{G} = \mathbf{G}_a + \mathbf{G}_b, \tag{123.15}$$

one has from (123.12), by inspection,

$$\boldsymbol{\delta}_{s1} = s\mathbf{G}, \quad \boldsymbol{\delta}_{m1} = m\mathbf{G}. \tag{123.16}$$

Since

$$\Delta\boldsymbol{\delta}_s = -(s-m)\,\mathbf{g}_b, \quad \Delta\boldsymbol{\delta}_m = (s-m)\,\mathbf{g}_a, \tag{123.17}$$

$\boldsymbol{\delta}_{s2}, \boldsymbol{\delta}_{s3}, \ldots, \boldsymbol{\delta}_{m2}, \ldots$ are then obtained successively, starting from the first surface.

So far we have contemplated in this section the variables $\mathbf{V}_j$ and $\mathbf{V}'_j$ and those which are naturally associated with them, such as $\mathbf{Y}_j$, $\mathbf{H}_j$, etc. We could equally well have focused our attention on $\boldsymbol{\beta}_j$ and $\boldsymbol{\beta}'_j$; and then, of course, $\mathbf{Y}^*_j\,(\equiv \alpha_j \mathbf{Y}_j)$ must appear where $\mathbf{Y}_j$ appeared before, whilst $\mathbf{H}^*_j$ similarly replaces $\mathbf{H}_j$, and so on. The formal modifications of the various equations are of a trivial nature, and one typical example will suffice to show what is involved. Thus, in place of (123.11) we shall have

$$\boldsymbol{\beta}_j = v_{aj}(\boldsymbol{\sigma} + \boldsymbol{\delta}_{\sigma j}) + v_{bj}(\boldsymbol{\mu} + \boldsymbol{\delta}_{\mu j}), \tag{123.18}$$

where
$$\left.\begin{aligned}\boldsymbol{\delta}_{\sigma j} &= m\mathbf{G}^*_{b;j} + s\mathbf{G}^*_{b:j} + s\mathbf{G}^*_a,\\ \boldsymbol{\delta}_{\mu j} &= s\mathbf{G}^*_{a;j} + m\mathbf{G}^*_{a:j} + m\mathbf{G}^*_b,\end{aligned}\right\} \tag{123.19}$$

with
$$\mathbf{G}^*_{a;j} = \sum_{i=1}^{j-1} \mathbf{g}^*_{ai}, \tag{123.20}$$

and so on.

It is rather striking that one can also write down equations for the intermediate variables in terms of the (reduced) ray-coordinates $\mathbf{y}'$ ($\equiv N'_k\mathbf{H}'_{Ek}$) and $\mathbf{y}_1$ ($\equiv N_1 m\mathbf{H}_1$), appropriate to the point characteristic, which closely correspond to (123.11), but are in fact formally somewhat simpler than these. Thus the first and fourth members of (123.8) read

$$\left.\begin{aligned}N_j v_{aj}\mathbf{H}_j &= m^{-1}v_{a1}\mathbf{y}_1 - \mathbf{G}_{a;j},\\ N_j v_{bj}\mathbf{H}_{Ej} &= v'_{bk}\mathbf{y}' + \mathbf{G}_{b:j}.\end{aligned}\right\} \tag{123.21}$$

Now

$$Nv_a v_b(\mathbf{H}_E - \mathbf{H}) = N(y_a v_b - y_b v_a)\,\mathbf{V} = (s-m)^{-1} f_K\,\mathbf{V}, \tag{123.22}$$

because of (118.19). (123.21) then gives straight away the relation

$$f_K\mathbf{V}_j = v_{aj}[\mathbf{y}' + (s-m)\,\mathbf{G}_{b:j}] + v_{bj}[\mathbf{y}_1 + (s-m)\,\mathbf{G}_{a;j}]. \tag{123.23}$$

This result is far from being of merely academic interest, for the following reason. We shall see later how $T$ may be computed, and, if it is to be a characteristic function in the proper sense of the term, it must be exhibited as a function of the appropriate variables, say $\boldsymbol{\sigma}$ and $\boldsymbol{\mu}$. The present method, however, has inherently the kind of flexibility to allow just as easily the computation of $T$—regarded merely as an optical distance—as a function of $\mathbf{y}'$ and $\mathbf{y}_1$; and in the course of this (123.23) will play a crucial part. On the other hand, we have the general relation

$$V = T - d'\alpha' + \boldsymbol{\beta}'.\mathbf{y}' - \boldsymbol{\beta}.\mathbf{y}_1/m \tag{123.24}$$

(where the second term on the right represents the movement of the posterior base-point from $O'_0$ to $E'$) so that the knowledge of $T(\mathbf{y}', \mathbf{y}_1)$ implies that of $V(\mathbf{y}', \mathbf{y}_1)$. The difficulties outlined in Section 117, at any rate, evidently do not arise here, because the *point* characteristics of the *individual* surfaces are never contemplated.

## 124. On the direct computation of the third-order characteristic coefficients

We return to the general question of the *direct* computation of the coefficients of $t$, meaning thereby any method which does not go through the displacement $\boldsymbol{\epsilon}'$. In particular, we shall consider the special case of $t^{(2)}$ in detail, in view of the apparent difficulty which we encountered at the end of Section 121. To deal with the general principles involved, let

$$T = \sum_{j=1}^{k} T_j(\bar{\xi}_j, \bar{\eta}_j, \bar{\zeta}_j) \qquad (124.1)$$

be given, the base-points $O_{0j}$, $O'_{0j}$ being understood. Now, for any ray through $K$, $\mathbf{H}_{j+1} = \mathbf{H}'_j$, i.e.

$$\frac{\partial T_{j+1}}{\partial \boldsymbol{\beta}_{j+1}} + \frac{\partial T_j}{\partial \boldsymbol{\beta}'_j} = 0 \quad (j = 2, \ldots, k). \qquad (124.2)$$

However, $\boldsymbol{\beta}_{j+1} \equiv \boldsymbol{\beta}'_j$, so that, in view of (124.1), we have the $k-1$ relations

$$\frac{\partial T}{\partial \boldsymbol{\beta}_j} = 0 \quad (j = 2, \ldots, k). \qquad (124.3)$$

These evidently express the fact that $T$ is stationary with respect to small variations of the intermediate variables: which, the initial and final rays being fixed, is simply a direct reflection of Fermat's Principle. Moreover, they constitute a set of $2k-2$ equations, which allow us, in principle, to express $\boldsymbol{\beta}_2, \ldots, \boldsymbol{\beta}_k$ as functions of $\boldsymbol{\beta}_1$ and $\boldsymbol{\beta}'_k$, or what comes to the same thing, of $\boldsymbol{\sigma}$ and $\boldsymbol{\mu}$; and the *exact* solution of these equations must of necessity just have the form (121.6). We thus have

$$\boldsymbol{\beta}_j = v_{aj}\,\boldsymbol{\sigma} + v_{bj}\,\boldsymbol{\mu} + \mathbf{B}_j, \qquad (124.4)$$

where the $\mathbf{B}_j$ are certain expressions which contain no linear terms:

$$\mathbf{B}_j = O(3). \qquad (124.5)$$

$T$, as given by (124.1), is as yet a function of $\boldsymbol{\sigma}$, $\boldsymbol{\mu}$ and $\boldsymbol{\beta}_2, \boldsymbol{\beta}_3, \ldots, \boldsymbol{\beta}_k$; and it is convenient to write it simply as

$$T = J(\boldsymbol{\sigma}, \boldsymbol{\mu}; \boldsymbol{\beta}_2, \ldots, \boldsymbol{\beta}_k) = J(\boldsymbol{\sigma}, \boldsymbol{\mu}; \boldsymbol{\beta}), \qquad (124.6)$$

so that $\boldsymbol{\beta}$ stands here for the whole set of intermediate variables. (124.4) is now to be inserted into (124.6). There comes

$$\left.\begin{aligned} T = T(\boldsymbol{\sigma},\boldsymbol{\mu}) &= J(\boldsymbol{\sigma},\boldsymbol{\mu};v_a\boldsymbol{\sigma}+v_b\boldsymbol{\mu}+\mathbf{B}) \\ &= J(\boldsymbol{\sigma},\boldsymbol{\mu};v_a\boldsymbol{\sigma}+v_b\boldsymbol{\mu})+\sum_{j=2}^{k}\left(\frac{\partial J}{\partial\beta_j}B_{yj}+\frac{\partial J}{\partial\gamma_j}B_{zj}\right)+\delta J, \end{aligned}\right\} \qquad (124.7)$$

where $\delta J$ involves second and higher derivatives of $J$. (All derivatives are evaluated at $\mathbf{B} = 0$.) Now, according to (124.3) the first derivatives on the right of (124.7) vanish. Further, all terms of $\delta J$ are at least quadratic in the $\mathbf{B}_j$, so that in view of (124.5), $J = O(6)$. It follows that

$$T = J(\boldsymbol{\sigma},\boldsymbol{\mu},v_a\boldsymbol{\sigma}+v_b\boldsymbol{\mu})+O(6). \qquad (124.8)$$

We therefore conclude that, upon substituting in (124.1) for the $\boldsymbol{\beta}_j$ their *paraxial* expressions in terms of $\boldsymbol{\sigma}$ and $\boldsymbol{\mu}$, the resulting function differs from the correct angle characteristic by terms which are of degree not less than *six*. On the other hand, the substitution of linear expressions in $T_j^{(1)}$ leaves the latter linear in $\xi$, $\eta$, $\zeta$, so that we are indeed justified in taking

$$t^{(2)} = \sum_{j=1}^{k} t_j^{(2)}(\bar{\xi}_j,\bar{\eta}_j,\bar{\zeta}_j), \qquad (124.9)$$

and simply using (118.5) to eliminate the intermediate variables. In short, the effect of the $\mathbf{B}_i$ induced by $\sum T_j^{(1)}$ in $t^{(2)}$ is nugatory.

This conclusion, though elementary in character, is so striking that it will not come amiss to confirm it by detailed calculation, in the light of the equations of Section 123. This may be done as follows. From (117.17) and (119.8)

$$T_j^{(1)} = \tfrac{1}{2}[f(\bar{\xi}-2\bar{\eta}+\bar{\zeta})-N'(l_0'-r)\bar{\xi}+N(l_0-r)\bar{\zeta}], \quad (124.10)$$

the index $j$ being suppressed on the right. By means of (119.6–7) we easily convince ourselves that the factors multiplying $\frac{1}{2}\bar{\xi}$ and $\frac{1}{2}\bar{\zeta}$ are $fv_a/v_a'$ and $fv_a'/v_a$ respectively, so that (124.10) becomes

$$T_j^{(1)} = \tfrac{1}{2}fv_av_a'\{[\Delta(\beta/v_a)]^2+[\Delta(\gamma/v_a)]^2\}. \qquad (124.11)$$

Now, from (123.18),

$$\Delta(\boldsymbol{\beta}/v_a) = \boldsymbol{\mu}\Delta\bar{\theta}+\Delta(\boldsymbol{\delta}_\sigma+\bar{\theta}\boldsymbol{\delta}_\mu) = \boldsymbol{\mu}\Delta\bar{\theta}+\mathbf{D}, \qquad (124.12)$$

say, where we have set $\qquad \bar{\theta} = v_b/v_a. \qquad (124.13)$

Accordingly (124.11) becomes

$$T_j^{(1)}(\bar{\xi}_j, \bar{\eta}_j, \bar{\zeta}_j) = \tfrac{1}{2} f v_a v'_a [\zeta(\Delta\bar{\theta})^2 + 2(\Delta\bar{\theta})\,\boldsymbol{\mu}.\mathbf{D} + \mathbf{D}.\mathbf{D}], \quad (124.14)$$

in a notation suggested just after equation (14.4). Summing over $j$ from 1 to $k$, we evaluate the various terms in turn. To begin with,

$$f v_a v'_a \Delta\bar{\theta} = f\phi(1-k)\, i_a = (s-m)^{-1}, \quad (124.15)$$

by (118.20–21) and (119.8). Hence

$$\sum f v_a v'_a (\Delta\bar{\theta})^2 = (s-m)^{-1}(\bar{\theta}'_k - \bar{\theta}_1) = 1/m, \quad (124.16)$$

since $\bar{\theta}'_k = -1$ and $\bar{\theta}_1 = -s/m$, according to (118.6). As regards the second term on the right of (124.14) we have, with (124.15),

$$(s-m)^{-1}\boldsymbol{\mu}.\sum\mathbf{D} = (s-m)^{-1}\boldsymbol{\mu}.(\mathbf{D}'_k - \mathbf{D}_1).$$

In view of (123.16) $\mathbf{D}_1$ vanishes. Again, $\boldsymbol{\delta}'_{\sigma k} = \boldsymbol{\delta}'_{\mu k} = s\mathbf{G}_a + m\mathbf{G}_b$, and so $\mathbf{D}'_k$ also vanishes. Thus we have finally

$$\sum T_j^{(1)}(\bar{\xi}_j, \bar{\eta}_j, \bar{\zeta}_j) = \zeta/2m + \tfrac{1}{2}\sum_j f v_a v'_a \mathbf{D}.\mathbf{D}. \quad (124.17)$$

Here (121.5) may be recalled: we see that there are indeed no terms of the fourth degree on the right of (124.17) at all. To this extent we have therefore confirmed explicitly the result we got earlier in this section. Here, however, we have more, in as far as the second term on the right of (124.17) is an explicit expression for the contribution by $\sum T_j^{(1)}$ to the terms of the sixth and higher degrees of $T$; see Section 130.

Before going on to consider the question of higher-order coefficients it will be wise to return briefly to the results of Section 121, with particular reference to the coefficient $p_6$ which was not included in (121.3).

## 125. The coefficient $p_6$. Incidental remarks concerning duality

To begin with, consider the determination of $p_6$ by the direct method described in the preceding section. From (117.17) one has at once

$$T = T^{(0)} + r\kappa(\Delta N)(\chi - \tfrac{1}{2}\kappa\chi^2) - \tfrac{1}{2}N'(l'_0 - r)(\xi + \tfrac{1}{4}\xi^2)$$
$$+ \tfrac{1}{2}N(l_0 - r)(\bar{\zeta} + \tfrac{1}{4}\bar{\zeta}^2) + O(6), \quad (125.1)$$

the index $j$ having been suppressed throughout. Since

$$\chi = \tfrac{1}{2}(\bar{\xi} - 2\bar{\eta} + \bar{\zeta}) + \tfrac{1}{8}(\bar{\xi} - \bar{\zeta})^2 + O(6), \qquad (125.2)$$

we then get from (125.1)

$$t^{(2)} = \tfrac{1}{8}Nr\{(k-1)^{-1}[\kappa(\bar{\xi} - 2\bar{\eta} + \bar{\zeta})^2 - (\bar{\xi} - \bar{\zeta})^2] + i_a[(\bar{\xi}^2/v'_a) - (\bar{\zeta}^2/v_a)]\}. \qquad (125.3)$$

In principle it only remains to make the usual paraxial substitutions for $\boldsymbol{\beta}$ and $\boldsymbol{\beta}'$. In this way one does indeed recover (121.3) exactly. On the other hand it appears that the labour involved in this procedure exceeds that required to go through the steps which lead from (120.5) to (120.15). This is largely because, whereas the expression (125.3) is quartic in the intermediate variables, (120.5) is only cubic; whilst this cubic has, at the same time, a linear factor. These simplifying features, taken together, probably make the method of Section 120 much the most convenient as far as the third-order aberrations are concerned. It is deficient only in as far as it does not provide us with the value of the coefficient $p_6$ which multiplies $\zeta^2$ in $t^{(2)}$, a coefficient required both in the context of object shifts and of the exact higher-order displacement.

To remedy this situation it suffices to make the appropriate paraxial substitutions, taking $\xi = \eta = 0$, i.e. one sets

$$\bar{\xi} \to v_b'^2\zeta, \quad \bar{\eta} \to v_b v_b' \zeta, \quad \bar{\zeta} \to v_b^2 \zeta, \qquad (125.4)$$

using (118.21) at the same time. Then

$$p_6 = \tfrac{1}{8}Nr\{(k-1)^{-1}[k(k-1)^2\theta^4 i^4 - (\theta^2\Delta v^2 + 2\phi\theta\Delta v)^2] + i\Delta[(\theta v + \phi)^4/v]\}, \qquad (125.5)$$

the subscript $a$ on the paraxial coefficients being understood, as in (120.14). The factors multiplying the various powers of $\theta$ may be simplified in the usual way. Upon finally summing over $j$ one gets the desired result,

$$p_6 = \tfrac{1}{4}\Sigma[\theta^4\mathfrak{p} + (s-m)^{-2}\tilde{\omega}(\theta^2 - \tfrac{1}{2}\phi^2/vv')]. \qquad (125.6)$$

It is instructive to consider (125.6) from another point of view. We have already introduced $T_b$ alongside $T_a$ as a device designed towards being able to write down formally explicit equations such as (123.11), representing the solution of (124.3). The relation between $T_b$ and $T_a$, or equivalently between $T_{bj}$ and $T_{aj}$, is very simple, and given the one, the other need not be calculated separately. At any rate, supplying the coefficients of the power series for $T_b$ and $T_a$ with additional subscripts $b$ and $a$ respectively, a moment's reflection shows that $p_{b6j}$ must be the same function of the paraxial $b$-coefficients as $p_{a1j}$ is of the $a$-coefficients. Thus, recalling (120.14) and the first member of (121.3),

$$p_{b6} = \tfrac{1}{8}N(k-1)\,i_b^2\,y_b\,(i_b'+v_b). \tag{125.7}$$

Writing $i_b = \theta i_a$, etc., as usual, and leaving the index $a$ understood, (125.7) becomes

$$p_{b6} = \tfrac{1}{4}\theta^4\mathfrak{p} - \tfrac{1}{8}\theta^3(s-m)^{-1}(\Delta v^2) - \tfrac{1}{8}\theta^2(s-m)^{-2}\tilde{\omega}. \tag{125.8}$$

On the other hand, selecting the quadratic terms of (123.1), we have

$$t_a^{(2)} - t_b^{(2)} = \tfrac{1}{8}(s-m)^{-1}\left[(\bar{\xi}^2/v_a'v_b') - (\bar{\xi}^2/v_a v_b)\right]. \tag{125.9}$$

The substitutions (125.4) in this at once give, with (118.19),

$$\begin{aligned} p_{a6} - p_{b6} &= \tfrac{1}{8}(s-m)^{-1}\,\Delta(v_b^3/v_a) \\ &= \tfrac{1}{8}(s-m)^{-1}\left[\theta^3\Delta(v^2) + 3\theta^2\phi\Delta v + \phi^3\Delta(1/v)\right]. \end{aligned} \tag{125.10}$$

Upon combining this with (125.8) and summing over $j$, one exactly recovers (125.6).

One easily convinces oneself that if $t_b^{(2)}$ be obtained by going through the displacement after the fashion of Section 120—one is then considering the displacement associated with the pupil planes —it is $p_{b1}$, and not $p_{b6}$, which remains undetermined. In this sense, therefore, $p_{a6}$ *can* be obtained by dealing with displacements alone. This conclusion involves no contradiction with what was said earlier, since we have now been contemplating displacements relating to two distinct pairs of conjugate planes, and the respective coefficients are related to each other by equations given essentially by (45.9): and they involve $p_6$, that is to say, $p_{a6}$, in their right-hand members.

In practice one may well decide to calculate the coefficients of both $T_a$ and $T_b$ to whatever order may be required. In the first place one then has an excellent check upon one's calculations, in as far as one's results must conform with (123.3). Further, thinking in terms of digital computers, the set of *instructions* which have to be fed into the machine need virtually be no greater than if $T_a$ alone were being considered, provided one does *not* make use of relations such as (118.21); that is to say, the use of $\theta$ is to be abandoned. In that case the same set of instructions will yield $p_{a\alpha}$ $(\alpha = 1, 2, 3, 4, 5, 6)$ on the one hand and $p_{b\alpha}$ $(\alpha = 6, 5, 3, 4, 2, 1)$ on the other; only the numerical input data differing in the two cases. It is true that there will be redundancy in the output, but—numerical checks quite apart—it is very advantageous to have the various *a*- and *b*-quantities available independently of each other at the intermediate stages of higher-order calculations. In conclusion it should be mentioned that the possibility of using one set of instructions to compute two distinct sets of coefficients is the result of a formally 'symmetric' construction of the whole theory with respect to *a*-quantities and *b*-quantities: at the most rudimentary level even the initial *a*- and *b*-rays are calculated according to the *same* programme. At any rate, the situation just described is exactly that dealt with at length under the heading of *duality* in Section 6 of OAC XII, the details of which may be translated into the present context without undue difficulty.

## 126. The idea of iteration

The time has come to contemplate the computation of coefficients of higher order. Here we can do little more than to adumbrate the general principles involved, though enough detail will be provided so that, not forgetting the availability of the very explicit details of the Lagrangian method, it should not be too difficult to translate theory into practice.

For the purpose of discussing the *modus operandi* of the process of iteration about to be described, we concern ourselves for the present with the displacement, regarded as a function of **S** and **M**. Suppose, then, that at each surface the quantities $\mathbf{g}_{aj}$ and $\mathbf{g}_{bj}$, as defined by (119.9), have been written down as power series in the 'local' variables $\mathbf{V}_j$ and $\mathbf{V}'_j$:

$$\mathbf{g}_{aj} = \mathbf{g}_{aj}^{(2)} + \mathbf{g}_{aj}^{(3)} + \dots, \quad \mathbf{g}_{bj} = \mathbf{g}_{bj}^{(2)} + \mathbf{g}_{bj}^{(3)} + \dots, \tag{126.1}$$

where $\mathbf{g}_{aj}^{(n)}$ and $\mathbf{g}_{bj}^{(n)}$ are each homogeneous of degree $2n-1$. With the usual *paraxial* substitutions one then gets at once $\mathbf{g}_{ai}^{(2)}$ and $\mathbf{g}_{bi}^{(2)}$ as functions of **S** and **M**, since the non-paraxial terms of (123.11) are irrelevant to the dominant terms of the $\mathbf{g}_j$. Correctly to the third order, the increments $\boldsymbol{\delta}_{sj}$ and $\boldsymbol{\delta}_{mj}$, as given by (123.12), follow immediately by mere summation. Indeed, as suggested just after equation (123.15) we obtain them most easily from

$$\left.\begin{aligned} \boldsymbol{\delta}_{sj}^{(2)} &= s\sum_{i=1}^{k}(\mathbf{g}_{ai}^{(2)}+\mathbf{g}_{bi}^{(2)})-(s-m)\sum_{i=1}^{j-1}\mathbf{g}_{bi}^{(2)},\\ \boldsymbol{\delta}_{mj}^{(2)} &= m\sum_{i=1}^{k}(\mathbf{g}_{ai}^{(2)}+\mathbf{g}_{bi}^{(2)})+(s-m)\sum_{i=1}^{j-1}\mathbf{g}_{ai}^{(2)},\end{aligned}\right\} \qquad (126.2)$$

the first sum on the right, which is independent of $j$, being the same in each case. In view of (123.11) we now have at this stage expressions for $\mathbf{V}_j$ and $\mathbf{V}_j'$ in terms of $\mathbf{S}$ and $\mathbf{M}$ which are correct to the *third* order, e.g.

$$\mathbf{V}_j = v_{aj}(\mathbf{S}+\boldsymbol{\delta}_{sj}^{(2)})+v_{bj}(\mathbf{M}+\boldsymbol{\delta}_{mj}^{(2)})+O(5). \qquad (126.3)$$

Now return to (126.1), and let $\mathbf{g}_j$ stand alternatively for $\mathbf{g}_{aj}$ and $\mathbf{g}_{bj}$. Substitute the expression (126.3) for $\mathbf{V}_j$ and the corresponding expression for $\mathbf{V}_j'$ in $\mathbf{g}_j$, rejecting all terms of degree greater than five. This means that every $\mathbf{g}_j^{(n)}$ which has $n > 3$ is to be ignored, whilst in $\mathbf{g}_j^{(3)}$ one may content oneself with the usual paraxial substitutions. As regards $\mathbf{g}_j^{(2)}$ one recovers the previous third-order terms, but in addition they will contain fifth-order terms which arise from the *known* third-order expressions for $\boldsymbol{\delta}_{sj}^{(2)}, \ldots, \boldsymbol{\delta}_{mj}^{(2)\prime}$. Altogether we now have $\mathbf{g}_j$ correctly to the *fifth* order. In the same way as before, simple summation immediately yields the increments correctly to the fifth order, so that $\mathbf{V}_j$ and $\mathbf{V}_j'$ are now known as functions of $\mathbf{S}$ and $\mathbf{M}$ with an error which is $O(7)$. These are now used in (126.1); and upon selecting all the terms of degree seven one has, when these are taken together with the lower-order terms already determined, $\mathbf{g}_j$ correctly to the *seventh* order. It should be obvious by now how one can proceed systematically, step by step, to whatever order desired; and this process is called *iteration*. Note that if $\boldsymbol{\epsilon}'$ is required to order $2n-1$, $\mathbf{g}_b^{(n)}$ need not be calculated, since according to (120.3)

$$\boldsymbol{\epsilon}' = (s-m)\sum_{j=1}^{k}\sum_{s=1}^{n}\mathbf{g}_{aj}^{(s)}+O(2n+1), \qquad (126.4)$$

and in this only the $\mathbf{g}_{bj}^{(s)}$ with $s < n$ are required in the course of iteration.

Suppose, then, that $\boldsymbol{\epsilon}'$ has been obtained, to the required order, as a function of **S** and **M**. As far as the displacement is concerned the presence of these variables, rather than of $\boldsymbol{\sigma}$ and $\boldsymbol{\mu}$, does no harm, for in the paraxial approximation **S** and **M** have the same significance as $\boldsymbol{\sigma}$ and $\boldsymbol{\mu}$, and we may simply consider the pseudo-displacement now to be defined with respect to **S** and **M**. In any event, one has for the variables $\mathbf{y}'_e (\equiv \mathbf{H}'_{Ek})$ and $\mathbf{y}_1 (\equiv m\mathbf{H}_1)$, which occur in the series (23.1) defining the effective coefficients, the expressions

$$\mathbf{y}'_e = \mathbf{S} + \boldsymbol{\delta}'_{sk}, \quad \mathbf{y}_1 = \mathbf{M} + \boldsymbol{\delta}_{m1}, \tag{126.5}$$

and here everything is known to the required order. The effective coefficients are therefore obtainable in a manner entirely analogous to that set out in Section 24.

If the angle characteristic is to be found from $\boldsymbol{\epsilon}'$ by integration one must first eliminate **S** and **M** in favour of $\boldsymbol{\sigma}$ and $\boldsymbol{\mu}$. Trivial though this task may be in principle, it is somewhat cumbersome and unattractive in practice, and we would like to avoid it. In other words, we should proceed in terms of $\boldsymbol{\sigma}$ and $\boldsymbol{\mu}$, and therefore in terms of $\mathbf{g}_j^*$ rather than $\mathbf{g}_j$ from the outset, as we shall indeed do in Sections 127–9. The general process of iteration is, of course, exactly the same as before, in view of the generic form of equations (123.18) and (123.20). On the other hand, one now unfortunately ends up with $\boldsymbol{\epsilon}^{*\prime} (\equiv \mathbf{H}_k^{*\prime} - m\mathbf{H}_1^*)$, and this is not the displacement. Rather than discuss this point now, it is preferable to come back to it in Section 130, where we resume the investigation of the whole problem of computing the coefficients of $T$.

## 127. Modified increments. The variables $\boldsymbol{\sigma}_j$ and $\boldsymbol{\mu}_j$

As soon as one attempts actually to iterate in the manner described in the preceding section one becomes aware of a most irksome complication. It is a consequence of the fact that the increments appropriate to $\boldsymbol{\beta}_j$ differ from those appropriate to $\boldsymbol{\beta}'_j$. Were the increments the same, as they are in O(9.3) or O(12.9), any series in powers of $\boldsymbol{\beta}_j$, $\boldsymbol{\beta}'_j$ could be rewritten *exactly* as a series in powers of $\boldsymbol{\sigma}$ and $\boldsymbol{\mu}$ by first using the *paraxial* relations and then replacing $\boldsymbol{\sigma}$ by $\boldsymbol{\sigma} + \boldsymbol{\delta}_{\sigma j}$ and $\boldsymbol{\mu}$ by $\boldsymbol{\mu} + \boldsymbol{\delta}_{\mu j}$, such a two-step procedure being very convenient in practice; cf. Sections 11, 81 and 84 of OAC. Here, however, one could at best use, in the first step just described, relations of the kind

$$\boldsymbol{\beta}_j = v_{aj}\boldsymbol{\sigma} + v_{bj}\boldsymbol{\mu}, \quad \boldsymbol{\beta}'_j = v'_{aj}\dot{\boldsymbol{\sigma}} + v'_{bj}\dot{\boldsymbol{\mu}}, \tag{127.1}$$

the additional accents being intended to keep track of the fact that in the resulting '*pseudo-expansion*' one must eventually replace $\boldsymbol{\sigma}$ by $\boldsymbol{\sigma}+\boldsymbol{\delta}_{\sigma j}$, but $\dot{\boldsymbol{\sigma}}$ by $\boldsymbol{\sigma}+\boldsymbol{\delta}'_{\sigma j}$, and so on. Thus, already in the context of the third-order terms of $\mathbf{g}^*_{aj}$, say, one is temporarily concerned with forty instead of the usual six coefficients, and likewise in fifth order with 220 instead of the usual twelve. This is obviously a very unhappy situation and we must seek to avoid it. This may, indeed, be done as follows.

We *demand* that $\boldsymbol{\beta}_j$ and $\boldsymbol{\beta}'_j$ be given by expressions of the form

$$\left.\begin{aligned} \boldsymbol{\beta}_j &= v_{aj}(\boldsymbol{\sigma}+\mathbf{d}_{\sigma j}) + v_{bj}(\boldsymbol{\mu}+\mathbf{d}_{\mu j}), \\ \boldsymbol{\beta}'_j &= v'_{aj}(\boldsymbol{\sigma}+\mathbf{d}_{\sigma j}) + v'_{bj}(\boldsymbol{\mu}+\mathbf{d}_{\mu j}), \end{aligned}\right\} \tag{127.2}$$

the same *modified increments* $\mathbf{d}_{\sigma j}$, $\mathbf{d}_{\mu j}$ appearing in both equations. Then we must have

$$v'_a\mathbf{d}_\sigma + v'_b\mathbf{d}_\mu = v'_a\boldsymbol{\delta}'_\sigma + v'_b\boldsymbol{\delta}'_\mu, \quad v_a\mathbf{d}_\sigma + v_b\mathbf{d}_\mu = v_a\boldsymbol{\delta}_\sigma + v_b\boldsymbol{\delta}_\mu. \tag{127.3}$$

These equations are easily solved for $\mathbf{d}_\sigma$ and $\mathbf{d}_\mu$, and it turns out that the solution may be written compactly as

$$\mathbf{d}_\sigma = \boldsymbol{\delta}_\sigma + v_b\boldsymbol{\delta}, \quad \mathbf{d}_\mu = \boldsymbol{\delta}_\mu - v_a\boldsymbol{\delta}, \tag{127.4}$$

where
$$\begin{aligned} \boldsymbol{\delta} &= (v'_a v_b - v_a v'_b)^{-1}(v'_a\Delta\boldsymbol{\delta}_\sigma + v'_b\Delta\boldsymbol{\delta}_\mu) \\ &= (s-m)(v'_a v_b - v_a v'_b)^{-1}(v'_b\mathbf{g}^*_a - v'_a\mathbf{g}^*_b). \end{aligned} \tag{127.5}$$

It may be remarked that the same device may be employed in the context of equation (123.23). When the surface is spherical

$$\mathbf{g}^*_b = \theta\mathbf{g}^*_a,$$

as we know from (119.12); and then (127.5) reduces to

$$\boldsymbol{\delta} = (s-m)\,\tilde{\omega}^{-1}\mathbf{w}, \tag{127.6}$$

in view of (119.14). The expression on the right makes no explicit reference to the *a*- or *b*-rays whatsoever.

A moment's reflection will show that the device just described is exactly equivalent to introducing in place of $\boldsymbol{\beta}_j$ and $\boldsymbol{\beta}'_j$ local variables $\boldsymbol{\sigma}_j$ and $\boldsymbol{\mu}_j$, defined as

$$\left.\begin{aligned} \boldsymbol{\sigma}_j &= -(v'_{aj}v_{bj}-v_{aj}v'_{bj})^{-1}(v'_{bj}\boldsymbol{\beta}_j - v_{bj}\boldsymbol{\beta}'_j), \\ \boldsymbol{\mu}_j &= (v'_{aj}v_{bj}-v_{aj}v'_{bj})^{-1}(v'_{aj}\boldsymbol{\beta}_j - v_{aj}\boldsymbol{\beta}'_j), \end{aligned}\right\} \tag{127.7}$$

a step already suggested by the form of (124.11). Note that $\boldsymbol{\sigma}$ and $\boldsymbol{\mu}$ retain their previous significance. Using (127.2), there comes

$$\boldsymbol{\sigma}_j = \boldsymbol{\sigma} + \mathbf{d}_{\sigma j}, \quad \boldsymbol{\mu}_j = \boldsymbol{\mu} + \mathbf{d}_{\mu j}. \tag{127.8}$$

The formally extremely simple structure of these equations implies a corresponding degree of formal simplification with regard to iteration. This may be seen as follows. Given, as before, some series in ascending powers of $\boldsymbol{\beta}_j$ and $\boldsymbol{\beta}'_j$, we put

$$\boldsymbol{\beta}_j = v_{aj}\boldsymbol{\sigma}_j + v_{bj}\boldsymbol{\mu}_j, \quad \boldsymbol{\beta}'_j = v'_{aj}\boldsymbol{\sigma}_j + v'_{bj}\boldsymbol{\mu}_j, \tag{127.9}$$

according to (127.7). The corresponding pseudo-expansion arises by merely omitting from every $\boldsymbol{\sigma}_j$ and $\boldsymbol{\mu}_j$ the index $j$. The correct series is restored by subsequently replacing $\boldsymbol{\sigma}$ by $\boldsymbol{\sigma}+\mathbf{d}_{\sigma j}$ and $\boldsymbol{\mu}$ by $\boldsymbol{\mu}+\mathbf{d}_{\mu j}$. Of course, it must not be thought that the labour involved in obtaining the pseudo-expansion has been reduced in this way, for (127.9) represents, in effect, just the usual paraxial substitutions which lead to it.

Two small points are worthy of remark at this stage. The first concerns equation (119.4). Because of (127.7) we can now write

$$\mathbf{g}_{aj} = \frac{\partial T_{aj}}{\partial \boldsymbol{\sigma}_j}, \quad \mathbf{g}_{bj} = \frac{\partial T_{bj}}{\partial \boldsymbol{\mu}_j}, \tag{127.10}$$

but these seem to have no virtue other than formal elegance. Moreover, they revolve about $\mathbf{g}_j$, whereas there are no corresponding equations for the more important quantities $\mathbf{g}^*_j$. The second point relates to the meaning of the expression $y_a(\boldsymbol{\sigma}+\mathbf{d}_\sigma)+y_b(\boldsymbol{\mu}+\mathbf{d}_\mu)$. It cannot stand for either $\mathbf{Y}^*$ or $\mathbf{Y}^{*\prime}$ on account of its symmetry with respect to primed and unprimed variables, and we suspect that it is closely related to the coordinates of the point of incidence, $\mathbf{y}$. Indeed, using (127.7–8), one finds after a certain amount of manipulation involving (117.4) and (118.18–19) that for a spherical surface

$$y_a(\boldsymbol{\sigma}+\mathbf{d}_\sigma)+y_b(\boldsymbol{\mu}+\mathbf{d}_\mu) = (\sigma/\Delta N)\,\mathbf{y}, \tag{127.11}$$

which confirms our expectations.

## 128. Third- and fifth-order pseudo-coefficients

It may not be entirely out of place to set out in somewhat greater detail those steps of the iterative process which are required to obtain *fifth*-order coefficients. This may serve as a guide towards what has to be done in practice when proceeding to higher orders, for which reason the notation will on the whole be already adapted to the general case. Again, it will suffice to consider only spherical surfaces explicitly, to avoid a number of complications which may well be regarded as irrelevant to the immediate purpose in hand. Nevertheless, here also the details of the notation will be so arranged as to allow readily for the presence of aspherical surfaces. We operate with the variables $\boldsymbol{\beta}$ and $\boldsymbol{\beta}'$, bearing in mind that the formal changes required to go over to $\mathbf{V}$ and $\mathbf{V}'$ instead are of an elementary nature; and they revolve mainly about the detailed form of (128.14) and (128.17). This situation is to a large extent the result of the flexibility of the whole method.

At every surface we must first expand $\mathbf{w}$, as defined by (119.14) in ascending powers of $\boldsymbol{\beta}$ and $\boldsymbol{\beta}'$, rejecting all terms of degree greater than five. We therefore require only $\mathbf{w}^{(2)}$ and $\mathbf{w}^{(3)}$:

$$\mathbf{w}^{(2)} = \tfrac{1}{2}[\kappa(\bar{\xi}-2\bar{\eta}+\bar{\zeta})+\bar{\zeta}]\,\boldsymbol{\beta}'-\tfrac{1}{2}[\kappa(\bar{\xi}-2\bar{\eta}+\bar{\zeta})+\bar{\xi}]\,\boldsymbol{\beta}, \qquad (128.1)$$

$$\begin{aligned}\mathbf{w}^{(3)} = {} & \tfrac{1}{8}[\bar{\zeta}^2+\kappa(\bar{\xi}-\bar{\zeta})^2-2\kappa(\bar{\xi}-2\bar{\eta}+\bar{\zeta})\,\bar{\zeta}-3\kappa^2(\bar{\xi}-2\bar{\eta}+\bar{\zeta})^2]\,\boldsymbol{\beta}' \\ & -\tfrac{1}{8}[\bar{\xi}^2+\kappa(\bar{\xi}-\bar{\zeta})^2-2\kappa(\bar{\xi}-2\bar{\eta}+\bar{\zeta})\,\bar{\xi}-3\kappa^2(\bar{\xi}-2\bar{\eta}+\bar{\zeta})^2]\,\boldsymbol{\beta}.\end{aligned} \qquad (128.2)$$

The remark following equation (124.4) should here be recalled.

Next we use (127.9) in (128.1–2), and, upon omitting the subscript $j$ from $\boldsymbol{\sigma}_j$ and $\boldsymbol{\mu}_j$, we obtain the corresponding terms of the pseudo-expansion of $\mathbf{w}$, as explained after equation (127.9). We henceforth distinguish pseudo-expansions, and individual terms thereof, by angular brackets. Thus, quite generally, restoring the index $j$ now to avoid confusion

$$\langle \mathbf{w}_j \rangle = \sum_{n} \langle \mathbf{w}_j^{(n)} \rangle = \sum_{n} \sum_{\alpha,\,\beta} (\mathfrak{w}^{(n)}_{\alpha\beta j}\,\boldsymbol{\sigma} + \overline{\mathfrak{w}}^{(n)}_{\alpha\beta j}\,\boldsymbol{\mu})\,\xi^{n-\alpha}\eta^{\alpha-\beta}\zeta^{\beta}. \qquad (128.3)$$

The coefficients which appear on the right, and more generally those in pseudo-expansions of any other quantity, will be called *pseudo-coefficients*; and they will always be distinguished by the use of german type. In practice, to avoid having constantly to write down the superscripts $(n)$ and the double subscripts $\alpha\beta$, it is useful to use a special notation for the lower orders, say for $n = 2$, 3 and 4 at least. Thus,

$$\langle \mathbf{w}_j^{(2)} \rangle = (`\mathfrak{p}_{1j}\xi + `\mathfrak{p}_{2j}\eta + `\mathfrak{p}_{3j}\zeta)\,\boldsymbol{\sigma} + (`\bar{\mathfrak{p}}_{1j}\xi + `\bar{\mathfrak{p}}_{2j}\eta + `\bar{\mathfrak{p}}_{3j}\zeta)\,\boldsymbol{\mu}, \quad (128.4)$$

$$\langle \mathbf{w}_j^{(3)} \rangle = (`\mathfrak{s}_{1j}\xi^2 + \ldots + `\mathfrak{s}_{6j}\zeta^2)\,\boldsymbol{\sigma} + (`\bar{\mathfrak{s}}_{1j}\xi^2 + \ldots `\bar{\mathfrak{s}}_{6j}\zeta^2)\,\boldsymbol{\mu}, \quad (128.5)$$

and so on.

The notation for $\langle \mathbf{g}_j^* \rangle$, i.e. $\langle \mathbf{g}_{aj}^* \rangle$ and $\langle \mathbf{g}_{bj}^* \rangle$, is entirely similar. Thus $\langle \mathbf{g}_j^* \rangle$ is represented by a series just like that in (128.3), except that $\mathfrak{g}_{\alpha\beta j}^{(n)}$, $\bar{\mathfrak{g}}_{\alpha\beta j}^{(n)}$ replace $\mathfrak{w}_{\alpha\beta j}^{(n)}$ and $\bar{\mathfrak{w}}_{\alpha\beta j}^{(n)}$ respectively. Strictly speaking we should write $\mathfrak{g}_{\alpha\beta j}^{*(n)}$ instead of $\mathfrak{g}_{\alpha\beta j}^{(n)}$, and so on, since $\mathfrak{g}_{\alpha\beta j}^{(n)}$ is naturally associated with $\langle \mathbf{g}_j \rangle$. However, it is best at this point not to encumber the notation too much: one simply has to remember that an asterisk is missing. Further, $\mathfrak{g}_{\alpha\beta j}^{(n)}$ stands for $\mathfrak{g}_{a\alpha\beta j}^{(n)}$ or $\mathfrak{g}_{b\alpha\beta j}^{(n)}$ according as $\langle \mathbf{g}_{aj}^* \rangle$ or $\langle \mathbf{g}_{bj}^* \rangle$ is being contemplated. Then, for spherical surfaces,

$$\mathfrak{g}_{a\alpha\beta j}^{(n)} = e_{aj}\,\mathfrak{w}_{\alpha\beta j}^{(n)}, \quad \mathfrak{g}_{b\alpha\beta j}^{(n)} = e_{bj}\,\mathfrak{w}_{\alpha\beta j}^{(n)}, \quad (128.6)$$

and similarly for the barred coefficients. These equations are not valid, of course, if they refer to $\langle \mathbf{g}_j \rangle$ rather than $\langle \mathbf{g}_j^* \rangle$; nor do they apply to aspherical surfaces, for then there are additional terms on the right.

For the lower orders one will again use a notation like that of equations (128.4, 5), i.e.

$$\langle \mathbf{g}_j^{*(2)} \rangle = (\mathfrak{p}_{1j}\xi + \ldots)\,\boldsymbol{\sigma} + (\bar{\mathfrak{p}}_{1j}\xi + \ldots)\,\boldsymbol{\mu}, \quad (128.7)$$

and so on. (Here also every coefficient should have an additional asterisk.) Our immediate task is now to obtain expressions for the pseudo-coefficients of orders three and five in a form resembling, if possible, the primary coefficients of Section 120.

It is advisable to be as systematic as possible, to which end we introduce three auxiliary quantities defined as follows:

$$P_j = \xi + 2\theta_j\eta + \theta_j^2\zeta, \quad Q_j = 2\phi_j(\eta + \theta_j\zeta), \quad R_j = \phi_j^2\zeta. \quad (128.8)$$

Then, for example, $\bar{\zeta} = v_{aj}^2 P_j + v_{aj} Q_j + R_j$. (128.9)

The index $j$ may now be safely omitted; and at the same time the index $a$ may also be left understood. Thus

$$\kappa(\bar{\xi} - 2\bar{\eta} + \bar{\zeta}) = \ddot{u}' P, \quad (128.10)$$

and then there comes from (128.1)

$$\langle \mathbf{w}^{(2)} \rangle = \tfrac{1}{2}\{[(ii' + v^2)\,v' - (ii' + v'^2)\,v]\,P + (\Delta v)\,R\}\,(\boldsymbol{\sigma} + \theta\boldsymbol{\mu}) - \tfrac{1}{2}\phi[(\Delta v^2)\,P + (\Delta v)\,Q]\,\boldsymbol{\mu}. \quad (128.11)$$

Upon multiplying throughout by $e\,(\equiv e_a)$ one naturally finds that the factor multiplying $P\boldsymbol{\sigma}$ is just $\mathfrak{p}$ as defined by (120.14). Also set

$$\mathfrak{a} = \tfrac{1}{2}(s-m)^{-1}(\Delta v^2), \quad `\tilde{\omega} = (s-m)^{-2}\tilde{\omega}, \quad (128.12)$$

and then

$$\langle \mathbf{g}^{*(2)} \rangle = (\mathfrak{p}P + \tfrac{1}{2}`\tilde{\omega}\zeta)(\boldsymbol{\sigma} + \theta\boldsymbol{\mu}) - (\mathfrak{a}P + \tfrac{1}{2}`\tilde{\omega}\phi^{-1}Q)\,\boldsymbol{\mu}, \quad (128.13)$$

whence, recalling (128.7, 8), the required coefficients are given by

$$\begin{aligned} &\mathfrak{p}_1 = \mathfrak{p}, && \mathfrak{p}_2 = 2\theta\mathfrak{p}, && \mathfrak{p}_3 = \theta^2\mathfrak{p} + \tfrac{1}{2}`\tilde{\omega}, \\ &\bar{\mathfrak{p}}_1 = \theta\mathfrak{p}_1 - \mathfrak{a}, && \bar{\mathfrak{p}}_2 = \theta\mathfrak{p}_2 - 2\theta\mathfrak{a} - `\tilde{\omega}, && \bar{\mathfrak{p}}_3 = \theta\mathfrak{p}_3 - \theta^2\mathfrak{a} - \theta`\tilde{\omega}. \end{aligned} \quad (128.14)$$

The coefficients of $\langle \mathbf{g}_b^{*(2)} \rangle$ follow from these by multiplying each of them by $\theta$. It must not be forgotten that $f_K$ has been given the value unity throughout. (If $f_K \neq 1$ one has to include factors $f_K$ and $f_K^2$ respectively on the right of the equations (128.12) for $\mathfrak{a}$ and $`\tilde{\omega}$, since every $\mathfrak{p}_\alpha$ must be dimensionally a length.)

We go on in like manner to the fifth-order coefficients. From (128.2) we get in the first place

$$\begin{aligned} 8\langle \mathbf{w}^{(3)} \rangle = {} & \{(v^2P + vQ + R)^2 + ii'[(v' + v)\,P + Q]^2 \\ & - 2ii'P(v^2P + vQ + R) - 3i^2i'^2P^2\}\,[v'(\boldsymbol{\sigma} + \theta\boldsymbol{\mu}) + \phi\boldsymbol{\mu}] \\ & - \{(v'^2P + v'Q + R)^2 + ii'[(v' + v)\,P + Q]^2 \\ & - 2ii'P(v'^2P + v'Q + R) - 3i^2i'^2P^2\}\,[v(\boldsymbol{\sigma} + \theta\boldsymbol{\mu}) + \phi\boldsymbol{\mu}]. \end{aligned} \quad (128.15)$$

Multiplying throughout by $\frac{1}{8}e_a$ one obtains for $\langle \mathbf{g}_a^{*(3)} \rangle$ an expression of the form

$$\begin{aligned} \langle \mathbf{g}^{*(3)} \rangle = {} & (\mathfrak{z}_1P^2 + \tfrac{1}{2}\mathfrak{z}_2\phi^{-1}PQ + \mathfrak{z}_3\phi^{-2}PR + \tfrac{1}{4}\mathfrak{z}_4\phi^{-2}Q^2 + \mathfrak{z}_5\phi^{-4}R^2) \\ & \times (\boldsymbol{\sigma} + \theta\boldsymbol{\mu}) + (\bar{\mathfrak{z}}_1P^2 + \tfrac{1}{2}\bar{\mathfrak{z}}_2\phi^{-1}PQ + \bar{\mathfrak{z}}_3\phi^{-2}PR + \tfrac{1}{4}\bar{\mathfrak{z}}_4\phi^{-2}Q^2 \\ & + \tfrac{1}{2}\bar{\mathfrak{z}}_5\phi^{-3}QR)\,\boldsymbol{\mu}. \end{aligned} \quad (128.16)$$

To evaluate the $\mathfrak{z}_\alpha$ and $\bar{\mathfrak{z}}_\alpha$ one essentially needs only to read off the factors multiplying $P^2, PQ, \ldots$ in (128.15). Still, this is a somewhat tedious business, and it will suffice to quote the results:

$$\left.\begin{aligned}
\mathfrak{z}_1 &= \tfrac{1}{4}[(k^2-5k+1)i^2+3(k-1)iv+3v^2]\mathfrak{p},\\
\mathfrak{z}_2 &= (v'+v)\,\phi\mathfrak{p},\\
\mathfrak{z}_3 &= \tfrac{1}{2}\mathfrak{z}_4-\tfrac{1}{2}{}^{\backprime}\tilde{\omega}ii',\\
\mathfrak{z}_4 &= \phi^2\mathfrak{p},\\
\mathfrak{z}_5 &= \tfrac{1}{8}\phi^{2\backprime}\tilde{\omega},\\
\bar{\mathfrak{z}}_1 &= \tfrac{1}{2}\mathfrak{z}_2-\tfrac{1}{4}(k-1)^2 i^2\mathfrak{a},\\
\bar{\mathfrak{z}}_2 &= \phi^2\mathfrak{p}-\tfrac{1}{2}{}^{\backprime}\tilde{\omega}\nabla v^2,\\
\bar{\mathfrak{z}}_3 &= -\tfrac{1}{2}\phi^2\,\mathfrak{a},\\
\bar{\mathfrak{z}}_4 &= 2\bar{\mathfrak{z}}_3,\\
\bar{\mathfrak{z}}_5 &= -4\mathfrak{z}_5.
\end{aligned}\right\}\qquad(128.17)$$

Returning now to $\xi$, $\eta$, $\zeta$ by using (128.8) in (128.16), there comes finally

$$\left.\begin{aligned}
\mathfrak{S}_1 &= \mathfrak{z}_1,\\
\mathfrak{S}_2 &= 4\theta\mathfrak{z}_1+\mathfrak{z}_2,\\
\mathfrak{S}_3 &= 2\theta^2\mathfrak{z}_1+\theta\mathfrak{z}_2+\mathfrak{z}_3,\\
\mathfrak{S}_4 &= 4\theta^2\mathfrak{z}_1+2\theta\mathfrak{z}_2+\mathfrak{z}_4,\\
\mathfrak{S}_5 &= 4\theta^3\mathfrak{z}_1+3\theta^2\,\mathfrak{z}_2+2\theta(\mathfrak{z}_3+\mathfrak{z}_4),\\
\mathfrak{S}_6 &= \theta^4\mathfrak{z}_1+\theta^3\mathfrak{z}_2+\theta^2(\mathfrak{z}_3+\mathfrak{z}_4)+\mathfrak{z}_5,\\
\bar{\mathfrak{S}}_1 &= \theta\mathfrak{S}_1+\bar{\mathfrak{z}}_1,\\
\bar{\mathfrak{S}}_2 &= \theta\mathfrak{S}_2+4\theta\bar{\mathfrak{z}}_1+\bar{\mathfrak{z}}_2,\\
\bar{\mathfrak{S}}_3 &= \theta\mathfrak{S}_3+2\theta^2\bar{\mathfrak{z}}_1+\theta\bar{\mathfrak{z}}_2+\bar{\mathfrak{z}}_3,\\
\bar{\mathfrak{S}}_4 &= \theta\mathfrak{S}_4+4\theta^2\bar{\mathfrak{z}}_1+2\theta\bar{\mathfrak{z}}_2+\bar{\mathfrak{z}}_4,\\
\bar{\mathfrak{S}}_5 &= \theta\mathfrak{S}_5+4\theta^3\bar{\mathfrak{z}}_1+3\theta^2\bar{\mathfrak{z}}_2+2\theta(\bar{\mathfrak{z}}_3+\bar{\mathfrak{z}}_4),\\
\bar{\mathfrak{S}}_6 &= \theta\mathfrak{S}_6+\theta^4\,\bar{\mathfrak{z}}_1+\theta^3\bar{\mathfrak{z}}_2+\theta^2(\bar{\mathfrak{z}}_3+\bar{\mathfrak{z}}_4)+\theta\bar{\mathfrak{z}}_5.
\end{aligned}\right\}\qquad(128.18)$$

We note in passing that the particular structure of equations such as (128.14) in which the same quantities tend to occur over and over again is no 'accident'. Thus, contemplate the quantity $N(Y\gamma-Z\beta)$ which was shown to be an optical invariant on general grounds in Section 8. We write it here as $N\alpha^{-1}(Y^*\gamma-Z^*\beta)$, and the fact that this has the same value in the image space as in the object space evidently implies identities between the coefficients which occur in the expansions of the increments. On the other hand we may also use the identity

$$\Delta[N\alpha^{-1}(Y^*\gamma - Z^*\beta)] = 0 \qquad (128.19)$$

at any particular surface. If we use (123.18), etc., in this, and restrict ourselves, for the sake of illustration, to the third order we get at once the relation

$$(\sigma_y \Delta\delta_{\mu z} - \sigma_z \Delta\delta_{\mu y}) - (\mu_y \Delta\delta_{\sigma z} - \mu_z \Delta\delta_{\sigma y})$$
$$= (\sigma_y \mu_z - \sigma_z \mu_y)\,[(\alpha'/\alpha) - 1] + O(6). \qquad (128.20)$$

Here $\Delta\boldsymbol{\delta}_\sigma = -(s-m)\,e_b\mathbf{w}$ and $\Delta\boldsymbol{\delta}_\mu = (s-m)\,e_a\mathbf{w}$, so that if we write

$$\mathbf{w} = w\boldsymbol{\sigma} + \overline{w}\boldsymbol{\mu} \qquad (128.21)$$

for the moment, there comes

$$\overline{w}^{(2)} = \theta w^{(2)} - \tfrac{1}{2}\phi\Delta(\beta^2 + \gamma^2). \qquad (128.22)$$

Then, upon multiplying throughout by $e_a$, one reads off the relations

$$\bar{\mathfrak{p}}_1 = \theta\mathfrak{p}_1 - \tfrac{1}{2}e_a\phi\Delta v_a^2,\quad \bar{\mathfrak{p}}_2 = \theta\mathfrak{p}_2 - e_a\phi\Delta v_a v_b,\quad \bar{\mathfrak{p}}_3 = \theta\mathfrak{p}_3 - \tfrac{1}{2}e_a\phi\Delta v_b^2, \qquad (128.23)$$

and these will be seen at once to be in complete harmony with the last three members of (128.14).

## 129. The fifth-order iteration equations

Having considered the various third- and fifth-order pseudo-coefficients in reasonable detail, it remains to be a little more explicit with regard to the iterative equations which yield the coefficients of the exact series. It will be recalled that the latter result from the pseudo-expansions by replacing in these $\boldsymbol{\sigma}$ by $\boldsymbol{\sigma} + \mathbf{d}_{\sigma j}$ and $\boldsymbol{\mu}$ by $\boldsymbol{\mu} + \mathbf{d}_{\mu j}$ throughout. Consequently it is most convenient from several points of view to denote the coefficients of the exact series by the same kernel-letters as those of the corresponding pseudo-expansions, except that italic type replaces german type. For example,

$$\mathbf{w}_j^{(n)} = \sum_{\alpha,\,\beta} (w_{\alpha\beta j}^{(n)}\boldsymbol{\sigma} + \overline{w}_{\alpha\beta j}^{(n)}\boldsymbol{\mu})\,\xi^{n-\alpha}\eta^{\alpha-\beta}\zeta^{\beta}, \qquad (129.1)$$

and $$\mathbf{w}_j^{(2)} = ({}'p_{1j}\xi+\ldots)\,\boldsymbol{\sigma}+({}'\bar{p}_{1j}\xi+\ldots)\,\boldsymbol{\mu}; \tag{129.2}$$

and so on. Finally it is desirable to introduce a further set of coefficients of an auxiliary character, namely those which occur in the expansions of $\mathbf{d}_{\sigma j}$ and $\mathbf{d}_{\mu j}$. We naturally write $d^{(n)}_{\sigma\alpha\beta j}$ and $d^{(n)}_{\mu\alpha\beta j}$ for these, together with their barred counterparts. For the lower orders one will of course write $p_{\sigma 1j}, \ldots, \bar{p}_{\mu 3j}$, and so on. (It should be borne in mind that strictly speaking every coefficient relating to $\mathbf{d}_\sigma$ and $\mathbf{d}_\mu$ should have an additional asterisk; and the latter will then be absent when one is dealing with $\mathbf{d}_s$ and $\mathbf{d}_m$.)

With our system of notation complete, we can now write down the *identity*

$$\sum_n \sum_{\alpha,\beta} [\mathfrak{g}^{(n)}_{\alpha\beta j}(\boldsymbol{\sigma}+\mathbf{d}_{\sigma j})+\bar{\mathfrak{g}}^{(n)}_{\alpha\beta j}(\boldsymbol{\mu}+\mathbf{d}_{\mu j})]\,{}^*\xi^{n-\alpha}\,{}^*\eta^{\alpha-\beta}\,{}^*\zeta^{\beta}$$
$$= \sum_n \sum_{\alpha,\beta} (g^{(n)}_{\alpha\beta j}\,\boldsymbol{\sigma}+\bar{g}^{(n)}_{\alpha\beta j}\,\boldsymbol{\mu})\,\xi^{n-\alpha}\eta^{\alpha-\beta}\zeta^{\beta}, \tag{129.3}$$

where $$\left.\begin{aligned} {}^*\xi &= \xi+2\boldsymbol{\sigma}\,.\,\mathbf{d}_{\sigma j}+\mathbf{d}_{\sigma j}\,.\,\mathbf{d}_{\sigma j},\\ {}^*\eta &= \eta+(\boldsymbol{\sigma}\,.\,\mathbf{d}_{\mu j}+\boldsymbol{\mu}\,.\,\mathbf{d}_{\sigma j})+\mathbf{d}_{\sigma j}\,.\,\mathbf{d}_{\mu j},\\ {}^*\zeta &= \zeta+2\boldsymbol{\mu}\,.\,\mathbf{d}_{\mu j}+\mathbf{d}_{\mu j}\,.\,\mathbf{d}_{\mu j}.\end{aligned}\right\} \tag{129.4}$$

(129.3) simply reflects the idea of the pseudo-expansion, and both its members equally represent $\mathbf{g}_j^*$, that is to say, $\mathbf{g}_{aj}^*$ or $\mathbf{g}_{bj}^*$, if every coefficient is given an additional index *a* or *b* respectively.

The modified increments are now to be written as power series. It is best to do so in two steps, first writing down the series for $^*\xi$, $^*\eta$, $^*\zeta$ to the required order. Thus one has, for example,

$$^*\xi = \xi+2[p_{\sigma 1j}\xi^2+(\bar{p}_{\sigma 1j}+p_{\sigma 2j})\,\xi\eta+p_{\sigma 3j}\xi\zeta+\bar{p}_{\sigma 2j}\eta^2$$
$$+\bar{p}_{\sigma 3j}\eta\zeta]+O(6), \tag{129.5}$$

with analogous equations for $^*\eta$ and $^*\zeta$. Selecting now the terms of degree 3 in (129.3) one has, of course,

$$p_\alpha = \mathfrak{p}_\alpha, \quad \bar{p}_\alpha = \bar{\mathfrak{p}}_\alpha \quad (\alpha = 1, 2, 3), \tag{129.6}$$

where the index $j$ has been left understood, since in equations of this kind *every* coefficient has this index. Next, selecting the terms of degree five, we find the following *fifth-order iteration equations*:

$$
\left.
\begin{aligned}
s_1 &= \mathfrak{s}_1 + 3\mathfrak{p}_1 p_{\sigma 1} + (\bar{\mathfrak{p}}_1 + \mathfrak{p}_2) p_{\mu 1},\\
\bar{s}_1 &= \bar{\mathfrak{s}}_1 + \mathfrak{p}_1 \bar{p}_{\sigma 1} + \bar{\mathfrak{p}}_1(\bar{p}_{\mu 1} + 2p_{\sigma 1}) + \bar{\mathfrak{p}}_2 p_{\mu 1},\\
s_2 &= \mathfrak{s}_2 + \mathfrak{p}_1(2\bar{p}_{\sigma 1} + 3p_{\sigma 2}) + \bar{\mathfrak{p}}_1 p_{\mu 2} + \mathfrak{p}_2(\bar{p}_{\mu 1} + 2p_{\sigma 1} + p_{\mu 2})\\
&\qquad + (\bar{\mathfrak{p}}_2 + 2\mathfrak{p}_3) p_{\mu 1},\\
\bar{s}_2 &= \bar{\mathfrak{s}}_2 + \mathfrak{p}_1 \bar{p}_{\sigma 2} + \bar{\mathfrak{p}}_1(2\bar{p}_{\sigma 1} + \bar{p}_{\mu 2} + 2p_{\sigma 2}) + \mathfrak{p}_2 \bar{p}_{\sigma 1}\\
&\qquad + \bar{\mathfrak{p}}_2(p_{\sigma 1} + 2\bar{p}_{\mu 1} + p_{\mu 2}) + 2\bar{\mathfrak{p}}_3 p_{\mu 1},\\
s_3 &= \mathfrak{s}_3 + 3\mathfrak{p}_1 p_{\sigma 3} + \bar{\mathfrak{p}}_1 p_{\mu 3} + \mathfrak{p}_2(\bar{p}_{\sigma 1} + p_{\mu 3})\\
&\qquad + \mathfrak{p}_3(2\bar{p}_{\mu 1} + p_{\sigma 1}) + \bar{\mathfrak{p}}_3 p_{\mu 1},\\
\bar{s}_3 &= \bar{\mathfrak{s}}_3 + \mathfrak{p}_1 \bar{p}_{\sigma 3} + \bar{\mathfrak{p}}_1(2p_{\sigma 3} + \bar{p}_{\mu 3}) + \bar{\mathfrak{p}}_2(\bar{p}_{\sigma 1} + p_{\mu 3})\\
&\qquad + \mathfrak{p}_3 \bar{p}_{\sigma 1} + 3\bar{\mathfrak{p}}_3 \bar{p}_{\mu 1},\\
s_4 &= \mathfrak{s}_4 + 2\mathfrak{p}_1 \bar{p}_{\sigma 2} + \mathfrak{p}_2(2p_{\sigma 2} + \bar{p}_{\mu 2}) + (\bar{\mathfrak{p}}_2 + 2\mathfrak{p}_3) p_{\mu 2},\\
\bar{s}_4 &= \bar{\mathfrak{s}}_4 + (2\bar{\mathfrak{p}}_1 + \mathfrak{p}_2) \bar{p}_{\sigma 2} + \bar{\mathfrak{p}}_2(2\bar{p}_{\mu 2} + p_{\sigma 2}) + 2\bar{\mathfrak{p}}_3 p_{\mu 2},\\
s_5 &= \mathfrak{s}_5 + 2\mathfrak{p}_1 \bar{p}_{\sigma 3} + \mathfrak{p}_2(\bar{p}_{\sigma 2} + 2p_{\sigma 3} + \bar{p}_{\mu 3}) + \bar{\mathfrak{p}}_2 p_{\mu 3}\\
&\qquad + \mathfrak{p}_3(2\bar{p}_{\mu 2} + p_{\sigma 2} + 2p_{\mu 3}) + \bar{\mathfrak{p}}_3 p_{\mu 2},\\
\bar{s}_5 &= \bar{\mathfrak{s}}_5 + (2\bar{\mathfrak{p}}_1 + \mathfrak{p}_2) \bar{p}_{\sigma 3} + \bar{\mathfrak{p}}_2(\bar{p}_{\sigma 2} + p_{\sigma 3} + 2\bar{p}_{\mu 3})\\
&\qquad + \mathfrak{p}_3 \bar{p}_{\sigma 2} + \bar{\mathfrak{p}}_3(3\bar{p}_{\mu 2} + 2p_{\mu 3}),\\
s_6 &= \mathfrak{s}_6 + \mathfrak{p}_2 \bar{p}_{\sigma 3} + \mathfrak{p}_3(2\bar{p}_{\mu 3} + p_{\sigma 3}) + \bar{\mathfrak{p}}_3 p_{\mu 3},\\
\bar{s}_6 &= \bar{\mathfrak{s}}_6 + (\bar{\mathfrak{p}}_2 + \mathfrak{p}_3) \bar{p}_{\sigma 3} + 3\bar{\mathfrak{p}}_3 \bar{p}_{\mu 3}.
\end{aligned}
\right\}
\tag{129.7}
$$

The equations are quite general, in the sense that no use has been made of any relations which may exist between the coefficients which occur on the right. They are valid whether the surface is spherical or not. When it is, they have to be used only once, i.e. for $\mathbf{g}_a^{*(2)}$ (every pseudo-coefficient having then the additional index $a$) since $\mathbf{g}_b^{*(2)} = \theta \mathbf{g}_a^{*(2)}$; otherwise they are used twice, once for the coefficients of $\mathbf{g}_a^{*(2)}$ and once for those of $\mathbf{g}_b^{*(2)}$. As regards the use of digital computers, one requires, however, only one set of instructions. One can presumably do even better than that, since six of the fifth-order equations are the duals of the remaining six, in the sense explained at the end of Section 125. Thus $s_1$ goes into $\bar{s}_6$, $\bar{s}_1$ into $s_6$, and so on, upon mutually interchanging the indices 1 and 3, and also the indices $\sigma$ and $\mu$ of the primary coefficients, at the same time replacing barred by unbarred symbols and vice versa;

whilst the duals of the fifth-order coefficients $\mathfrak{s}_1, \bar{\mathfrak{s}}_1, \ldots$ are $\bar{\mathfrak{s}}_6, \mathfrak{s}_6, \ldots$ as already set out in Section 125.

Comparison of (129.7) with O(11.3) shows that the one may be obtained from the other by straightforward transcription according to the scheme

$$s_\alpha \to s_\alpha, \quad \mathfrak{s}_\alpha \to \mathfrak{s}_\alpha \quad (\alpha = 1, \ldots, 6); \quad \mathfrak{a} \to \mathfrak{p}_1, \quad \mathfrak{b} \to \mathfrak{p}_2, \quad \mathfrak{c} \to \mathfrak{p}_3; \tag{129.8}$$

$$\left.\begin{aligned} `A_p &\to p_{\mu 1}, & `B_p &\to p_{\mu 2}, & `C_p &\to p_{\mu 3}, \\ `A_q &\to -p_{\sigma 1}, & `B_q &\to -p_{\sigma 2}, & `C_q &\to -p_{\sigma 3}; \end{aligned}\right\} \tag{129.9}$$

and analogously for the barred coefficients. Therefore, if one wishes to go to the seventh order, one does not need to work out the required iteration equations *ab initio*. Instead, one can directly use O(81.3), with the appropriate transcription of the symbols which occur in it.

All the third- and fifth-order coefficients $p_\alpha, \bar{p}_\alpha, s_\alpha, \bar{s}_\alpha$ can now be calculated from (129.6–7), together with (128.14) and (128.17–18); and for the sake of convenience we recall here the connection between the $p_{\sigma\alpha}, \ldots, \bar{p}_{\mu\alpha}$ on the one hand and the $p_\alpha, \bar{p}_\alpha (\alpha = 1, 2, 3)$ on the other. This is quite generally provided by equations (123.19) and (127.5). For spherical surfaces in particular,

$$\left.\begin{aligned} \mathbf{d}_{\sigma j} &= s \sum_{i=1}^{k} (1 + \theta_i)\, \mathbf{g}_i^* - (s - m) \sum_{i=1}^{j-1} \theta_i \mathbf{g}_i^* + `\tilde{\omega}_j^{-1} v_{bj}\, \phi_j\, \mathbf{g}_j^*, \\ \mathbf{d}_{\mu j} &= m \sum_{i=1}^{k} (1 + \theta_i)\, \mathbf{g}_i^* + (s - m) \sum_{i=1}^{j-1} \mathbf{g}_i^* - `\tilde{\omega}_j^{-1} v_{aj}\, \phi_j\, \mathbf{g}_j^*, \end{aligned}\right\} \tag{129.10}$$

where $\mathbf{g}_i^*$ stands for $\mathbf{g}_{ai}^*$ throughout. Then, for example, in view of (129.6),

$$p_{\sigma 1 j} = s \sum_{i=1}^{k} (1 + \theta_i)\, \mathfrak{p}_{1i} - (s - m) \sum_{i=1}^{j-1} \theta_i \mathfrak{p}_{1i} + `\tilde{\omega}_j^{-1} v_{bj}\, \phi_j \mathfrak{p}_{1j}, \tag{129.11}$$

and so on, in an obvious way, for the remaining coefficients.

The form of (129.11) is characteristic of the comparative lack of neatness one gets in Hamiltonian calculations. In the corresponding relation of the Lagrangian method only the middle term (supplied with the appropriate constant factor) remains.

## 130. On the computation of the higher-order characteristic coefficients. Fifth-order pseudo-coefficients and iteration equations

We resume the discussion concerning the problem of the coefficients of $T$ where we left off at the end of Section 126. As we know, in principle two broad alternatives present themselves from the point of view of practice, namely, either $T$ is computed by integration of the displacement, or else directly. With regard to the first of these, there are again several alternatives, and we first deal with these in turn.

If, on account of (120.3), one proceeds exclusively in terms of $\mathbf{g}_j$, $\boldsymbol{\epsilon}'$ will appear as a function of the 'wrong' variables $\mathbf{S}$ and $\mathbf{M}$, and these must be eliminated in favour of $\boldsymbol{\sigma}$ and $\boldsymbol{\mu}$ before we can integrate. As already explained at the end of Section 126, this is an unattractive proposition, and we do not contemplate it. If, on the other hand, we deal with $\mathbf{g}_j^*$ alone, we end up with $\boldsymbol{\epsilon}^{*\prime}$ and this is not the displacement. This defect can be remedied, for example, as follows. In the course of iteration one makes use of $\mathbf{g}_b^*$ as well as of $\mathbf{g}_a^*$, so that we also have

$$\boldsymbol{\epsilon}_b^{*\prime} \equiv \alpha_k' \mathbf{H}_{Ek}' - s\alpha_1 \mathbf{H}_{E1} = -(s-m) \sum_{j=1}^{k} \mathbf{g}_{bj}^* = -(s-m)\,\mathbf{G}_b^* \qquad (130.1)$$

available to a certain order. We can write

$$\boldsymbol{\epsilon}_b^{*\prime} = \alpha'\mathbf{H}' - s\alpha\mathbf{H} + (s-m)\,m^{-1}\boldsymbol{\mu},$$

where the subscripts $k$ and 1 have been suppressed, without risk of confusion. Using this relation, we readily convince ourselves that, identically,

$$\boldsymbol{\epsilon}' = \left(\frac{s}{\alpha'} - \frac{m}{\alpha}\right) \mathbf{G}_a^* + \left(\frac{1}{\alpha'} - \frac{1}{\alpha}\right)(m\mathbf{G}_b^* + \boldsymbol{\mu}). \qquad (130.2)$$

Now, restricting ourselves again to the fifth order, suppose that $\mathbf{G}_a^*$ is known to this order: we have already seen in detail how to calculate it. In so doing we shall have obtained $\mathbf{G}_b^*$ correctly to the third order, and this is all that is required, since $(1/\alpha') - (1/\alpha) = O(2)$. Still, even here the factors arising from $\alpha$ and $\alpha'$ are not as simple as one might like them to be, since they have to be written as series in powers of $\xi$, $\eta$, $\zeta$, and not of $\bar{\xi}$ and $\bar{\zeta}$; but at least we now have $\boldsymbol{\epsilon}_3'$ and $\boldsymbol{\epsilon}_5'$ as functions of $\boldsymbol{\sigma}$ and $\boldsymbol{\mu}$, as desired.

So far we have explored only procedures based on either **g** to the exclusion of **g*** or vice versa. A little reflection shows that perhaps we should compromise and use both at the same time. In the context of fifth-order aberrations in particular one then has a very simple state of affairs which may be described as follows. One first proceeds as set out in Sections 128 and 129, but *only to the third order*, the coefficients of $\mathbf{d}_{\sigma j}$ and $\mathbf{d}_{\mu j}$ being included in this third-order computation. Now let $\mathbf{g}_j (\equiv \mathbf{g}_{aj})$—not $\mathbf{g}_j^*$—be expanded in powers of $\boldsymbol{\beta}_j'$ and $\boldsymbol{\beta}_j$. The third-order coefficients are already contained in (120.5), but those of the fifth order must be found *ab initio* from (119.9). The usual paraxial substitutions then lead to $\langle \mathbf{g}_j^{(2)} \rangle$ and $\langle \mathbf{g}_j^{(3)} \rangle$, the first of which is just the right-hand member of (120.15). To get the correct expression for $\mathbf{g}_j^{(3)}$ we now have to replace $\boldsymbol{\sigma}$ by $\boldsymbol{\sigma} + \mathbf{d}_{\sigma j}$ and $\boldsymbol{\mu}$ by $\boldsymbol{\mu} + \mathbf{d}_{\mu j}$ in $\langle \mathbf{g}_j^{(2)} \rangle$. To this end equations (129.7) may be used as they stand, provided we understand the coefficients on the left, and all pseudo-coefficients on the right to refer to $\mathbf{g}_j$ rather than $\mathbf{g}_j^*$; whilst the coefficients $p_{\sigma 1}, \ldots, \bar{p}_{\mu 3}$ are already known. In view of (120.3) we thus obtain $\boldsymbol{\epsilon}'$, correct to the fifth order, as a function of the desired variables $\boldsymbol{\sigma}$ and $\boldsymbol{\mu}$.

We note the following point in passing. **G** has the generic form

$$\mathbf{G} = G\boldsymbol{\sigma} + \bar{G}\boldsymbol{\mu}, \tag{130.3}$$

say, where $G$ and $\bar{G}$ are functions of $\xi$, $\eta$, $\zeta$. (24.4) therefore entails that the integrability condition

$$2\partial\bar{G}/\partial\xi = \partial G/\partial\eta, \tag{130.4}$$

must be satisfied. Thus, if, consistently with (123.7), we temporarily write $\sum p_{1aj} = P_1$, and so on, there is one third-order condition:

$$2\bar{P}_1 = P_2, \tag{130.5}$$

and three fifth-order conditions:

$$4\bar{S}_1 = S_2, \quad \bar{S}_2 = S_4, \quad 2\bar{S}_3 = S_5. \tag{130.6}$$

Having calculated the various coefficients independently of each other, one's confidence in the numerical results will be greatly enhanced if the identities (130.6) are in fact satisfied.

Finally, we must consider the direct calculation of $T$. For this purpose one may again use to advantage a kind of hybrid method: 'hybrid', that is to say, only in the very limited sense that for practical convenience $\mathbf{g}_j^*$ and $T_j$ are separately expanded in powers of $\boldsymbol{\beta}'$ and $\boldsymbol{\beta}$. With regard to $T$, one has in the first place, from (121.4),

$$T = \sum_{j=1}^{k} [T_j^{(0)} + T_j^{(1)}(\boldsymbol{\beta}_j', \boldsymbol{\beta}_j) + t_j(\boldsymbol{\beta}_j', \boldsymbol{\beta}_j)]. \qquad (130.7)$$

The first sum on the right is a constant, which is, in fact, the optical distance between $O_{01}$ and $O'_{0k}$. We henceforth bring this constant over to the left and absorb it in $T$. The usual paraxial substitutions then yield the pseudo-expansion

$$\langle T\rangle = \sum_{j=1}^{k} [\langle T_j^{(1)}(\boldsymbol{\sigma}, \boldsymbol{\mu})\rangle + \langle t_j(\boldsymbol{\sigma}, \boldsymbol{\mu})\rangle]. \qquad (130.8)$$

It should be carefully noted that the meaning of $T_j^{(1)}$ and $t_j$, regarded as *functional* symbols, is not the same now as it was in (130.7). The exact equation which is generated by the pseudo-expansion (130.8) is

$$T = \sum_{j=1}^{k} [T_j^{(1)}(\boldsymbol{\sigma} + \mathbf{d}_{\sigma j}, \boldsymbol{\mu} + \mathbf{d}_{\mu j}) + t_j(\boldsymbol{\sigma} + \mathbf{d}_{\sigma j}, \boldsymbol{\mu} + \mathbf{d}_{\mu j})]. \quad (130.9)$$

Using (127.2) in (124.12), one sees at once that $\mathbf{D} = \mathbf{d}_\mu \Delta\bar{\theta}$, and since according to (124.15), $fv_a v_a' \Delta\bar{\theta} = (s-m)^{-1}$ (when $f_K = 1$), we can write (124.17) in the form

$$\sum_{j=1}^{k} T_j^{(1)}(\boldsymbol{\sigma} + \mathbf{d}_{\sigma j}, \boldsymbol{\mu} + \mathbf{d}_{\mu j}) = \zeta/2m + \sum_{j=1}^{k} b_j \mathbf{d}_{\mu j} \cdot \mathbf{d}_{\mu j}, \quad (130.10)$$

where $$b_j = \tfrac{1}{2}(s-m)^{-1}\Delta\bar{\theta}_j.$$

For the aberration function we thus finally get the equation

$$t(\xi, \eta, \zeta) \equiv \sum_{j=1}^{k} [t_j(\boldsymbol{\sigma} + \mathbf{d}_{\sigma j}, \boldsymbol{\mu} + \mathbf{d}_{\mu j}) + b_j \mathbf{d}_{\mu j} \cdot \mathbf{d}_{\mu j}]. \quad (130.11)$$

Let us suppose once again that only the third- and fifth-order aberration functions are to be found. The first of these, i.e. $t^{(2)}$, was already obtained previously. At any rate, we write

$$t_j^{(2)} = \langle t_j^{(2)}\rangle = \mathfrak{p}_{1j}\xi^2 + \mathfrak{p}_{2j}\xi\eta + \ldots + \mathfrak{p}_{6j}\zeta^2. \qquad (130.12)$$

The somewhat overworked coefficients on the right are to be regarded as defined by this equation. They are evidently given by (121.3) and (125.6), the summation signs being understood to have been omitted from the right-hand members of these six equations. We also require

$$\langle t^{(3)} \rangle = \mathfrak{s}_1 \xi^3 + \mathfrak{s}_2 \xi^2 \eta + \ldots + \mathfrak{s}_{10} \zeta^3, \tag{130.13}$$

the index $j$ having been suppressed throughout. To get the coefficients $\mathfrak{s}_\alpha$ we must first find the terms of degree six of $T$. There is a considerable resemblance between these and (125.1):

$$16 t^{(3)}(\bar{\xi}, \bar{\eta}, \bar{\zeta}) = f[(\bar{\xi} - \bar{\zeta})^2 (\bar{\xi} + \bar{\zeta}) - \kappa(\bar{\xi} - 2\bar{\eta} + \bar{\zeta})(\bar{\xi} - \bar{\zeta})^2$$
$$+ \kappa^2 (\bar{\xi} - 2\bar{\eta} + \bar{\zeta})^3] + N r i_a \left( \frac{\xi^3}{v'_a} - \frac{\zeta^3}{v_a} \right), \tag{130.14}$$

all variables referring to the surface in question. Using (128.8), this becomes

$$16 \langle t^{(3)}(\xi, \eta, \zeta) \rangle = f\{(P \Delta v^2 + Q \Delta v)^2 [P(\nabla v^2 - ii') + Q \nabla v + 2R]$$
$$+ \kappa^{-1} (ii')^3 P^3\} + N r i \Delta[(v^2 P + vQ + R)^3 / v]. \tag{130.15}$$

From here we proceed as after (128.15). The factors multiplying $QR^2$ and $Q^3$ turn out to be zero. Set

$$\left.\begin{aligned}
\mathfrak{z}_1 &= \tfrac{1}{8}\mathfrak{p}[(k^2 - 3k + 1) i^2 + 3(k-1) iv + 3v^2], \\
\mathfrak{z}_2 &= \tfrac{1}{2}\phi(v' + v)\mathfrak{p}, \\
\mathfrak{z}_3 &= \tfrac{1}{16}(v'^2 - vv' + v^2)\,{}^{\backprime}\tilde{\omega}, \\
\mathfrak{z}_4 &= \tfrac{1}{2}\phi^2 \mathfrak{p}, \\
\mathfrak{z}_5 &= \tfrac{1}{2}\phi^2 \mathfrak{a}, \\
\mathfrak{z}_6 &= \tfrac{3}{16}\phi^2\,{}^{\backprime}\tilde{\omega}, \\
\mathfrak{z}_7 &= \tfrac{4}{3}\mathfrak{z}_6, \\
\mathfrak{z}_8 &= -\tfrac{1}{16}\phi^4({}^{\backprime}\tilde{\omega}/vv'),
\end{aligned}\right\} \tag{130.16}$$

where $\mathfrak{p}$ is given by (120.14), and $\mathfrak{a}$ and ${}^{\backprime}\tilde{\omega}$ by (128.12). The $\mathfrak{z}_\alpha$ are of course quite distinct from those which appear in (128.17), though they are distantly related to them. Then

$$
\begin{aligned}
\mathfrak{s}_1 &= \mathfrak{z}_1,\\
\mathfrak{s}_2 &= 6\theta\mathfrak{z}_1+\mathfrak{z}_2,\\
\mathfrak{s}_3 &= 3\theta^2\mathfrak{z}_1+\theta\mathfrak{z}_2+\mathfrak{z}_3,\\
\mathfrak{s}_4 &= 12\theta^2\mathfrak{z}_1+4\theta\mathfrak{z}_2+\mathfrak{z}_4,\\
\mathfrak{s}_5 &= 12\theta^3\mathfrak{z}_1+6\theta^2\mathfrak{z}_2+2\theta(2\mathfrak{z}_3+\mathfrak{z}_4)+\mathfrak{z}_5,\\
\mathfrak{s}_6 &= 3\theta^4\mathfrak{z}_1+2\theta^3\mathfrak{z}_2+\theta^2(2\mathfrak{z}_3+\mathfrak{z}_4)+\theta\mathfrak{z}_5+\mathfrak{z}_6,\\
\mathfrak{s}_7 &= 8\theta^3\mathfrak{z}_1+4\theta^2\mathfrak{z}_2+2\theta\mathfrak{z}_4,\\
\mathfrak{s}_8 &= 12\theta^4\mathfrak{z}_1+8\theta^3\mathfrak{z}_2+\theta^2(4\mathfrak{z}_3+5\mathfrak{z}_4)+2\theta\mathfrak{z}_5+\mathfrak{z}_7,\\
\mathfrak{s}_9 &= 6\theta^5\mathfrak{z}_1+5\theta^4\mathfrak{z}_2+4\theta^3(\mathfrak{z}_3+\mathfrak{z}_4)+3\theta^2\mathfrak{z}_5+2\theta(\mathfrak{z}_6+\mathfrak{z}_7),\\
\mathfrak{s}_{10} &= \theta^6\mathfrak{z}_1+\theta^5\mathfrak{z}_2+\theta^4(\mathfrak{z}_3+\mathfrak{z}_4)+\theta^3\mathfrak{z}_5+\theta^2(\mathfrak{z}_6+\mathfrak{z}_7)+\mathfrak{z}_8.
\end{aligned}
\tag{130.17}
$$

It remains to write down the iteration equations for the coefficients $s_\alpha$ of the contribution by the $j$th surface to the fifth-order aberration function $t^{(3)}(\xi,\eta,\zeta)$, i.e. to the terms of degree six of the expression within the square brackets on the right of equation (130.11). They are, with $b=\frac{1}{2}(s-m)^{-1}f_K$,

$$
\begin{aligned}
s_1 &= \mathfrak{s}_1+4\mathfrak{p}_1p_{\sigma1}+\mathfrak{p}_2p_{\mu1}+bp_{\mu1}^2,\\
s_2 &= \mathfrak{s}_2+4\mathfrak{p}_1(\bar{p}_{\sigma1}+p_{\sigma2})+\mathfrak{p}_2(3p_{\sigma1}+\bar{p}_{\mu1}+p_{\mu2})\\
&\qquad +2(\mathfrak{p}_3+\mathfrak{p}_4)p_{\mu1}+2bp_{\mu1}(\bar{p}_{\mu1}+p_{\mu2}),\\
s_3 &= \mathfrak{s}_3+4\mathfrak{p}_1p_{\sigma3}+\mathfrak{p}_2(\bar{p}_{\sigma1}+p_{\mu3})+2\mathfrak{p}_3(p_{\sigma1}+\bar{p}_{\mu1})\\
&\qquad +\mathfrak{p}_5p_{\mu1}+b(2p_{\mu1}p_{\mu3}+\bar{p}_{\mu1}^2),\\
s_4 &= \mathfrak{s}_4+4\mathfrak{p}_1\bar{p}_{\sigma2}+\mathfrak{p}_2(2\bar{p}_{\sigma1}+3p_{\sigma2}+\bar{p}_{\mu2})+2\mathfrak{p}_3p_{\mu2}+2\mathfrak{p}_4(p_{\sigma1}\\
&\qquad +\bar{p}_{\mu1}+p_{\mu2})+2\mathfrak{p}_5p_{\mu1}+b(p_{\mu2}^2+2p_{\mu1}\bar{p}_{\mu2}+2\bar{p}_{\mu1}p_{\mu2}),\\
s_5 &= \mathfrak{s}_5+4\mathfrak{p}_1\bar{p}_{\sigma3}+\mathfrak{p}_2(\bar{p}_{\sigma2}+3p_{\sigma3}+\bar{p}_{\mu3})+2\mathfrak{p}_3(\bar{p}_{\sigma1}+p_{\sigma2}+\bar{p}_{\mu2}+p_{\mu3})\\
&\qquad +2\mathfrak{p}_4(\bar{p}_{\sigma1}+p_{\mu3})+\mathfrak{p}_5(p_{\sigma1}+3\bar{p}_{\mu1}+p_{\mu2})+4\mathfrak{p}_6p_{\mu1}\\
&\qquad +2b(p_{\mu2}p_{\mu3}+p_{\mu1}\bar{p}_{\mu3}+\bar{p}_{\mu1}p_{\mu3}+\bar{p}_{\mu1}\bar{p}_{\mu2}),\\
s_6 &= \mathfrak{s}_6+\mathfrak{p}_2\bar{p}_{\sigma3}+2\mathfrak{p}_3(p_{\sigma3}+\bar{p}_{\mu3})+\mathfrak{p}_5(\bar{p}_{\sigma1}+p_{\mu3})\\
&\qquad +4\mathfrak{p}_6\bar{p}_{\mu1}+b(p_{\mu3}^2+2\bar{p}_{\mu1}\bar{p}_{\mu3}),\\
s_7 &= \mathfrak{s}_7+2\mathfrak{p}_2\bar{p}_{\sigma2}+2\mathfrak{p}_4(p_{\sigma2}+\bar{p}_{\mu2})+2\mathfrak{p}_5p_{\mu2}+2bp_{\mu2}\bar{p}_{\mu2},\\
s_8 &= \mathfrak{s}_8+\mathfrak{p}_2\bar{p}_{\sigma3}+2\mathfrak{p}_3\bar{p}_{\sigma2}+2\mathfrak{p}_4(\bar{p}_{\sigma2}+p_{\sigma3}+\bar{p}_{\mu3})+\mathfrak{p}_5(p_{\sigma2}+3\bar{p}_{\mu2}\\
&\qquad +2p_{\mu3})+4\mathfrak{p}_6p_{\mu2}+b(2p_{\mu2}\bar{p}_{\mu3}+2\bar{p}_{\mu2}p_{\mu3}+\bar{p}_{\mu2}^2),\\
s_9 &= \mathfrak{s}_9+2(\mathfrak{p}_3+\mathfrak{p}_4)\bar{p}_{\sigma3}+\mathfrak{p}_5(\bar{p}_{\sigma2}+p_{\sigma3}+3\bar{p}_{\mu3})\\
&\qquad +4\mathfrak{p}_6(\bar{p}_{\mu2}+p_{\mu3})+2b(\bar{p}_{\mu2}+p_{\mu3})\bar{p}_{\mu3},\\
s_{10} &= \mathfrak{s}_{10}+\mathfrak{p}_5\bar{p}_{\sigma3}+4\mathfrak{p}_6\bar{p}_{\mu3}+b\bar{p}_{\mu3}^2.
\end{aligned}
\tag{130.18}
$$

The operation of the duality principle is clearly visible.

The coefficients $s_{\alpha j}$ ($\alpha = 1, \ldots, 10$) may now be calculated from these equations for each of the surfaces, for all the third- and fifth-order pseudo-coefficients are known from equations (121.3) and (125.6) (without the signs of summation) and from (130.16–17); whilst the coefficients of the third-order increments are known from Section 129 (see, in particular, (129.11)). The coefficients of $t^{(3)}$ finally follow by mere summation over the whole system:

$$s_\alpha = \sum_{j=1}^{k} s_{\alpha j}. \tag{130.19}$$

We note that the results embodied in the lengthy equations (128.17–18) and (129.7) were not used in obtaining the $s_\alpha$ in the manner just described. To this extent they are indeed redundant. They have, however, been included since they will be required as soon as one wishes to use the present method to calculate the coefficients of the *seventh*-order aberration function.

## Problems

**P.12 (i).** Find the shape of the most general surface which has a pair of (non-coincident) conjugate points $B$, $B'$ entirely free from spherical aberration. (Such a surface is a 'Cartesian ovoid'.)

**P.12 (ii).** Show that the surface of the preceding question can be a sphere, and find the locations of $B$ and $B'$ when this is the case.

**P.12 (iii).** Show by considering the angle characteristic that the conjugate points of the sphere defined in the two preceding problems are such that rays through them satisfy the sine-condition. (Remark: The sphere thus has a pair of non-coincident conjugate planes such that not only spherical aberration but also circular coma of all orders is absent. The axial points of these planes are therefore called the *aplanatic points* of the sphere; cf. Section 33 (i).)

**P.12 (iv).** By inspection of the third-order aberrations show that no surface other than a sphere possesses a pair of aplanatic points.

**P.12 (v).** Obtain an expression for $\mathbf{g}^{(3)}(\mathbf{V}', \mathbf{V})$ and hence show that the coefficient of $\mathbf{S}(S_y^2 + S_z^2)^2$ in $\langle \mathbf{g}^{(3)} \rangle$ is

$$-\tfrac{3}{4}\mathfrak{p}[(k^2 - k + 1)i^2 + 3(k-1)iv + 3v^2].$$

**P.12 (vi).** Show that (with $f_K = 1$)

$$2\phi^2\mathfrak{p} = `\tilde{\omega}(ii' - vv'),$$

and hence that in (130.16)

$$16\mathfrak{z}_3 + 4\mathfrak{z}_4 = `\tilde{\omega}(k^2 - k + 1)\, i^2.$$

**P.12 (vii).** Writing $\boldsymbol{\delta}_\sigma = \boldsymbol{\sigma}\delta_\sigma + \boldsymbol{\mu}\bar{\delta}_\sigma$, etc., show from equation (127.19) that

$$\bar{\delta}_{\mu j} + \delta_{\sigma j} = \bar{\delta}_{\mu 1} + \delta_{\sigma 1} - \tfrac{1}{2}[(\beta_j^2 + \gamma_j^2) - (\beta_1^2 + \gamma_1^2)] + O(4).$$

**P.12 (viii).** A set of variables $x_1, x_2, \ldots, x_k$ is related to a variable $x$ through the set of equations

$$x_j = a_j x + \sum_{i=1}^{j} f_i(x_i) \quad (j = 1, 2, \ldots, k),$$

where $$f_j(x_j) = a_{2j} x_j^3 + a_{3j} x_j^5 + a_{4j} x_j^7 + \ldots.$$

The various $a$'s are given constants. Develop in detail an iterative method of solving these equations, i.e. for finding the $x_j$ as power series in $x$. Proceed at least to the seventh degree.

# SOLUTIONS OF THE PROBLEMS

## Chapter 1

**P.1 (i).** We have $dx/dt = 2x^{\frac{1}{2}}$, $dy/dt = yx^{-\frac{1}{2}}$, $dz/dt = zx^{-\frac{1}{2}}$. These derivatives are proportional to the direction cosines $\alpha$, $\beta$, $\gamma$ of the tangent to the ray at the point $x, y, z$, and so are the derivatives of $F$, if $F(x, y, z) =$ constant is the equation of the normal surface through the point in question. There therefore must exist a function $\theta$ such that

$$dF = x^{-\frac{1}{2}}\theta(2x\,dx + y\,dy + z\,dz), \qquad \text{(P.1 (i), 1)}$$

granted that the given congruence of rays is normal. By inspection $\theta = x^{\frac{1}{2}}$ is such a function. The required normal surfaces therefore have the equation

$$2x^2 + y^2 + z^2 = \text{const.}; \qquad \text{(P.1 (i), 2)}$$

and they are evidently ellipsoids of revolution.

**P.1 (ii).** By the argument of the preceding solution, there must exist a function $\theta$ such that

$$\frac{\partial F}{\partial x} = \theta\alpha, \quad \frac{\partial F}{\partial y} = \theta\beta, \quad \frac{\partial F}{\partial z} = \theta\gamma,$$

and these require that

$$\frac{\partial(\theta\gamma)}{\partial y} - \frac{\partial(\theta\beta)}{\partial z} = 0, \quad \frac{\partial(\theta\alpha)}{\partial z} - \frac{\partial(\theta\gamma)}{\partial x} = 0, \quad \frac{\partial(\theta\beta)}{\partial x} - \frac{\partial(\theta\alpha)}{\partial y} = 0.$$

Multiply these equations by $\alpha$, $\beta$, $\gamma$ respectively and add them. The required necessary condition follows at once.

**P.1 (iii).** By inspection, taking $N = 1$,

$$V^*(APA') = [(a-ky^2)^2 + (b-y)^2]^{\frac{1}{2}} + [(a-ky^2) + (b+y)^2]^{\frac{1}{2}}.$$

Expanding this in powers of $y^2$,

$$V^* = 2(a^2+b^2)^{\frac{1}{2}} + a(a^2+b^2)^{-\frac{3}{2}}[a - 2(a^2+b^2)k]y^2 + O(y^4), \qquad \text{(P.1 (iii), 1)}$$

from which the conclusion stated in $(a)$ follows by inspection.

Next if the reflecting surface is such that $\tilde{V}$ is constant, its equation must be

$$[(a-x)^2+(b-y)^2]^{\frac{1}{2}}+[(a-x)^2+(b+y)^2]^{\frac{1}{2}} = 2(a^2+b^2)^{\frac{1}{2}}.$$

From this one obtains by repeated squaring the required equation

$$\frac{(a-x)^2}{a^2}+\frac{y^2}{a^2+b^2} = 1,$$

which is that of an ellipse with $A$ and $A'$ as foci. It may be written as

$$x = \tfrac{1}{2}[a/(a^2+b^2)]\,y^2+O(y^4).$$

Thus, near the origin one has a parabola for which $k$ has just the value which makes the coefficient of $y^2$ in (P.1 (iii), 1) zero.

**P.1 (iv).** One may think of the medium as stratified, the equation of the boundary of any layer being $N$ = constant. Then, if neighbouring points on the ray define the displacement $d\mathsf{s}$,

$$d(N\mathsf{e}) = \kappa \operatorname{grad} N\,ds,$$

since in (2.6) the normal $\mathsf{p}$ is in the direction of grad $N$, $\kappa$ being a scalar factor. Scalar multiplication throughout by $\mathsf{e}$ gives

$$\mathsf{e}\,.\,d(N\mathsf{e}) = N\mathsf{e}\,.\,d\mathsf{e}+\mathsf{e}\,.\,\mathsf{e}\,dN \equiv dN = \kappa\, d\mathsf{s}\,.\operatorname{grad} N = \kappa\, dN,$$

bearing in mind that $\mathsf{e}\,.\,\mathsf{e} = 1$. Hence $\kappa = 1$, so that one has the required equation.

## Chapter 2

**P.2 (i).** The system is invariant under rotations about the $\bar{x}$-axis, and also invariant under translations along the $\bar{y}$- and $\bar{z}$-axes. Its point characteristic must therefore have the generic form

$$V = f(x', x, u), \qquad \text{(P.2 (i), 1)}$$

where $u = (y'-y)^2+(z'-z)^2$; see also Sections 13 and 84. By rotation and translation we can always arrange the coordinates of the initial point $A$ and the final point $A'$ to become $(x, 0, 0)$ and $(x', \hat{y}', 0)$ respectively, relative to the new coordinates. Then

$$V = f(x', x, \hat{y}'^2). \qquad \text{(P.2 (i), 2)}$$

In view of the elementary laws of refraction the ray through $A$ and $A'$ will lie in the plane $\bar{z} = 0$, and we need to consider this alone.

Then if $P(a, y_0, 0)$ is the point of incidence

$$V^*(APA') = N'(PA') + N(AP)$$
$$= N'[(x'-a)^2 + (\hat{y}'-y_0)^2]^{\frac{1}{2}} + N[(a-x)^2 + y_0^2]^{\frac{1}{2}}. \quad (\text{P.2 (i)}, 3)$$

This must be stationary with respect to small variations of $y_0$, and so

$$ky_0[(a-x)^2 + y_0^2]^{-\frac{1}{2}} = (\hat{y}'-y_0)[(x'-a)^2 + (\hat{y}'-y_0)^2]^{-\frac{1}{2}},$$

with $k = N/N'$. This is a quartic equation for $y_0$ which may be written as

$$(y_0^2 + A)(y_0 - \hat{y}')^2 - By_0^2 = 0, \quad (\text{P.2 (i)}, 4)$$

where $A = (a-x)^2/(1-k^2)$, $B = k^2(x'-a)^2/(1-k^2)$. Let the appropriate root of (P. 2 (i), 4) be

$$y_0 = \chi(x', x, \hat{y}').$$

Then $V$ is the function which is obtained when $\chi(x', x, \hat{y}')$ is substituted for $y_0$ on the right of (P.2 (i), 3) and $\hat{y}'$ then replaced by $[(y'-y)^2 + (z'-z)^2]^{\frac{1}{2}}$ in the function which results from the first substitution.

**P.2 (ii).** Because in this limit the system is telescopic, and the ray-coordinates $\beta', \gamma', \beta, \gamma$ can no longer be given values independently of each other.

**P.2 (iii).** Since the points $B'$, $B$ have the coordinates $(q', 0, 0)$ and $(q, 0, 0)$ respectively, we have by an elementary geometrical argument

$$T = \Delta\{N[(q - x_P)\alpha - (y_P\beta + z_P\gamma)]\}, \quad (\text{P.2 (iii)}, 1)$$

where $x_P$, $y_P$, $z_P$ are the coordinates of the point of incidence $P$, and, whatever $X$ may be, $\Delta X \equiv X' - X$. Note that

$$N'(y'_P\beta' + z'_P\gamma'),$$

for instance, is the optical length of the part of the refracted ray, produced backwards, between the foot of the normal from the origin on to the ray and $P$. Now when the surface is spherical one has in (2.6)

$$\mathsf{p} = (1 - cx_P, -cy_P, -cz_P), \quad (\text{P.2 (iii)}, 2)$$

where $c = 1/r$. Hence

$$\sigma\mathsf{p}.\mathsf{p} = \sigma = \Delta(N\mathsf{e}.\mathsf{p}) = \Delta\{N[\alpha - c(x_P\alpha + y_P\beta + z_P\gamma)]\}.$$

Using this on the right of (P.2(iii), 1), there comes

$$T = \Delta[N(q-r)\alpha]+r\sigma. \qquad \text{(P.2(iii), 3)}$$

On the other hand, taking the scalar product of each member of (2.6) with itself, one has

$$\begin{aligned}\sigma^2 &= [\Delta(N\mathrm{e})].[\Delta(N\mathrm{e})]\\ &= N'^2+N^2-2NN'\mathrm{e}'.\mathrm{e}\\ &= N'^2+N^2-2NN'(\alpha\alpha'+\beta\beta'+\gamma\gamma')\\ &= N'^2+N^2-2NN'[(1-\xi)^{\frac{1}{2}}(1-\zeta)^{\frac{1}{2}}+\eta].\end{aligned}$$

With this result, (P.2(iii), 3) will be seen at once to be equivalent to (4.5). (See also equations (117.2–17).)

**P.2 (iv).** The required optical distance is, by inspection,

$$W_1 = N'[x'\alpha'+(y'\beta'+z'\gamma')-(y_P\beta'+z_P\gamma')]+N(y_P\beta+z_P\gamma). \qquad \text{(P.2(iv), 1)}$$

In view of (2.6), $N'\beta' = N\beta, \quad N'\gamma' = N\gamma,$ (P.2(iv), 2)

since $p_y = p_z = 0$ here, and so (P.2(iv), 1) reduces to

$$W_1 = N'x'\alpha'+N(y'\beta+z'\gamma). \qquad \text{(P.2(iv), 3)}$$

But $N'^2\alpha'^2 = N'^2-N'^2(\beta'^2+\gamma'^2) = N'^2-N^2(\beta^2+\gamma^2),$

by (P.2(iv), 2), so that (P.2(iv), 3) is just the stated result.

**P.2 (v).** The coordinate axes having been suitably chosen, equations (7.2), taken together with (6.2), require that

$$\frac{\partial T}{\partial \beta'}+m\frac{\partial T}{\partial \beta} = 0, \quad \frac{\partial T}{\partial \gamma'}+m\frac{\partial T}{\partial \gamma} = 0, \qquad \text{(P.2(v), 1)}$$

for all rays. The first of these shows that $\beta$ and $\beta'$ can occur in $T$ only in the combination $\beta-m\beta'$, and the second likewise shows that $\gamma$ and $\gamma'$ can only occur in the combination $\gamma-m\gamma'$. Hence the required result is

$$T = g(\beta-m\beta', \gamma-m\gamma'), \qquad \text{(P.2(v), 2)}$$

where $g$ is some function of two arguments.

## Chapter 3

**P.3(i).** In the notation of equation (117.2), the equation of the paraboloid is

$$s = 2k^{-\frac{1}{2}}u^{\frac{1}{4}}. \tag{P.3 (i), 1}$$

Then in (117.15)

$$\psi = u^{-\frac{1}{2}}/4k, \tag{P.3 (i), 2}$$

whence in (117.12)

$$u\dot{s} - \tfrac{1}{2}s = -\tfrac{1}{2}k^{-\frac{1}{2}}u^{\frac{1}{4}} = -\psi^{-\frac{1}{2}}/4k. \tag{P.3 (i), 3}$$

Equation (117.12) therefore becomes

$$T = \Delta(Nq\alpha) - (1/4k)(\Delta N\alpha)^2 [(\Delta N)^2(1 + 2\kappa\chi) - (\Delta N\alpha)^2]^{-\frac{1}{2}}. \tag{P.3 (i), 4}$$

Since the expression in the square brackets vanishes when

$$\beta' = \gamma' = \beta = \gamma = 0$$

it follows at once that $T$ cannot be written as a power series in the ray-coordinates.

**P.3(ii).** $\mathscr{W}$ has the property in question for example when it is the surface of an anchor-ring, i.e. a circular torus (see equation (93.1)). Such a surface can be thought of as generated by rotating a circle of radius $R_2$, centre $C$, about a line $\mathscr{L}$ in the plane of the circle, where the perpendicular distance $R_1$ from $C$ to $\mathscr{L}$ satisfies the inequality $R_1 > R_2$. In the course of the rotation $C$ itself generates a circle $\mathscr{C}$ of radius $R_1$. It is obvious that every normal to $\mathscr{W}$, i.e. every ray, passes through both $\mathscr{C}$ and $\mathscr{L}$.

Take one such ray, and suppose that the angle $\theta$ which it makes with $\mathscr{L}$ is not a right angle: this will be the case for every ray $\mathscr{R}$ which does not happen to lie in the plane of $\mathscr{C}$. Next contemplate a narrow bundle of rays, i.e. a general parabasal bundle, about $\mathscr{R}$. These will cover small segments of both $\mathscr{C}$ and $\mathscr{L}$, and these segments are indeed the parabasal focal lines in this case. The first of these is normal to $\mathscr{R}$, but the second is not.

**P.3(iii).** By substitution from (10.3) one finds that

$$\lambda_y' - \lambda_y = b_2(\hat{y}'z' - \hat{z}'y') + b_4(\hat{z}y' - \hat{y}z') + b_6(\hat{y}z' - \hat{z}y') + b_9(\hat{z}y - \hat{y}z).$$

$\lambda'_z - \lambda_z$ must reduce to a similar expression except that $y$ and $z$ will be interchanged throughout. Hence $\lambda'_y - \lambda_y = -(\lambda'_z - \lambda_z)$, which is the required result.

**P.3 (iv).** In the first place one has to have $b_1 = b_5$ and then we must take $d' = -1/b_1$. Set $y = \rho\cos\theta$, $z = \rho\sin\theta$ in (10.15). Eliminating $\theta$ one gets

$$\rho^2 d'^2 (b_3 b_7 - b_4 b_6)^2 = (b_6^2 + b_7^2) Y'^2 - 2(b_3 b_6 + b_4 b_7) Y' Z' + (b_3^2 + b_4^2) Z'^2. \quad \text{(P.3 (iv), 1)}$$

The discriminant of the quadratic form on the right is negative, and (P.3 (iv), 1) therefore represents an ellipse. The condition that it must not shrink to a point is

$$b_3 b_7 - b_4 b_6 \neq 0, \quad \text{(P.3 (iv), 2)}$$

which is just (10.4).

**P.3 (v).** The required equations come from the first member of (3.6):

$$[\dot{a}_0 + (\tfrac{1}{2}\dot{b}_1 y'^2 + \dot{b}_2 y' z' + \ldots + \tfrac{1}{2}\dot{b}_{10} z^2)]^2 + (b_1 y' + b_3 y + b_4 z)^2 + (b_2 y' + b_5 z' + b_6 y + b_7 z)^2 = 1, \quad \text{(P.3 (v), 1)}$$

where a dot denotes differentiation with respect to $x'$. Hence the term independent of the ray-coordinates requires that $\dot{a}_0 = \pm 1$, and since on the axis $V$ is an increasing function of $x'$,

$$a_0 = x'. \quad \text{(P.3 (v), 2)}$$

By inspection of the quadratic terms of (P.3 (v), 1) one then has the differential equations

$$\left.\begin{aligned} \dot{b}_1 + b_1^2 = 0, \quad \dot{b}_2 = 0, \quad \dot{b}_3 + b_1 b_3 = 0, \quad \dot{b}_4 + b_1 b_4 = 0,\\ \dot{b}_5 + b_5^2 = 0, \quad \dot{b}_6 + b_5 b_6 = 0, \quad \dot{b}_7 + b_5 b_7 = 0,\\ \dot{b}_8 + b_3^2 + b_6^2 = 0, \quad \dot{b}_9 + b_3 b_4 + b_6 b_7 = 0, \quad \dot{b}_{10} + b_4^2 + b_7^2 = 0. \end{aligned}\right\} \quad \text{(P.3 (v), 3)}$$

Let the coordinates be so taken that the initial values of the various functions relate to $x' = 0$; and let these values be distinguished by bars. Then, from the first of (P.3 (v), 3),

$$b_1(x') = \bar{b}_1/(1 + \bar{b}_1 x'). \quad \text{(P.3 (v), 4)}$$

The second equation is trivial, and the third gives

$$b_3(x') = \bar{b}_3/(1+\bar{b}_1 x'), \qquad \text{(P.3(v), 5)}$$

and one can easily continue in this way.

**P.3 (vi).** Take the generators of the cylinder to be parallel to the $z$-axis. When the whole cylinder is displaced in this direction one ends up with the same optical system, and so one must have

$$V(y', z'+a, y, z+a) \equiv V(y', z', y, z),$$

for any $a$. It follows that $V$ can depend upon $z$ and $z'$ only in the combination $z'-z$. Moreover, the positive $z$-direction is here not preferred over the negative $z$-direction, and so $V$ must be an even function of $z'-z$:

$$V = V[y', y, (z'-z)^2], \qquad \text{(P.3(vi), 1)}$$

and this function can be written as a power series in the three variables which occur as arguments. (See also Section 84.)

**P.3 (vii).** Yes. There are five for a fixed position of the object, namely the coefficients of $y'^3, y'^2 y, y'y^2, y'(z'-z)^2, y(z'-z)^2$.

## Chapter 4

**P.4 (i).** Using (14.23), with $\hat{s}$ replaced by $\hat{m}$, one has, in view of (14.29), the pair of equations

$$mq' = \hat{m}(d'+smq), \quad q' = d'+m\hat{m}q.$$

These are very easily solved for $q$ and $q'$, and recalling (14.27) one has at once

$$q' = (s-\hat{m})f, \quad q = \left(\frac{1}{\hat{m}}-\frac{1}{m}\right)f. \qquad \text{(P.4(i), 1)}$$

**P.4 (ii).** From the basic equations $\boldsymbol{\beta} = -\partial V/\partial \mathbf{y} = -\mathbf{y}'V_\eta - 2\mathbf{y}V_\zeta$ and $\boldsymbol{\beta}' = \partial V/\partial \mathbf{y}' = 2\mathbf{y}'V_\xi + \mathbf{y}V_\eta$. These can be written in the form (14.5), with

$$A = -2V_\zeta/V_\eta, \; B = -1/V_\eta, \; C = (V_\eta^2 - 4V_\xi V_\zeta)/V_\eta, \; D = -2V_\xi/V_\eta,$$

and (14.7) follows from these by inspection.

**P.4 (iii).** We argue exactly as in Section 15, except that since the ratios $Y'/y$ and $Z'/z$ need no longer be constant we have to write

$$\mathbf{Y}' = mD(\zeta)\,\mathbf{y}, \qquad \text{(P.4(iii), 1)}$$

where $D(0) = 1$. The generic form of (P.4(iii), 1) is required by the symmetry of $K$, and by the fact that $\mathbf{Y}'$ must be independent of $\mathbf{y}'$. Using the same rotational invariants as those which occur in (15.6), the required result follows at once. $D(\zeta)$ is a 'distortion function' whose meaning is defined by (P.4(iii), 1).

**P.4 (iv).** All rays from an object point $O$ unite in a single point $O'(X', Y', Z')$ where, by hypothesis,

$$X' = C(\zeta), \quad \mathbf{Y}' = D(\zeta)\,\mathbf{y}_1. \qquad \text{(P.4(iv), 1)}$$

Here $C$ and $D$ are functions of $\zeta$ such that $C(0) = 0$ and $D(0) = 1$. Using the usual argument, one gets

$$V = g(\zeta) - [(d'+C)^2 + \xi - 2D\eta + D^2\zeta]^{\frac{1}{2}}. \qquad \text{(P.4(iv), 2)}$$

This has the required form if one sets

$$K = 2d'C + C^2 + \zeta D^2. \qquad \text{(P.4(iv), 3)}$$

Then, writing $C(\zeta) = c_1\zeta + \ldots, \quad D(\zeta) = 1 + d_1\zeta + \ldots,$

$$K(\zeta) = k_1\zeta + \ldots = (1 + 2d'c_1)\,\zeta + \ldots,$$

so that

$$c_1 = (k_1 - 1)/2d'. \qquad \text{(P.4(iv), 4)}$$

Again, from $X' = c_1\zeta + \ldots$, $Y'^2 + Z'^2 = \zeta + \ldots$ one has

$$X' = c_1(Y'^2 + Z'^2) + \ldots.$$

The required curvature is therefore $2c_1 = (k_1 - 1)/d'$.

**P.4 (v).** By differentiation of (P.4(iv), 2) we have

$$\boldsymbol{\beta}' = -(\mathbf{y}' - D\mathbf{y}_1)/R, \qquad \text{(P.4(v), 1)}$$

where $R$ is the square root on the right of that equation, whence

$$\alpha' = (d' + C)/R. \qquad \text{(P.4(v), 2)}$$

Therefore

$$\begin{aligned} \boldsymbol{\epsilon}' &= \mathbf{y}' - \mathbf{y}_1 + d'\boldsymbol{\beta}'/\alpha' \\ &= (d'+C)^{-1}[d'(D-1)\,\mathbf{y}_1 + C(\mathbf{y}' - \mathbf{y}_1)]. \end{aligned} \qquad \text{(P.4(v), 3)}$$

When $C = 0$ this becomes $\boldsymbol{\epsilon}' = (D-1)\,\mathbf{y}_1$, as it must.

**P.4 (vi).** If one, say, doubles the size of the object at infinity one evidently doubles the slope of the initial rays, all of which are mutually parallel. Therefore $\mathbf{y}'$ is proportional to $\boldsymbol{\beta}/\alpha$. From (14.35) one has, in the paraxial limit, $\mathbf{y}' = f(\boldsymbol{\beta} - m\boldsymbol{\beta}')$, and taking the limit $m \to 0$, the required constant of proportionality is seen to have the value $f$.

**P.4 (vii).** Consider the plane through $E$ normal to a family of initial rays. Let a particular such ray $\mathscr{R}$ intersect this plane in the point $D$ and $\mathscr{E}'$ in the point $D'$. Then $\tilde{V}(D, O')$ is a function of $\boldsymbol{\beta}$ alone, say

$$\tilde{V}(D, O') = g(\zeta). \qquad \text{(P.4 (vii), 1)}$$

Moreover, $\tilde{V}(D, O') = \tilde{V}(D, D') + \tilde{V}(D', O')$,

i.e. $$g(\zeta) = W_{10}(y', z', \beta, \gamma) + [d'^2 + (f\beta/\alpha - y')^2 + (f\gamma/\alpha - z')^2]^{\frac{1}{2}} \qquad \text{(P.4 (vii), 2)}$$

where (15.14) has been used. Equation (P.4 (vii), 2) can be rewritten immediately in the required form (15.20).

**P.4 (viii).** Allowing for the fact that the functions $g(\zeta)$ which occur in $V$ and $V_0$ respectively need not be the same, we have

$$v = V - V_0 = (1 + \xi - 2\eta + \zeta)^{\frac{1}{2}} - (1 + \xi - 2D\eta + D^2\zeta)^{\frac{1}{2}} + g^*(\zeta). \qquad \text{(P.4 (viii), 1)}$$

It is best to write this first as

$$v = (1 + u)^{\frac{1}{2}} - [(1 + u) - 2(D - 1)\eta + (D^2 - 1)\zeta]^{\frac{1}{2}} + g^*(\zeta). \qquad \text{(P.4 (viii), 2)}$$

Using the power series for $D$, this becomes

$$v = (1 + u)^{\frac{1}{2}} - \{1 + u - 2d_1(\eta\zeta - \zeta^2) - [2d_2\eta\zeta^2 - (2d_2 + d_1^2)\zeta^3]\ldots\}^{\frac{1}{2}} + g^*(\zeta).$$

Expanding this in the usual way one finds, on rejecting terms of degree exceeding six and terms depending on $\zeta$ alone, that

$$v^{(2)} = d_1\eta\zeta, \quad v^{(3)} = \tfrac{1}{2}[-d_1\xi\eta\zeta + d_1\xi\zeta^2 + 2d_1\eta^2\zeta + (2d_2 - 3d_1)\eta\zeta^2]. \qquad \text{(P.4 (viii), 3)}$$

Note that in (23.2–3) the non-zero characteristic coefficients are:

$$p_5 = d_1, \quad s_5 = -\tfrac{1}{2}d_1, \quad s_6 = \tfrac{1}{2}d_1, \quad s_8 = d_1, \quad s_9 = d_2 - \tfrac{3}{2}d_1, \qquad \text{(P.4 (viii), 4)}$$

and one then finds that the only non-vanishing effective coefficients are

$$\bar{p}_3^* = d_1, \quad \bar{s}_6^* = d_2, \qquad \text{(P.4(viii), 5)}$$

consistently with (P.4(v), 3).

When $V$ is given by (P.4(iv), 2) one finds after the same fashion that

$$v^{(2)} = \tfrac{1}{2}c_1\xi\zeta + (d_1 - c_1)\eta\zeta. \qquad \text{(P.4(viii), 6)}$$

The only non-vanishing coefficients in (19.6) are

$$\sigma_4 = c_1, \quad \sigma_5 = d_1 - c_1. \qquad \text{(P.4(viii), 7)}$$

**P.4(ix).** Relative to coordinates with origin at $I'$ (Fig. 4.2) the equation of a meridional ray is (with $d' = 1$)

$$y = -\rho x + \sigma_1\rho^3, \qquad \text{(P.4(ix), 1)}$$

where $\sigma_1\rho^3$ is neglected compared with $\rho$. Differentiating (P.4(ix), 1) with respect to $\rho$,

$$3\sigma_1\rho^2 = x, \qquad \text{(P.4(ix), 2)}$$

and upon eliminating $\rho$ between the two preceding equations there comes

$$27\sigma_1 y^2 = 4x^3. \qquad \text{(P.4(ix), 3)}$$

In view of the axial symmetry of the caustic surface, this result implies (19.11).

**P.4(x).** We proceed after the fashion of the work following equation (19.9). Write $2nv_{00}^{(n)} = b$, $\bar{\rho} = -c\rho_0$, $\chi = k\rho_0^{2n-2}$, so that

$$\hat{y} = b\rho^{2n-1} - k\rho_0^{2n-2}\rho. \qquad \text{(P.4(x), 1)}$$

Differentiation with respect to $\rho$ leads to the equation

$$(2n-1)\,b\bar{\rho}^{2n-2} = k\rho_0^{2n-2},$$

or

$$k = (2n-1)\,bc^{2n-2}. \qquad \text{(P.4(x), 2)}$$

(19.9) then becomes

$$2(n-1)\,c^{2n-1} + (2n-1)\,c^{2n-2} - 1 = 0. \qquad \text{(P.4(x), 3)}$$

Removing a factor $(c+1)^2$ on the left,

$$\sum_{s=0}^{2n-3} (-1)^s (s+1)\,c^s = 0, \qquad \text{(P.4(x), 4)}$$

which is identical with (21.7).

**P.4(xi).** By inspection of (19.7) and (25.1) the conditions are

$$\text{(i) } p_2 = 0, \quad \text{(ii) } s_2 = \tfrac{8}{3}p_1, \quad \text{(iii) } s_5 + \tfrac{3}{4}s_7 = 2p_3 + \tfrac{3}{2}p_4 - \tfrac{1}{2}p_5. \qquad \text{(P.4(xi), 1)}$$

**P.4(xii).** The displacement in $\mathscr{I}'$ is given by (P.4(v), 3). The effective distortion is the displacement for the ray through $E'$. Hence, setting $\mathbf{y}' = 0$, and bearing in mind that $\boldsymbol{\epsilon}'$ then vanishes by hypothesis, we must have

$$D = C + 1, \qquad \text{(P.4(xii), 1)}$$

which is the required result, in view of (P.4(iv), 3).

**P.4(xiii).** From (P.4(iv), 1) and (P.4(xii), 1),

$$X' = D - 1, \quad Y'^2 + Z'^2 = \zeta D^2. \qquad \text{(P.4(xiii), 1)}$$

Setting $D = 1 + d_1\zeta + d_2\zeta^2 + \ldots$, we therefore have

$$X' = d_1\zeta + d_2\zeta^2 + \ldots, \quad Y'^2 + Z'^2 = \zeta + 2d_1\zeta^2 + \ldots.$$

Eliminating $\zeta$ between these equations the required result follows immediately.

**P.4(xiv).** We have $\boldsymbol{\beta}' = (\mathbf{y}' - \mathbf{y}_1)\dot{f}(u)$, where a dot denotes differentiation with respect to $u$. Hence $\alpha'^2 = 1 - u\dot{f}^2$, and therefore

$$\boldsymbol{\epsilon}' = (\mathbf{y}' - \mathbf{y}_1)\,[1 + (1 - u\dot{f}^2)^{-\frac{1}{2}}\dot{f}]. \qquad \text{(P.4(xiv), 1)}$$

One sees immediately that $\epsilon_y'^2 + \epsilon_z'^2$ is a function of $u$ only, and $\alpha'$ likewise depends upon $u$ alone; and this implies the stated result. Taking $z' = 0$ as usual, one has

$$\epsilon_y'/\epsilon_z' = (y' - y_1)/z',$$

so that the circle is not concentric with $I'$.

## Chapter 5

**P.5(i).** Since coma of all orders is definitely present, third-order coma must be present and for this, and therefore for all orders, the comatic ratio has the value 3, i.e.

$$\bar{K} = 3K. \qquad \text{(P.5(i), 1)}$$

Equations (33.5–6) become, since $\delta = 0$,

$$3K = \frac{d(\rho K)}{d\rho} - \tan^3\phi' \frac{dK}{d\rho},$$

and, in view of (32.6), i.e. $\tan\phi' = -\rho$, one has the differential equation

$$\rho(1+\rho^2)\frac{dK}{d\rho} = 2K, \qquad \text{(P.5 (i), 2)}$$

which has the solution $$K = \frac{\text{const.}\,\rho^2}{1+\rho^2}. \qquad \text{(P. 5 (i), 3)}$$

**P.5 (ii).** We are concerned with rays through $O$ and $E'$, and for these $\rho = 0$, i.e. $\xi = \eta = 0$. Let

$$v = \eta D(\zeta) + \ldots, \qquad \text{(P.5 (ii), 1)}$$

where the dots stand for terms at least quadratic in $\rho$. Then

$$\boldsymbol{\beta}' = \mathbf{y}_1[(1+\zeta)^{-\frac{1}{2}} + D(\zeta)], \qquad \text{(P.5 (ii), 2)}$$

since $\mathbf{y}' = 0$. Let $\overline{\psi}'$ be the angle between the axis of $K$ and the line $E'I'$, and take $z_1 = 0$ as usual. Then we define as the analogue of $\Delta(\rho)$ the quantity

$$\Delta_D(h') = \sin\phi' - \sin\overline{\psi}', \qquad \text{(P.5 (ii), 3)}$$

and so

$$\Delta_D = h'D(\zeta).$$

**P.5 (iii).** Set $n_1 = m$ in (40.5) and use (40.11).

**P.5 (iv).** The result is $4(n-2)(n-3)k_5 - 2(n-3)k_3 + 3k_{11}$, but one should remember the powers of $c$ absorbed in $\hat{k}_\alpha$, cf. (51.2).

**P.5 (v).** Setting $\hat{p}_1 = \hat{p}_2 = 0$, and $p_\alpha = 0$ ($\alpha = 1, \ldots, 5$) in the first two members of (45.9), one has the pair of equations

$$p^4 p_6 + j_1(1 - \hat{m}^3/m) = 0, \quad bp^3 p_6 - j_1(1 - s\hat{m}^2/m) = 0.$$

Their mutual compatibility requires that

$$p(1 - s\hat{m}^2/m) + b(1 - \hat{m}^3/m) = 0.$$

With (45.5) this reduces to $\hat{m}^2 = 1$.

**P.5 (vi).** One here has

$$q' - r = Nr/(N'-N), \quad q - r = -N'r/(N'-N),$$

and $f = (NN'r)/(N'-N)$, so that incidentally $f = 1$ implies $(N'-N)r = \kappa^{-1}$. Use of these equations in (4.5) gives the required result at once.

**P.5 (vii).** By the usual argument (cf. P.4 (iii))

$$V^\dagger = g^\dagger(\zeta) - [1 - 2(D-1)\eta + \zeta D^2]^{\frac{1}{2}},$$

where $D$ is defined by the equation $\mathbf{Y}' = D(\zeta)\mathbf{y}_1$.

## Chapter 6

**P.6(i).** Yes. In the limit $k \to -1$ equation (63.12) becomes

$$x = \tfrac{1}{4}(m^2 - 1)(y^2 + z^2). \qquad \text{(P.6(i), 1)}$$

Then (63.13) becomes at the same time

$$X' = \tfrac{1}{4}(m^{-2} - 1)(Y'^2 + Z'^2), \qquad \text{(P.6(i), 2)}$$

which is also the equation of a paraboloid.

**P.6(ii).** Let $z = 0$ in (63.12) and let $|x|$ and $|y|$ become large. The equation becomes that of a pair of straight lines, that is, of the asymptotes, and one reads off the result

$$\tan\psi = k^{-1}(k^2 - 1)^{\frac{1}{2}}(m^2 - 1)^{-\frac{1}{2}}. \qquad \text{(P.6(ii), 1)}$$

In the same way from (63.13)

$$\tan\psi' = m(k^2 - 1)^{\frac{1}{2}}(m^2 - 1)^{-\frac{1}{2}}. \qquad \text{(P.6(ii), 2)}$$

Replacing $k$ by $\tan\psi'/m\tan\psi$ in (P.6(ii), 1), the stated result follows.

**P.6(iii).** It is advisable to use $\mathbf{y}$ rather than $\mathbf{y}_1$ as coordinates here. Then

$$V = g(\zeta) - (1 + \xi - 2m\eta + m^2\zeta)^{\frac{1}{2}} + v(\xi, \eta, \zeta), \qquad \text{(P.6(iii), 1)}$$

and since, when $s = 1/m$, $O_0C = CE'$, this function must be invariant under the mutual interchange of $\mathbf{y}$ and $\mathbf{y}'$, i.e. of $\xi$ and $\zeta$. To the required order we therefore have to consider the invariance of

$$-\tfrac{1}{2}(\xi - 2m\eta + m^2\zeta) - \tfrac{1}{2}(m/f)\zeta + \tfrac{1}{8}(\xi - 2m\eta + m^2\zeta)^2 + v^{(2)}(\xi, \eta, \zeta).$$

Note that $d' = (s - m)f = (m^{-1} - m)f = 1$, so that the coefficient of $\zeta$ is $-\tfrac{1}{2}(m^2 + m/f) = -\tfrac{1}{2}$, as must be the case to ensure invariance of the paraxial part of $V$. We are left with

$$v^{(2)}(\xi, \eta, \zeta) + \tfrac{1}{8}(\xi - 2m\eta + m^2\zeta)^2 \equiv v^{(2)}(\zeta, \eta, \xi) + \tfrac{1}{8}(\zeta - 2m\eta + m^2\xi)^2. \qquad \text{(P.6(iii), 2)}$$

Rejecting the unwanted relation which involves $p_6$ we have the one remaining relation

$$p_2 - p_5 = \tfrac{1}{2}m(1 - m^2). \qquad \text{(P.6(iii), 3)}$$

With the present coordinates $h$ will appear in $\boldsymbol{\epsilon}'$ in place of $h'$.

Going over to $h'$ therefore we must replace $p_2$ by $mp_2$ and $p_5$ by $m^3p_5$, the new coefficients being those which occur in (19.1). Bearing (19.6) in mind we thus arrive at

$$(\sigma_2/m^2) - \sigma_5 = \tfrac{1}{2}(1-m^2)/m^2. \qquad \text{(P.6 (iii), 4)}$$

When $d' \neq 1$, we have to supply a factor $d'^{-2}$ on the right. But we have seen that $d' = (1-m^2)f/m$; so that when $f$ is now taken to be unity we have to supply a factor $m^2(1-m^2)^{-2}$ on the right of (P.6 (iii), 4); and the latter thereupon becomes identical with (56.12).

**P.6 (iv).** $V_1$ refers to a curved posterior base-surface, not $\mathscr{E}'$.

**P.6 (v).** Write
$$(1-t\xi)^{\frac{1}{2}}(1-t\zeta)^{\frac{1}{2}} = \Sigma\phi_n t^n. \qquad \text{(P.6 (v), 1)}$$

Differentiate with respect to $t$ and then multiply by $(1-t\xi)(1-t\zeta)$ throughout. There comes

$$-\tfrac{1}{2}[\zeta(1-t\xi)+\xi(1-t\zeta)]\,\Sigma\phi_n t^n = (1-t\xi)(1-t\zeta)\,\Sigma n\phi_n t^{n-1},$$

whence

$$-\tfrac{1}{2}(\xi+\zeta)\,\Sigma\phi_n t^n + \xi\zeta\Sigma\phi_n t^{n+1} - \Sigma n\phi_n t^{n-1} + (\xi+\zeta)\,\Sigma n\phi_n t^n - \xi\zeta\Sigma n\phi_n t^{n+1} \equiv 0. \qquad \text{(P.6 (v), 2)}$$

The factor multiplying $t^s$ $(s = 0, 1, 2, \ldots)$ must vanish, and this requires just that the stated recurrence relation be satisfied.

## Chapter 7

**P.7 (i).** Setting $P = 1$, $J = 1$ in (82.2) one has

$$\left.\begin{aligned} \epsilon'_y &= [(R-Q)\,\eta+(\tilde{R}-\tilde{Q})\,\tau]\,y'-(1-R)\,y_1, \\ \epsilon'_z &= [(R-Q)\,\eta+(\tilde{R}-\tilde{Q})\,\tau]\,z'+\tilde{R}y_1, \end{aligned}\right\} \qquad \text{(P.7(i), 1)}$$

with $z_1 = 0$, as usual. Introducing polar coordinates in the customary way, $\boldsymbol{\epsilon}'$ becomes a function of $2\theta$, and the terms independent of $\theta$ give the coordinates of the centres of the zonal circles. They are

$$\hat{y} = [\tfrac{1}{2}(R-Q)\,\rho^2-(1-R)\,]\,h', \quad \hat{z} = [\tfrac{1}{2}(\tilde{R}-\tilde{Q})\,\rho^2+\tilde{R}]\,h'. \qquad \text{(P.7 (i), 2)}$$

A straight line is generated if and only if $\hat{y} = k\hat{z}$ for some constant $k$, i.e.

$$\xi(2w^2\dot{C}+C)+2C = k[\xi(2w^2\dot{\tilde{C}}+\tilde{C})+2\tilde{C}]. \qquad \text{(P.7 (i), 3)}$$

Set $$\tilde{C} = k^{-1}C + \Gamma.$$
Then the equation for $\Gamma$ is
$$2\xi(1+\xi)\dot{\Gamma} + (\xi+2)\Gamma = 0,$$
the solution of which near $\xi = 0$ is const. $\xi^{-1}$, which means that $\Gamma$ must vanish, taking regularity into account. Hence $\tilde{C}$ and $C$ must stand in a constant ratio to one another.

**P.7 (ii).** The possibility of determining $\hat{s}$ and $\hat{m}$ is assured if the set of equations (79.3) is soluble. The discriminant of these equations turns out to be
$$[\bar{d}^{-1}bc(1-pq)]^3 = \bar{d}^{-6} \neq 0,$$
which was to be shown.

**P.7 (iii).** Since $\boldsymbol{\epsilon}' = -(\partial T/\partial \boldsymbol{\beta}') - m(\partial T/\partial \boldsymbol{\beta})$ by definition, one merely has to introduce the derivatives with respect to $\xi, \eta, \zeta, \tau$, which is easily done with the help of (15.11) and (72.4–5).

**P.7 (iv).** By (80.3), writing $\tilde{t}^{\#(5)} = \tilde{q}_1\xi^4 + \tilde{q}_2\xi^3\eta + \ldots + \tilde{q}_{15}\zeta^4$,
$$\begin{aligned}\tilde{\alpha}_9 &= (s-m)A^2\tilde{t}^{\#(5)} \\ &= (s-m)^5(16\partial_{\xi\xi\zeta\zeta} - 8\partial_{\xi\eta\eta\zeta} + \partial_{\eta\eta\eta\eta})(\tilde{q}_6\xi^2\zeta^2 + \tilde{q}_8\xi\eta^2\zeta + \tilde{q}_{11}\eta^4) \\ &= 8(s-m)^5(8\tilde{q}_6 - 2\tilde{q}_8 + 3\tilde{q}_{11}).\end{aligned}$$
The coefficients govern ninth-order quintic skew coma.

## Chapter 8

**P.8 (i).** As in P.6 (iii), one should use invariants defined in terms of $y, z$ rather than $m_1 y$, $m_2 z$. Taking $d' = 1$,
$$V = \text{const.} - \frac{m_1\zeta_1}{2f_1} - \frac{m_2\zeta_2}{2f_2} - (1 + \xi_1 - 2m_1\eta_1 + m_1^2\zeta_1 + \xi_2 - 2m_2\eta_2 + m_2^2\zeta_2)^{\frac{1}{2}} + v, \quad \text{(P.8 (i), 1)}$$
all non-linear terms of $g(\zeta_1, \zeta_2)$ having been absorbed in $v$. A sharp paraxial image is supposed to exist. $B'$, with which $E'$ will be taken to be coincident, is chosen to be so situated that $O_0C = CB'$ in the usual notation. Then $V$ must be invariant under the simultaneous interchange of $\xi_1$ with $\zeta_1$ and $\xi_2$ with $\zeta_2$. In the first place
$$\frac{(1-m_1^2)f_1}{m_1} = \frac{(1-m_2^2)f_2}{m_2} = 1. \quad \text{(P.8 (i), 2)}$$

Considering the quadratic terms of (P.8(i), 1) one may omit those varying as $\zeta_1^2, \zeta_1\zeta_2, \zeta_2^2$, $\xi_1^2, \xi_1\xi_2, \xi_2^2$, $\xi_1\zeta_1, \xi_2\zeta_2, \eta_1\eta_2$; the first three because they will give rise to relations involving irrelevant coefficients (i.e. those which do not enter into the displacement), the next three because they become redundant when the first three are rejected, whilst the last three are in any case not affected when $\xi_1$, $\xi_2$ are interchanged with $\zeta_1, \zeta_2$ respectively. One is left with the quadratic terms

$$\begin{aligned} &m_1p_4\xi_1\eta_1 + m_2p_5\xi_1\eta_2 + m_1p_6\eta_1\xi_2 + m_2p_7\xi_2\eta_2 + m_2^2p_9\xi_1\zeta_2 \\ &\quad + m_1^2p_{11}\zeta_1\xi_2 + m_1^3p_{13}\eta_1\zeta_1 + m_1m_2^2p_{14}\eta_1\zeta_2 + p_{15}m_1^2m_2\zeta_1\eta_2 \\ &\quad + p_{16}m_2^3\eta_2\zeta_2 + \tfrac{1}{8}(-4m_1\xi_1\eta_1 - 4m_2\xi_1\eta_2 - 4m_1\eta_1\xi_2 - 4m_2\xi_2\eta_2 \\ &\quad + 2m_2^2\xi_1\zeta_2 + 2m_1^2\zeta_1\xi_2 - 4m_1^3\eta_1\zeta_1 - 4m_1m_2^2\eta_1\zeta_2 - 4m_1^2m_2\zeta_1\eta_2 \\ &\quad - 4m_2^3\eta_2\zeta_2). \end{aligned} \qquad \text{(P.8(i), 3)}$$

The aberration function $v^{(2)}$ has been taken to be defined in terms of the more usual variables $y_1 (\equiv m_1y)$ and $z_1 (\equiv m_2z)$, and the $p_\alpha$ are therefore those in (88.6). Since $V$ is, however, contemplated at present as a function of $y$ and $z$ one has to supply the various powers of $m_1$ and $m_2$ which appear in (P.8(i), 3). By inspection one now reads off the *five* third-order relations

$$\left.\begin{aligned} p_4 - m_1^2p_{13} &= \tfrac{1}{2}(1-m_1^2), \\ p_5 - m_1^2p_{15} &= \tfrac{1}{2}(1-m_1^2), \\ m_2^2p_9 - m_1^2p_{11} &= \tfrac{1}{4}(m_1^2-m_2^2), \\ p_6 - m_2^2p_{14} &= \tfrac{1}{2}(1-m_2^2), \\ p_7 - m_2^2p_{16} &= \tfrac{1}{2}(1-m_2^2). \end{aligned}\right\} \qquad \text{(P.8(i), 4)}$$

**P.8 (ii).** The total number of coefficients of a polynomial of degree $n+1$ in four variables is $\frac{1}{6}(n+2)(n+3)(n+4)$. A typical term is of the form $c_{\lambda\mu\nu}y'^{n-\lambda+1}y^{\lambda-\mu}z'^{\mu-\nu}z^\nu$, which takes a factor $(-1)^\mu$ when the signs of $z'$ and $z$ are simultaneously reversed. Thus $c_{\lambda\mu\nu} = 0$ when $\mu$ is odd. When $n$ is even the number of coefficients which so vanish is $\frac{1}{12}(n+2)(n+3)(n+4)$, and there are still $\frac{1}{2}(n+2)$ coefficients multiplying powers of $y$ and $z$ alone. One is left with the number of coefficients given by (88.14). The case of odd $n$ may be dealt with in the same way.

**P.8(iii).** One possibility (which is particularly well adapted to straight line images) is to take the posterior base-surface to be a circular cylinder $\mathscr{W}_0$ through $E'$. The axis of the cylinder is normal to the meridional plane and passes through the point $(0, y_1, 0)$. The equation of $\mathscr{W}_0$ is evidently

$$x' = 1-(1-\xi+2\eta)^{\frac{1}{2}}. \qquad \text{(P.8(iii), 1)}$$

Then
$$V_1 = g(\zeta)-(1+\zeta+\tau)^{\frac{1}{2}}+v_1, \qquad \text{(P.8(iii), 2)}$$

in view of (86.1), and this defines $v_1$. Referring to Section 36, one has

$$\beta' = \frac{\partial V_1}{\partial y'}-(y'-y_1)/X', \quad \gamma' = \frac{\partial V_1}{\partial z'} \qquad \text{(P.8(iii), 3)}$$

where $X'$ is defined by (37.5). In (P.8(iii), 2) only $v_1$ depends upon $y'$. In the usual way one then finds

$$\epsilon'_y = X'\frac{\partial v_1}{\partial y'}, \quad \epsilon'_z = X'\frac{\partial v_1}{\partial z'}+(z'-z_1)[1-X'(1+\zeta+\tau)^{-\frac{1}{2}}].$$

**P.8 (iv).** We now have generically

$$Y' = y_1 A(\zeta_1, \zeta_2), \quad Z' = z_1 B(\zeta_1, \zeta_2). \qquad \text{(P.8(iv), 1)}$$

There can be no term of the form $z_1 C(\zeta_1, \zeta_2)$ in $Y'$, since when the $\bar{y}$- and $\bar{y}'$-axes are inverted $Y'$ and $y_1$ reverse sign independently of $z_1$; and so on. By the usual argument

$$V = g(\zeta_1, \zeta_2)-[1+(y_1 A-y')^2+(z_1 B-z')^2]^{\frac{1}{2}}. \qquad \text{(P.8(iv) 2)}$$

Write for the moment

$$A = 1+a_1\zeta_1+a_2\zeta_2+O(4), \quad B = 1+b_1\zeta_1+b_2\zeta_2+O(4).$$

Then the quadratic terms of (P.8(iv), 2) other than those which remain when $a_1 = a_2 = b_1 = b_2 = 0$ are

$$v^{(2)} = a_1\eta_1\zeta_1+a_2\eta_1\zeta_2+b_1\zeta_1\eta_2+b_2\eta_2\zeta_2, \qquad \text{(P.8(iv), 3)}$$

terms depending upon $\zeta_1$ and $\zeta_2$ alone having been rejected. Comparison with (88.6) then gives

$$A = 1+p_{13}\zeta_1+p_{14}\zeta_2+O(4), \; B = 1+p_{15}\zeta_1+p_{16}\zeta_2+O(4). \qquad \text{(P. 8(iv), 4)}$$

All other third-order coefficients vanish, as they must.

**P.8(v).** Not in general. Write (88.8) as

$$\hat{y} = p + a\cos 2\theta + b\sin 2\theta, \quad \hat{z} = q + c\cos 2\theta + d\sin 2\theta. \quad \text{(P.8(v), 1)}$$

Then, for example, when $ad - bc = 0$ a zonal curve degenerates into a segment of a straight line. No matter what the values of the coefficients may be, provided only that

$$\frac{(3p_4 - p_6)\,p_6}{(p_5 - 3p_7)\,p_5} \geqslant 0, \qquad \text{(P.8(v), 2)}$$

there will exist a value of $\psi$ such that this degeneration occurs. The zonal curves are circles if and only if

$$p_5 - 3p_7 = \pm 2p_5 \quad \textit{and} \quad 3p_4 - p_6 = \pm 2p_6. \quad \text{(P.8(v), 3)}$$

## Chapter 9

**P.9(i).** The number in question is the number of coefficients governing the terms of degree $n$ in $\chi, \tau$ in the function $G$ of equation (99.4), and this is $n+1$.

**P.9(ii).** When $\beta = \gamma = 0$

$$-y' = [2\alpha' J_\xi + (J_\chi - q')]\,\beta'/\alpha', \quad -z' = (J_\chi - q')\,\gamma'/\alpha'. \quad \text{(P.9(ii), 1)}$$

In the absence of spherical aberration $y'$ and $z'$ must vanish. Thus $J$ has to satisfy the two conditions

$$J_\xi = 0, \quad J_\chi = q' \quad \text{when} \quad \beta = \gamma = 0. \qquad \text{(P.9(ii), 2)}$$

Reference to (95.4) at once shows these to imply that

$$k_1 = 0, \quad k_4 = q', \quad p_1 = p_4 = p_8 = 0, \qquad \text{(P.9(ii), 3)}$$

consistently with (95.6–7).

**P.9(iii).** A circular patch obtains if and only if the factors multiplying $\beta'$ and $\gamma'$ respectively in (P.9(ii), 1) are equal. Therefore one must have

$$J_\xi = 0 \quad \text{when} \quad \beta = \gamma = 0. \qquad \text{(P.9(iii), 1)}$$

## Chapter 10

**P.10 (i).** Let $m$ tend to zero in (106.7). Then

$$\mathbf{y}'_e = (\kappa + sc_1)\,\boldsymbol{\sigma} + (1 - \kappa + sc_2)\,\boldsymbol{\mu}, \quad \mathbf{y}_1 = \boldsymbol{\mu}.$$

Also
$$\boldsymbol{\epsilon}' = s(c_1\boldsymbol{\sigma} + c_2\boldsymbol{\mu}). \qquad \text{(P.10(i), 1)}$$

From these equations one gets the desired result

$$\boldsymbol{\epsilon}'_1 = s(\kappa + sc_1)^{-1}\{c_1\mathbf{y}'_e + [\kappa c_2 + (\kappa - 1)\,c_1]\,\mathbf{y}_1\}. \qquad \text{(P.10(i), 2)}$$

This will be found to be in agreement with (106.11).

**P.10 (ii).** Taking $q' = q = r = 1$ and $N = 1$ in (4.5) one has

$$T = [(N' - 1)^2 + 2N'\chi]^{\frac{1}{2}}, \qquad \text{(P.10(ii), 1)}$$

where $\chi$ is given by (65.10). Setting $N' = \underset{0}{N'}(1 + \nu'_1\omega + \ldots)$ and expanding $T$ in ascending powers of $\omega$, one gets the stated result without difficulty.

**P.10 (iii).** Write (101.4) in the slightly more general form

$$\omega = \delta\lambda/(1 + \alpha\,\delta\lambda), \qquad \text{(P.10(iii), 1)}$$

whence
$$\delta\lambda = \omega/(1 - \alpha\omega). \qquad \text{(P.10(iii), 2)}$$

Now, if $\delta\lambda^* = \lambda - \underset{0}{\lambda}{}^*$, then

$$\begin{aligned}\omega^* &= \delta\lambda^*/(1 + \alpha^*\,\delta\lambda^*)\\ &= (\delta\lambda - \mu)/[1 + \alpha^*(\delta\lambda - \mu)],\end{aligned} \qquad \text{(P.10(iii), 3)}$$

where $\mu = \underset{0}{\lambda}{}^* - \underset{0}{\lambda}$. Inserting (P.10(iii), 1) into (P.10(iii), 3) one gets an expression for $\omega^*$ which is linear in $\omega$ provided one chooses

$$\alpha^* = \alpha/(1 + \alpha\mu). \qquad \text{(P.10(iii), 4)}$$

Then
$$\omega^* = (1 + \alpha\mu)\,[(1 + \alpha\mu)\,\omega - \mu]. \qquad \text{(P.10(iii), 5)}$$

**P.10 (iv).** In (107.3) $\underset{0}{K} = 0$, and therefore $\underset{0\,0}{JR} = 1$, in view of (31.1). Hence $\underset{1}{K} = 0$ if

$$\underset{1}{C} = 2\xi\underset{0\,0\,1}{JP\dot{S}}. \qquad \text{(P.10(iv), 1)}$$

However, $\underset{0\,0}{JP} = (1 + \underset{0}{\delta})^{-1}$ by (31.4) and it suffices to set $\underset{0\,0}{JP} = 1$.

## Chapter 11

**P.11 (i).** Consider $V = \int N ds$ as a function of the coordinates of points on the ray along which the integral is extended, so that some initial point on the ray is kept fixed. We denote them by $x, y, z$ although primed symbols would be more consistent. Then

$$dV/ds = N, \qquad \text{(P.11 (i), 1)}$$

bearing in mind that $d/ds$ denotes differentiation along the ray. Take the gradient of both members of this equation, so that

$$d(\text{grad } V)/ds = \text{grad } N. \qquad \text{(P.11 (i), 2)}$$

Equations (112.4) were derived directly from Fermat's Principle. From the *first* three we have

$$\text{grad } V = \text{Grad } N, \qquad \text{(P.11 (i), 3)}$$

and with this (P. 11 (i), 2) becomes just the required equation.

**P.11 (ii).** Since $\mathsf{p} = (1, 0, 0)$ the three equations (111.7) read

$$(b'\alpha'/N') - (b\alpha/N) = \sigma, \quad a'\beta'/N' = a\beta/N, \quad a'\gamma'/N' = a\gamma/N. \qquad \text{(P.11 (ii), 1)}$$

Squaring and adding the last two,

$$(a'/N')^2 \sin^2 \phi' = (a/N)^2 \sin^2 \phi. \qquad \text{(P.11 (ii), 2)}$$

On the other hand

$$N'^2 = a' + (b' - a') \cos^2 \phi', \quad N^2 = a + (b - a) \cos^2 \phi. \qquad \text{(P.11 (ii), 3)}$$

Then (P.11 (ii), 2) becomes $Q' = Q$, where

$$Q = \frac{a^2 \sin^2 \phi}{a + (b - a) \cos^2 \phi}.$$

Solving for $\sin^2 \phi'$, we get

$$\sin^2 \phi' = \frac{b'Q}{a'^2 + (b' - a')Q}, \qquad \text{(P.11 (ii), 4)}$$

which is, in effect, just the required equation.

## Chapter 12

**P.12 (i).** The surface $\mathcal{S}$ in question must obviously be a surface of revolution. Let $A(0, 0, 0)$ be its pole, $B(-q, 0, 0)$ and $B'(q', 0, 0)$ the required points ($q, q' > 0$), and $P(x, y, z)$ the point of incidence of some ray through $B$ and $B'$. Then $\tilde{V}(BPB')$ must have the same value for all such points, i.e.

$$N'[(q'-x)^2+(y^2+z^2)]^{\frac{1}{2}}+N[(q-x)^2+(y^2+z^2)]^{\frac{1}{2}} = N'q'+Nq. \tag{P.12(i), 1}$$

We had taken $q$ and $q'$ to be positive. If the object point is virtual, let the coordinates of $B$ be $(q, 0, 0)$ with $q > 0$. Then (P.12(i), 1) has to be replaced by

$$N'[(q'-x)^2+(y^2+z^2)]^{\frac{1}{2}}-N[(q-x)^2+(y^2+z^2)]^{\frac{1}{2}} = N'q'-Nq. \tag{P.12(i), 2}$$

In general $\mathscr{S}$ is therefore of the fourth degree.

**P.12 (ii).** Let $\mathscr{S}$ be a sphere of radius $r$, so that

$$y^2+z^2 = 2rx-x^2. \tag{P.12(ii), 1}$$

Using this in (P.12(i), 2), there comes

$$N'[q'^2+2(r-q')\,x]^{\frac{1}{2}}-N[q^2+2(r-q)\,x]^{\frac{1}{2}} = N'q'-Nq.$$

This is possible only if

$$N'q' = Nq \quad \textit{and} \quad N'^2(q'-r) = N^2(q-r). \tag{P.12(ii), 2}$$

(Were (P.12(ii), 1) used in (P.12(i), 1) one would not be able to satisfy the resulting equation.) (P.12(ii), 2) yield

$$N'q' = Nq = (N'+N)\,r. \tag{P.12(ii), 3}$$

The locations of $B$ and $B'$ are therefore determined, and both points lie on the same side of $\mathscr{S}$. Note that

$$q'/q = m = N/N', \tag{P.12(ii), 4}$$

so that the actual magnification associated with $B$ and $B'$ is $(N/N')^2$.

**P.12 (iii).** By differentiation of (117.17) we have

$$N\alpha\mathbf{y} = f(1+2\kappa\chi)^{-\frac{1}{2}}(\alpha'\boldsymbol{\beta}-\alpha\boldsymbol{\beta}')+N(q-r)\,\boldsymbol{\beta},$$

$$N'\alpha'\mathbf{y}' = f(1+2\kappa\chi)^{-\frac{1}{2}}(\alpha'\boldsymbol{\beta}-\alpha\boldsymbol{\beta}')+N'(q'-r)\,\boldsymbol{\beta}'. \tag{P.12(iii), 1}$$

If spherical aberration of all orders is absent, $\mathbf{y}$ and $\mathbf{y}'$ vanish together, and so

$$\Delta[N(q-r)\,\boldsymbol{\beta}] = 0. \tag{P.12(iii), 2}$$

Using (P.12(ii), 2) and (P.12(ii), 4), this reduces to

$$\boldsymbol{\beta} - m\boldsymbol{\beta}' = 0, \qquad \text{(P.12(iii), 3)}$$

i.e. the sine-condition is satisfied.

**P.12(iv).** According to (120.26) the vanishing of third-order spherical aberration and coma would require that the two equations

$$\mathfrak{p} + \mathfrak{c} = 0, \quad \theta\mathfrak{p} + \tilde{\theta}\mathfrak{c} = 0 \qquad \text{(P.12(iv), 1)}$$

be simultaneously satisfied. Since $\mathfrak{c} \neq 0$ by hypothesis ($y \neq 0$), one would have to have $\theta - \tilde{\theta} = 0$. However,

$$\theta - \tilde{\theta} = r\phi/y \neq 0. \qquad \text{(P.12(iv), 2)}$$

**P.12(v).** Write

$$a = V'^2 + W'^2, \quad b = VV' + WW', \quad c = V^2 + W^2. \quad \text{(P.12(v), 1)}$$

Then, from (119.9)

$$\mathbf{g} = \{f(1 - \kappa\chi + \tfrac{3}{2}\kappa^2\chi^2)\,[v'(1 - \tfrac{1}{2}c + \tfrac{3}{8}c^2) - v(1 - \tfrac{1}{2}a + \tfrac{3}{8}a^2)]$$
$$+ Nri\}\,\Delta\mathbf{V} + O(6). \quad \text{(P.12(v), 2)}$$

Here

$$\chi = 1 - (\alpha\alpha' + \beta\beta' + \gamma\gamma') = 1 - (1+c)^{-\frac{1}{2}}(1+a)^{-\frac{1}{2}}(1+b)$$
$$= \tfrac{1}{2}(a - 2b + c) - \tfrac{1}{8}(3a^2 - 4ab + 2ac - 4bc + 3c^2) + O(6). \text{ (P.12(v), 3)}$$

Inserting this in (P.12(v), 2) and selecting the terms of degree 5 in $\mathbf{V}'$, $\mathbf{V}$, there comes

$$\mathbf{g}^{(3)} = f\{-\tfrac{3}{8}(va^2 - v'c^2) - \tfrac{1}{4}\kappa(va - v'c)(a - 2b + c)$$
$$+ \tfrac{1}{8}(v' - v)\,[\kappa(3a^2 - 4ab + 2ac - 4bc + 3c^2) + 3\kappa^2(a - 2b + c)^2]\}\,\Delta\mathbf{V}.$$
$$\text{(P.12(v), 4)}$$

The factor multiplying $\mathbf{S}(S_y^2 + S_z^2)$ in $\langle\mathbf{g}^{(3)}\rangle$ is therefore

$$\tfrac{1}{8}f(k-1)\,i[-3vv'^4 + 3v'v^4 - 2\kappa(v'-v)^2(vv'^2 - v'v^2)$$
$$+ \kappa(v'-v)(3v'^4 - 4v'^3v + 2v'^2v^2 - 4v'v^3 + 3v^4) + 3\kappa^2(v'-v)^5].$$

Taking out a factor $v'-v$ from the expression in the square brackets and noting that $\kappa(v'-v)^2 = ii'$, this becomes

$$\tfrac{3}{8}f(k-1)^2 i^2[-vv'(v'^2+vv'+v^2)+ii'(v'^2+v^2)+i^2i'^2].$$

The expression in the square brackets clearly has a factor

$$ii'-vv' = cy(i'+v),$$

the other factor being $(k^2-k+1)i^2+3(k-1)iv+3v^2$. Recalling (120.14) and (119.8) the required result now follows at once.

**P.12 (vi).** The first relation follows immediately from (120.12), (120.14) and the second member of (128.12). Then

$$16\mathfrak{z}_3+4\mathfrak{z}_4 = `\tilde{\omega}(ii'+v'^2-2vv'+v^2) = `\tilde{\omega}(k^2-k+1)i^2. \quad \text{(P.12(vi), 1)}$$

**P.12 (vii).** The invariant $N\alpha^{-1}(Y^*\gamma-Z^*\beta)$ has the same value at the $j$th as at the first surface. As after equation (128.19), one has

$$(s-m)f_K^{-1}N\alpha^{-1}(Y^*\gamma-Z^*\beta)$$
$$= \alpha^{-1}\{(\sigma_y\mu_z-\sigma_z\mu_y)+(\sigma_y\delta_{\mu z}-\sigma_z\delta_{\mu y})-(\mu_y\delta_{\sigma z}-\mu_z\delta_{\sigma y})\}+O(6)$$
$$= \alpha^{-1}(\sigma_y\mu_z-\sigma_z\mu_y)[1+(\bar{\delta}_\mu+\delta_\sigma)]+O(6). \quad \text{(P.12(vii), 1)}$$

Hence $$\alpha^{-1}(1+\bar{\delta}_\mu+\delta_\sigma) = \alpha_1^{-1}(1+\bar{\delta}_{\mu 1}+\delta_{\sigma 1})+O(4),$$

or $$\bar{\delta}_\mu+\delta_\sigma = (\alpha/\alpha_1-1)+\bar{\delta}_{\mu 1}+\delta_{\sigma 1}+O(4), \quad \text{(P.12(vii), 2)}$$

which is, in effect, the required relation.

**P.12 (viii).** We have the given equations

$$x_j = a_j x+\sum_{i=1}^{j}(a_{2i}x_i^3+a_{3i}x_i^5+a_{4i}x_i^7+\ldots). \quad \text{(P.12(viii), 1)}$$

Let $$x_j = a_j x+A_{2j}x^3+A_{3j}x^5+A_{4j}x^7+\ldots. \quad \text{(P.12(viii), 2)}$$

Inserting this into both members of (P.12(viii), 1), there comes

$$a_j x+A_{2j}x^3+A_{3j}x^5+A_{4j}x^7+\ldots$$
$$\equiv a_j x+\sum_{i=1}^{j}[a_{2i}(a_i x+A_{2i}x^3+A_{3i}x^5+\ldots)^3$$
$$+a_{3i}(a_i x+A_{2i}x^3+\ldots)^5+a_{4i}(a_i x+\ldots)^7+\ldots]. \quad \text{(P.12(viii), 3)}$$

Suppressing the index $i$ on the right for the sake of clarity one reads off the relations

$$\left.\begin{aligned} A_{2j} &= \sum_{i=1}^{j} a^3 a_2, \\ A_{3j} &= \sum_{i=1}^{j} (3a^2 a_2 A_2 + a^5 a_3), \\ A_{4j} &= \sum_{i=1}^{j} (3a^2 a_2 A_3 + 3aa_2 A_2^2 + 5a^4 a_3 A_2 + a^7 a_4), \end{aligned}\right\} \quad \text{(P.12(viii), 4)}$$

and so on. The first of these gives $A_{2i}$ in terms of given constants, after which the second gives $A_{3i}$, and so on.

# LIST OF PRINCIPAL SYMBOLS†

*Roman symbols*

$a$ measure of primary asphericity $(= 4r^3s_2 - 1)$, 280

$\bar{a}$ coefficient of linear transformation of direction cosines, 115

$A$ axial point of surface of revolution, 14, 270

$A, A'$ initial and final points of ray, 5

$A$ the generator $(s-m)^2[4(\partial^2/\partial\xi\,\partial\zeta)-(\partial^2/\partial\eta^2)]$, 127

$\bar{A}$ generator adjoint to $A$, 127

$\mathfrak{a}$ $\frac{1}{2}(s-m)^{-1}\Delta(v^2)$, 299

$\mathscr{A}$ axis of symmetry of axially symmetric system, 22

$\mathscr{A}$ axis of system at least doubly plane-symmetric, 191

$b$ $1-p\ \big(=(s-m)/(s-\hat{m})\big)$, 119

$b$ $(1-p)/(1-pq)\ \big(=(\hat{s}-m)/(\hat{s}-\hat{m})\big)$, 122

$\bar{b}$ coefficient of linear transformation of direction cosines, 115

$B, B'$ base-points, i.e. axial points of $\mathscr{B}, \mathscr{B}'$, 39

$\hat{B}'$ moved posterior base-point, 95

$\mathsf{B}$ magnetic induction vector, 258

$\mathscr{B}, \mathscr{B}'$ anterior and posterior base-planes, 7

$\mathscr{B}^{*\prime}$ curved posterior base-surface, 98

$c$ $1-q\ \big(=(s-m)/(\hat{s}-m)\big)$, 97, 116

$c$ $(1-q)/(1-pq)\ \big(=(s-\hat{m})/(\hat{s}-\hat{m})\big)$, 122

$c$ axial curvature of $\mathscr{S}$, 272

$\bar{c}$ coefficient of linear transformation of direction cosines, 115

$c_n (n = 1, 2, \ldots)$ coefficients of $C(\xi)\ (= f_{10}^{(n+1)})$, 83, 121

$c_\alpha (\alpha = 1, 2)$ paraxial chromatic characteristic aberration coefficients, 230, 236

† The page numbers refer to the place where a symbol with the given meaning first occurs.

$\underset{m}{c_\alpha}(\alpha = 1, 2)$ chromatic $m$th-order coefficients of $c_\alpha$, 231

$c_1^*, \bar{c}_1^*$ effective paraxial chromatic aberration coefficients, 238

$C$ axial point of central plane, 137, 156

$C(\xi)$ $(\partial f/\partial \eta)_{\eta=\zeta=0}(f = v, t, \ldots)$, 83, 121, 239

$\tilde{C}(\xi)$ $(\partial v/\partial \tau)_{\eta=\zeta=0}$, 184

$C(\zeta)$ one of a pair of functions describing $\mathscr{I}^{*\prime}$, 151

$\mathscr{C}$ a curve, 4

$\mathscr{C}$ central plane of reversible or concentric system, 136, 156

$\mathscr{C}$ axis of symmetry of toroidally symmetric system, 212

$d$ length of straight segment of ray through $E'$, 244

$d$ distance between $O_0$ and $E$, 43

$d'$ distance between $B'$ and $O_0'$, or $E'$ and $O_0'$, 19, 42

$\hat{d}, \hat{d}'$ values of $d$, $d'$ constant upon shifts of $O_0$ or $E'$, 97

$d_j'$ distance between $A_{j+1}$ and $A_j$, 272

$\bar{d}$ coefficient of linear transformation of direction cosines, 115

$d^*$ $-m^{-1}(1-m^2)f$, 138

$d^{(n)}$ $(2n-1)$th-order part of $D = T - \tilde{T}$, 126

$d^{(n)}_{\sigma\alpha\beta j}, d^{(n)}_{\mu\alpha\beta j}$ $(2n-1)$th-order coefficients of $\mathbf{d}_\sigma, \mathbf{d}_\mu$, 302

$D$ deformation of the wavefront, 105

$D$ $T - \tilde{T}$, 126

$D$ length of straight segment of ray, 244

$D'$ point of intersection of $\mathscr{R}$ with posterior base-surface, 42, 98

$D(\zeta)$ one of a pair of functions describing $\mathscr{I}^{*\prime}$, 151

$D_k (k = 1, 2, 3)$ components of $\mathsf{D}$, 258

$\mathbf{d}_\sigma, \mathbf{d}_\mu$ modified increments, 295

$\mathbf{D}$ $\Delta(\boldsymbol{\beta}/v_a) - \boldsymbol{\mu}\Delta\bar{\theta}$, 288

$\mathsf{D}$ electric displacement vector, 258

$\mathsf{D}_0$ amplitude of $\mathsf{D}$, 258

$e$ *Nri*, 277

$e_i (i = 1, 2, 3)$ components of $\mathsf{e}$ $(= \alpha, \beta, \gamma)$, 251

$E, E'$ points of $\mathscr{E}, \mathscr{E}'$, 42

$E_j, E'_j$ points conjugate to $E$ before and after $\mathscr{S}_j$, 269

$E_k (k = 1, 2, 3)$ components of $\mathsf{E}$, 258

$\mathsf{e}$ unit tangent vector to ray, 3

$\mathsf{E}$ electric field vector, 258

$\mathsf{E}_0$ amplitude of $\mathsf{E}$, 258

$\mathscr{E}, \mathscr{E}'$ paraxial pupil planes, 42

$f$ mean focal length of $K$, 40

$f$ mean focal length of spherical surface, 276

$f$ aberration function associated with $F$, 19

$f_K$ (mean) focal length of $K$ as a whole, 274

$f_n$ $n$th-order part of the aberration function $f$, 31, 224

$f^{(n)}$ $(2n-1)$th-order aberration function, 50, 226

$f_1, f_2$ focal lengths of anamorphotic system, 193

$\hat{f}, \hat{f}'$ anterior and posterior focal lengths, 40

$\underset{m}{f}_n$ characteristic aberration function of coordinate order $n$ and chromatic order $m$, 224

$\underset{m}{f}{}^{(n)}_{\mu\nu}$ characteristic aberration coefficients of coordinate order $n$ and chromatic order $m$, 226

$f^{(n)}_{\mu\nu}$ characteristic aberration coefficients of order $2n-1$ (coefficients of $f^{(n)}$), 51

$f_{n\lambda\mu\nu}$ general $n$th-order characteristic aberration coefficient, 31

$F$ general characteristic function, 17

$F_0$ general ideal characteristic function, 19, 224

$F, F'$ principal focal points, 40

$g$ arbitrary additive function in $F$, 20, 47, 52

$G(\chi)$ angle characteristic of concentric system with central base-points, 158

$G_n$ $(2n-1)$th-order coefficient of $G(\chi)$, 158

$\mathbf{g}$ $\Delta\mathbf{\Lambda}$ $(\mathbf{g}_a, \mathbf{g}_b = \Delta\mathbf{\Lambda}_a, \Delta\mathbf{\Lambda}_b)$, 275

$\mathbf{g}^*$ $\Delta\mathbf{\Lambda}^*$, 277

$\mathbf{G}$ $\mathbf{G}_a+\mathbf{G}_b$, 285

$\mathbf{G}_{a;j}$ $\sum_{i=1}^{j-1}\mathbf{g}_{ai}$, 284

$\mathbf{G}_{a:j}$ $\sum_{i=j}^{k}\mathbf{g}_{ai}$, 284

$\mathbf{G}_a$ $\mathbf{G}_{a;j}+\mathbf{G}_{a:j}$, 284

$\mathfrak{g}_{\alpha\beta j}^{(n)}$ fifth-order pseudo-coefficients of $\mathbf{g}$ or $\mathbf{g}^*$, 298

$h$ object height, 196

$h'$ ideal image height, 55, 196

$\mathbf{H}_j$ coordinates of intersection point of $\mathscr{R}$ with normal plane through point before $\mathscr{S}_j$ conjugate to $B$, 275

$\mathbf{H}_j$ (reduced) coordinates of intersection point of $\mathscr{R}$ with normal plane through $O_{0j}$, 278

$\mathbf{H}_j^*$ $\alpha_j\mathbf{H}_j$, 277

$\mathbf{H}_{Ej}$ (reduced) coordinates of intersection point of $\mathscr{R}$ with normal plane through $E_j$, 283

$\mathsf{H}$ magnetic field vector, 258

$\mathsf{H}_0$ amplitude of $\mathsf{H}$, 258

$i_a, i_b$ paraxial constants relating to $\mathbf{i}$, 273

$I, I'$ angles of incidence and refraction, 3

$I'$ ideal image point, 46

$\mathbf{i}$ $c\mathbf{y}+\boldsymbol{\beta}$, 273

$\mathscr{I}, \mathscr{I}'$ conjugate object and image planes, 19

$\hat{\mathscr{I}}, \hat{\mathscr{I}}'$ shifted conjugate object and image planes, 114

$\mathscr{I}^*$ curved object surface, 153

$\mathscr{I}^{*\prime}$ curved image surface, 150

$j_1$ $\frac{1}{8}(\hat{m}-m)(s-\hat{m})^{-4}$, 120

$j_2$ $\frac{1}{16}(\hat{m}-m)(s-\hat{m})^{-6}$, 123

$J$ $(w^2-\xi P^2)^{-\frac{1}{2}}$, 83

$J$ the generator $(s-m)^2(\partial/\partial\eta)$, 129

$\bar{J}$ generator adjoint to $J$, 130

$k$ $N/N'$, 278

$k_\alpha$ $t_{\mu\nu}^{(n)}\left(\alpha = \frac{1}{2}\mu(\mu+1)+\nu+1\right)$, 117

$k_\alpha$ paraxial characteristic coefficients, 39, 172, 198

$K$ generic symbol for optical system, 8

$K$ offence against the sine-condition ($=\kappa_s/h'$), 88

$\bar{K}$ $\kappa_t/h'$, 90

$K_t, K_s, K_m$ tangential, sagittal, mean comatic asymmetry, 61

$l_0$ location of $O_0$, 273

$l_{0j}$ $-y_{aj}/v_{aj}$ (i.e. location of $O_{0j}$), 285

$L$ $(1-m^2)^{-2}(m^2\bar{\xi}-2m\bar{\eta}+\bar{\zeta})$, 139

$m$ reduced magnification associated with $\mathscr{I}, \mathscr{I}'$, 19, 41

$\hat{m}$ value of $m$ after $\mathscr{I}$ has been moved, 114

$M$ $(1-m^2)^{-2}[m\bar{\xi}-(1+m^2)\bar{\eta}+m\bar{\zeta}]$, 139

$M$ the generator $(s-m)^2(\partial/\partial\zeta)$, 127

$\bar{M}$ generator adjoint to $M$, 127

$\mathbf{M}$ $\mathbf{V}_1-m\mathbf{V}_k'$, 274

$n$ wave-index (magnitude of $\mathsf{n}$), 255

$N, N'$ refractive index in object and image space, 3

$N(x, y, z)$ refractive index function, 5

$N$ ratio of energy density to magnitude of energy flux, 260

$N$ $(1-m^2)^{-2}(\bar{\xi}-2m\bar{\eta}+m^2\bar{\zeta})$, 139

$\mathsf{n}$ the vector $(N_\alpha, N_\beta, N_\gamma)$, 250

$\mathsf{n}$ grad $V$, 259

$O, O'$ points of $\mathscr{I}, \mathscr{I}'$, 19

$O_0, O_0'$ axial points of $\mathscr{I}, \mathscr{I}'$, 42

$O_0^*$ $O_0^*$ and $O_0'$ are equidistant from central plane, 137

$O_{0j}, O_{0j}'$ points conjugate to $O_0$ before and after $\mathscr{S}_j$, 269

$O(n)$ order symbol (terms of order exceeding $n$), 26

$O(n_r)$ terms of coordinate order exceeding $n$ whose chromatic order exceeds $r$, 227

$p$ location of $E$, 273

$p$ $-(\hat{m}-m)/(s-\hat{m})$, 119

$p_i\,(i = 1, \ldots, 4)$ variable conjugate to ray-coordinate $q_i$, 18

$p_\alpha\,(\alpha = 1, \ldots, 6)$ third-order characteristic aberration coefficients of symmetric system, 55, 61, 281

$p_\alpha\,(\alpha = 1, \ldots, 6)$ third-order characteristic normal aberration coefficients of semi-symmetric system, 175

$p_\alpha\,(\alpha = 1, \ldots, 8)$ third-order characteristic aberration coefficients of $c$-symmetric or $t$-symmetric system, 196, 215

$p_\alpha\,(\alpha = 1, \ldots, 16)$ third-order characteristic aberration coefficients of doubly plane-symmetric system, 201

$\tilde{p}_\alpha\,(\alpha = 1, 2, 3)$ third-order characteristic skew aberration coefficients of semi-symmetric system, 175

$\hat{p}_\alpha\,(\alpha = 1, \ldots, 6)$ third-order characteristic aberration coefficients associated with shifted pupil or object, 97

$p_\alpha^\dagger\,(\alpha = 1, \ldots, 6)$ third-order spherical point characteristic aberration coefficients, 101

$p_\alpha^*, \bar{p}_\alpha^*\,(\alpha = 1, 2, 3)$ third-order effective aberration coefficients, 74

$p_\alpha^\ddagger, \bar{p}_\alpha^\ddagger\,(\alpha = 1, 2, 3)$ third-order effective aberration coefficients associated with posterior coordinates in $\mathscr{W}_0$, 101

$p_{\alpha j}, \bar{p}_{\alpha j}\,(\alpha = 1, 2, 3)$ third-order coefficients of $\mathbf{g}_j^*$ (or $\mathbf{g}_j$), 302

$`p_{\alpha j}, `\bar{p}_{\alpha j}\,(\alpha = 1, 2, 3)$ third-order coefficients of $\mathbf{w}_j$, 298

$p_{\sigma\alpha j}, \bar{p}_{\sigma\alpha j}$ $(\alpha = 1, 2, 3)$ third-order coefficients of $\mathbf{d}_{\sigma j}$, 302

$p_{\mu\alpha j}, \bar{p}_{\mu\alpha j}$ $(\alpha = 1, 2, 3)$ third-order coefficients of $\mathbf{d}_{\mu j}$, 302

$P$ point of incidence, 3

$P$ $1-2w(dS/d\xi)$, 83

$P_j$ $\xi+2\theta_j\eta+\theta_j^2\zeta$, 298

$\mathfrak{p}$ $\frac{1}{2}N(1-k)i_a^2 y_a(i'_a+v_a)$, 279

$\mathfrak{p}_{\alpha j}$ $(\alpha = 1, \ldots, 6)$ third-order pseudo-coefficients of $t_j$, 307

$\mathfrak{p}_{\alpha j}, \bar{\mathfrak{p}}_{\alpha j}$ $(\alpha = 1, 2, 3)$ third-order pseudo-coefficients of $\mathbf{g}_j^*$ (or $\mathbf{g}_j$), 298

$`\mathfrak{p}_{\alpha j}, `\bar{\mathfrak{p}}_{\alpha j}$ $(\alpha = 1, 2, 3)$ third-order pseudo-coefficients of $\mathbf{w}$, 298

$\mathsf{p}$ unit normal to refracting surface, 3

$q$ $-\hat{x}'/(1-\hat{x}') = (\hat{s}-s)/(\hat{s}-m)$, 96, 116

$q, q'$ location of $B$, $B'$ relative to centre of concentric system, 157

$q, q'$ location of local base-points relative to $A$, 270

$q_i$ $(i = 1, \ldots, 4)$ ray-coordinates, 17

$Q$ $1-2w^3(dC/d\xi)$, 83

$Q_j$ $2\phi_j(\eta+\theta_j\zeta)$, 298

$r$ radius of spherical surface, 14

$R$ $1+wC$, 83

$R'$ distance between $E'$ and $I'$, 100

$R_j$ $\phi^2\zeta$, 298

$R_{nm}$ circle polynomials, 110

$\mathscr{R}$ generic ray, 6

$\mathscr{R}_0$ base-ray, 26

$\underset{0}{\mathscr{R}}_0$ base-ray at colour $\underset{0}{\lambda}$, 224

$s$ reduced magnification associated with pupil planes, 42

$s$ $\frac{1}{2}s(u)$ is the $x$-coordinate of points on refracting surface, 270

$s_n$ coefficients of power series for $s(u)$, 272

$s_n$ coefficients of $S(=f_{00}^{(n+1)})$, 83

$\hat{s}$ value of $s$ after shift of pupil planes, 116

$s_\alpha\,(\alpha = 1, \ldots, 10)$ fifth-order characteristic aberration coefficients of symmetric system, 62, 74

$s_\alpha\,(\alpha = 1, \ldots, 10)$ fifth-order characteristic normal aberration coefficients of semi-symmetric system, 180

$\tilde{s}_\alpha\,(\alpha = 1, \ldots, 6)$ fifth-order characteristic skew aberration coefficients of semi-symmetric system, 181

$\hat{s}_\alpha\,(\alpha = 1, \ldots, 10)$ fifth-order characteristic aberration coefficient associated with shifted pupils or object, 100

$s_\alpha^\dagger\,(\alpha = 1, \ldots, 10)$ fifth-order spherical point characteristic aberration coefficients, 101

$s_\alpha^*, \bar{s}_\alpha^*\,(\alpha = 1, \ldots, 6)$ fifth-order effective aberration coefficients, 24

$s_\alpha^\ddagger, \bar{s}_\alpha^\ddagger\,(\alpha = 1, \ldots, 6)$ fifth-order effective aberration coefficients associated with posterior coordinates in $\mathscr{W}_0$, 101

$s_\alpha^*\,(\alpha = 1, \ldots, 6)$<br>
$\bar{s}_\alpha^*\,(\alpha = 1, \ldots, 6)$<br>
$\tilde{s}_\alpha^*\,(\alpha = 1, \ldots, 6)$<br>
${}^{\#}s_\alpha^*\,(\alpha = 1, \ldots, 3)$<br>
${}^{\#}\bar{s}_\alpha^*\,(\alpha = 1, \ldots, 3)$<br>
} fifth-order effective aberration coefficients of semi-symmetric system, 181

$s_{\alpha j}\,(\alpha = 1, \ldots, 10)$ fifth-order coefficients of $t_j$, 309

$s_{\alpha j}, \bar{s}_{\alpha j}\,(\alpha = 1, \ldots, 6)$ fifth-order coefficients of $\mathbf{g}_j^*$ or $\mathbf{g}_j$, 302

${}^{`}s_{\alpha j}, {}^{`}\bar{s}_{\alpha j}\,(\alpha = 1, \ldots, 6)$ fifth-order coefficients of $\mathbf{w}_j$, 302

$S$ the generator $(s-m)^2(\partial/\partial\xi)$, 127

$\bar{S}$ generator adjoint to $S$, 127

$S(\xi)$ $f(\xi, 0, 0)\,(f = v, t, \ldots)$, 83, 121, 239

**S** $\mathbf{V}_1 - s\mathbf{V}_k'$, 274

$\mathbf{s}$ vector from initial point to point of incidence, 4

S energy flux, 259

$\mathfrak{s}_{\alpha j}\,(\alpha = 1, \ldots, 10)$ fifth-order pseudo-coefficients of $t_j$, 308

$\mathfrak{s}_{\alpha j}, \bar{\mathfrak{s}}_{\alpha j}\,(\alpha = 1, \ldots, 6)$ fifth-order pseudo-coefficients of $\mathbf{g}_j^*$ or $\mathbf{g}_j$, 298

$`\mathfrak{s}_{\alpha j}, `\bar{\mathfrak{s}}_{\alpha j}\,(\alpha = 1, \ldots, 6)$ fifth-order pseudo-coefficients of $\mathbf{w}_j$, 298

$\mathscr{S}$ surface of revolution, usually spherical, 14, 269

$t$ angle characteristic aberration function, 52, 236

$t_j$ $t$ of $j$th surface, local base-points, 282

$t_\alpha\,(\alpha = 1, \ldots, 15)$ seventh-order characteristic aberration coefficients, 74

$t_\alpha^\dagger\,(\alpha = 1, \ldots, 15)$ seventh-order point characteristic aberration coefficients, 101

$t_\alpha^*, \bar{t}_\alpha^*\,(\alpha = 1, \ldots, 10)$ seventh-order effective aberration coefficients, 74

$t_\alpha^\ddagger, \bar{t}_\alpha^\ddagger\,(\alpha = 1, \ldots, 10)$ seventh-order effective aberration coefficients associated with posterior coordinates in $\mathscr{W}_0$, 101

$t_{\mu\nu}^{(n)}$ $(2n-1)$th-order angle characteristic aberration coefficients of symmetric system, 73

tr trace (of tensor): $\operatorname{tr} \epsilon_{kl} = \epsilon_{11} + \epsilon_{22} + \epsilon_{33}$, 261

$T$ angle characteristic, 11

$T_0$ ideal angle characteristic, 46

$T_a$ angle characteristic, base-points $O_0, O_0'$, 283

$T_b$ angle characteristic, base-points $E, E'$, 283

$T_{aj}, T_{bj}$ $T_a$, $T_b$ for $j$th refracting surface, local base-points, 283

$\hat{T}$ angle characteristic referred to shifted base-points, 119

$\tilde{T}$ focal angle characteristic, 125

$T^*$ $T$ of concentric system, base-points $O_0^*, O_0'$, 137

$T_j^{(n)}$ $(2n-1)$th-order part of $T_j\,(= t_j^{(n)}, n > 1)$, 282

$u$ $\xi - 2\eta + \zeta$, 47

$u$ $1 - sm$, 137

$u$ $y_P^2 + z_P^2$, 270

$u^{(n)}_{\mu\nu}, \bar{u}^{(n)}_{\mu\nu}$ (2$n$ – 1)th-order effective aberration coefficients of symmetric system, 73

$u^{(n)}_{\mu\nu}, \bar{u}^{(n)}_{\mu\nu}$, $\tilde{u}^{(n)}_{\mu\nu}$, ${}^{\#}u^{(n)}_{\mu\nu}, {}^{\#}\bar{u}^{(n)}_{\mu\nu}$ } (2$n$ – 1)th-order effective aberration coefficients of semi-symmetric system, 180

$v$ point characteristic aberration function, 52

$v$ $s-m$, 139

$v^{\dagger}$ spherical point characteristic aberration function, 100

$v^{\#}$ normal point characteristic aberration function of semi-symmetric system, 174

$\tilde{v}^{\#}$ skew point characteristic aberration function of semi-symmetric system, 174

$v_1$ point characteristic aberration function for arbitrary posterior base-surface, 99

$\tilde{v}_1$ modified point characteristic aberration function, 99

$\tilde{v}^{\dagger}$ modified spherical point characteristic aberration function, 102

$\hat{v}$ point characteristic aberration function associated with shifted posterior base-plane, 95

$\mathring{v}$ $v(0, y', z', y, z)$, 98

$v^{(n)}, v^{\dagger(n)}, \ldots$ (2$n$ – 1)th-order part of $v$, $v^{\dagger}$, ..., 10

$v^{(n)}_{\lambda}$ part of $v^{(n)}$ varying with $\rho$, $h'$ as $\rho^{2n-\lambda}h'^{\lambda}$, 67

$v^{(n)}_{\mu\nu}, \mathring{v}^{(n)}_{\mu\nu}$ (2$n$ – 1)th-order (point) characteristic aberration coefficients: coefficients of $v^{(n)}$, $\mathring{v}^{(n)}$, 67

$v^{(n)}_{\mu\nu}, \tilde{v}^{(n)}_{\mu\nu}$ (2$n$ – 1)th-order characteristic aberration coefficients: coefficients of $v^{\#(n)}$, $\tilde{v}^{\#(n)}$, 175

$v_{aj}, v_{bj}$ paraxial constants relation to $\boldsymbol{\beta}_j$ or $\mathbf{V}_j$, 273, 274

$V$ point characteristic, 8

$V$ effectively the phase of electromagnetic field vectors, 258

$V$ $\beta/\alpha$, 274

$V^*$ optical length, 4

$\tilde{V}$ optical distance, 5

$V^\dagger$ spherical point characteristic, 100

$V_1$ point characteristic associated with arbitrary posterior base-surface, 98

$\hat{V}$ point characteristic associated with shifted base-surfaces, 95

$V_0$ ideal point characteristic, 20, 46, 49

$w$ $(1+\xi)^{\frac{1}{2}}$, 83

$w$ $1-m^2$, 139

$w^{(n)}_{\alpha\beta}, \overline{w}^{(n)}_{\alpha\beta}$ $(2n-1)$th-order coefficients of $\mathbf{w}$, 301

$W$ energy density, 259

$W$ $\gamma/\alpha$, 274

$W_1, W_2$ mixed characteristic functions, 15

$W_{10}, W_{20}$ ideal mixed characteristic functions, 15

$\mathbf{w}$ $\mathbf{g}^*/e$, 277

$\mathfrak{w}^{(n)}_{\alpha\beta}, \overline{\mathfrak{w}}^{(n)}_{\alpha\beta}$ $(2n-1)$th-order pseudo-coefficients of $\mathbf{w}$, 297

$\mathscr{W}$ wavefront, 34

$\mathscr{W}_0$ posterior base-surface for $V^\dagger$: ideal spherical wavefront, 100, 103

$x$ $\bar{x}$-coordinates of initial point, 5, 10

$x'$ $\bar{x}'$-coordinate of final point, 5, 10

$\bar{x}, \bar{x}'$ current Cartesian coordinate, 5

$x_P$ $x$-coordinate of point of incidence, 270

$\hat{x}$ amount of shift of object surface, 114

$\hat{x}'$ amount of shift of posterior base-plane, 95

$X'$ $(1-x')/\alpha'$, 100

$X'$ $\bar{x}'$-coordinate of points of $\mathscr{I}^{*\prime}$, 150

$y$ $\bar{y}$-coordinate of initial point, 5, 10, 16

$y'$ $\bar{y}'$-coordinate of final point, or of $\mathscr{B}'$, 5, 10, 16

$\bar{y}, \bar{y}'$ current Cartesian coordinates in object and image space, 5

$y_1$ $my$, 47
$\hat{y}$ $\bar{y}'$-component of partial displacement, 56
$y_e, y_e'$ $\bar{y}$-, $\bar{y}'$-coordinates of points of $\mathscr{E}, \mathscr{E}'$, 61, 164
$\hat{y}'$ $\bar{y}'$-coordinate of points of shifted exit pupil, 97
$y^{\dagger\prime}$ $\bar{y}'$-coordinate of points in $\mathscr{W}_0$, 103
$y^{\#}$ $-z$, 172
$y_P$ $y$-coordinate of point of incidence, 270
$y_{aj}, y_{bj}$ paraxial constants relating to $\mathbf{Y}_j$, 273
$Y$ $y$-coordinate of points of polar tangent plane of $\mathscr{S}$, 285
$Y'$ $\bar{y}'$-coordinate of points of $\mathscr{I}'$ in the context of $V$, 19, 29, 40
$Y'$ $\bar{y}'$-coordinate of points $\mathscr{I}^{*\prime}$, 150
$Y^*$ $\alpha Y$, 285

$z$ $\bar{z}$-coordinate of initial point, 5, 10, 16
$z'$ $\bar{z}$-coordinate of final point, or of $\mathscr{B}'$, 5, 10, 16
$\bar{z}, \bar{z}'$ current Cartesian coordinates in object and image space, 5
$z_1$ $mz$, 47
$z_1$ $(N'/N)z$, 190
$\hat{z}$ $\bar{z}'$-component of partial displacement, 56
$z_e, z_e'$ $\bar{z}$-, $\bar{z}'$-coordinates of points in $\mathscr{E}, \mathscr{E}'$, 61, 164
$\hat{z}'$ $\bar{z}'$-coordinate of points of shifted exit pupil, 97
$z^{\dagger\prime}$ $\bar{z}'$-coordinate of points of $\mathscr{W}_0$, 103
$z^{\#}$ $y$, 172
$z_P$ $z$-coordinate of point of incidence, 270
$Z$ $z$-coordinate of points of polar tangent plane of $\mathscr{S}$, 285
$Z'$ $\bar{z}'$-coordinate of points of $\mathscr{I}'$ in context of $V$, 19, 29, 40
$Z'$ $\bar{z}'$-coordinate of points of $\mathscr{I}^{*\prime}$, 150
$Z^*$ $\alpha Z$, 285
$\mathfrak{z}_\alpha, \bar{\mathfrak{z}}_\alpha (\alpha = 1, \ldots, 5)$ coefficients ancillary to $\mathfrak{s}_\alpha, \bar{\mathfrak{s}}_\alpha (\alpha = 1, \ldots, 6)$, 300
$\mathfrak{z}_\alpha (\alpha = 1, \ldots, 8)$ coefficients ancillary to $\mathfrak{s}_\alpha (\alpha = 1, \ldots, 10)$, 308

*Greek symbols*

$\alpha$ $x$-direction cosine, 10
$\alpha_r$ absolute invariant of order $r$, 128

$\beta$ $y$-direction cosine, 10

$\gamma$ $z$-direction cosine, 10

$\delta$ variational symbol, 4
$\delta$ longitudinal spherical aberration, 88
$\Delta$ retardation of the wavefront, 106
$\Delta$ defined by $\Delta X = X' - X$ for any $X$, 270
$\boldsymbol{\delta}$ $(v'_a v_b - v'_b v_a)^{-1}(v'_a \Delta\boldsymbol{\delta}_\sigma + v'_b \Delta\boldsymbol{\delta}_\mu)$, 295
$\boldsymbol{\delta}_\sigma, \boldsymbol{\delta}_\mu$ increments relating to $\boldsymbol{\sigma}, \boldsymbol{\mu}$, 286
$\boldsymbol{\delta}_s, \boldsymbol{\delta}_m$ increments relating to $\mathbf{S}$, $\mathbf{M}$, 284

$\epsilon_{kl}$ dielectric tensor, 258
$\epsilon_m (m = 1, 2, 3)$ principal values of $\epsilon_{kl}$, 263
$\boldsymbol{\epsilon}'$ aberration of ray (displacement), 19
$\boldsymbol{\epsilon}'_n$ $n$th-order displacement, 32
$\underset{m}{\boldsymbol{\epsilon}'_n}$ $n$th-order displacement of chromatic order $m$, 225
$^*\boldsymbol{\epsilon}'$ pseudo-displacement, 85
$\hat{\boldsymbol{\epsilon}}'$ displacement after shift of $O_0$ or $E'$, 97
$\boldsymbol{\epsilon}'_n(\psi_n)$ $n$th-order displacement induced by $\psi_n$, 33

$\zeta$ rotational invariant (appropriate to context), 36, 38, 48
$\zeta$ reflection invariant, 191, 200, 214
$\bar{\zeta}$ $\beta^2 + \gamma^2$, 48
$\mathfrak{z}$ corresponds to $\zeta$ for non-standard reference planes, 95, 117

$\eta$ rotational invariant (appropriate to context), 36, 38, 48
$\bar{\eta}$ $\beta'\beta + \gamma'\gamma$, 48

$\eta$ reflection invariant, 191, 200, 214

$\hat{\eta}$ corresponds to $\eta$ for non-standard reference planes, 95, 117

$\eta_{kl}$ tensor inverse to $\epsilon_{kl}$, 261

$\theta$ angular polar coordinate in $\mathscr{E}'$, 55

$\theta$ $v_b/i_a$, 274

$\tilde{\theta}$ $y_b/y_a$, 281

$\bar{\theta}$ $v_b/v_a$, 288

$\kappa$ $NN'/(N'-N)^2$, 14

$\kappa$ $\nu/\nu'$, 235

$\kappa_t, \kappa_s$ tangential and sagittal circular coma, 59, 84

$\kappa_n$ $(-1)^n \binom{\frac{1}{2}}{n} d^*$, 138

$\lambda$ wavelength, 1

$\underset{0}{\lambda}$ base-colour, 223

$\boldsymbol{\lambda}$ Lagrange invariants, 29

$\boldsymbol{\Lambda}$ $-Nv\mathbf{H}$, 275

$\boldsymbol{\Lambda}^*$ $\alpha\boldsymbol{\Lambda}$, 275

$\mu_{2n-1,\alpha}$ $m$-invariants of order $2n-1$, 129

$\boldsymbol{\mu}$ $\boldsymbol{\beta}-m\boldsymbol{\beta}'$, 76, 273

$\boldsymbol{\mu}_j$ $\boldsymbol{\mu}+\mathbf{d}_{\mu j}$, 296

$\nu$ $N/\underset{0}{N}$, 230

$\nu'$ $N'/\underset{0}{N'}$, 230

$\nu_m$ dispersion coefficient of chromatic order $m$, 227

$\xi$ rotational invariant (appropriate to context), 36, 38, 48

$\xi$ reflection invariant, 191, 200, 214

$\bar{\xi}$ $\beta'^2+\gamma'^2$, 48

$\hat{\xi}$ corresponds to $\xi$ for non-standard reference planes, 95, 117

$\pi_\alpha(\alpha = 1, \ldots, 8)$ second-order characteristic aberration coefficients of singly plane-symmetric system, 204

$\rho$ radial polar coordinate in $\mathscr{E}'$, 55

$\rho_0$ largest value of $\rho$, 56

$\sigma$ $\Delta(N\cos I)$, 5

$\sigma$ transverse spherical aberration: $\epsilon'_y(h' = 0, \theta = 0)$, 84

$\sigma_\alpha(\alpha = 1, \ldots, 5)$ Seidel coefficients, 55

$\tilde{\sigma}_\alpha(\alpha = 2, 3, 5)$ skew Seidel coefficients, 176

$\sigma_{2n-1,\alpha}$ $s$-invariants of order $2n-1$, 118, 129

$\boldsymbol{\sigma}$ $\boldsymbol{\beta}-s\boldsymbol{\beta}'$, 76, 273

$\boldsymbol{\sigma}_j$ $\boldsymbol{\sigma}+\mathbf{d}_{\sigma j}$, 296

$\Sigma$ summation sign, 18

$\tau$ skew rotational invariant, 37, 171

T the generator $(s-m)(\partial/\partial\tau)$, 183

$\overline{\mathrm{T}}$ generator adjoint to T, 183

$\phi, \phi'$ angles between $\mathscr{A}$ and initial and final axial rays, 84

$\phi$ $f_K/[(s-m)Nri_a]$, 275

$\phi_\lambda^{(n)}$ factor multiplying $\rho^{2n-\lambda}h'^\lambda$ in $v^{(n)}$, 67

$\chi$ out-of-focus parameter, 54

$\chi$ $1-(\alpha\alpha'+\beta\beta'+\gamma\gamma')$, 158

$\psi$ angular polar coordinate in $\mathscr{I}$, 197

$\psi$ $u(ds/du)^2$, 271

$\psi_n$ any one $n$th-order term of aberration function, 33

$\omega$ chromatic coordinate, 224, 228

$\Omega$ primary chromatic part of $(D-d)$-sum, 244

*Special symbols*

$\nabla$ defined by $\nabla X = X' + X$ for any $X$, 270

$\tilde{\omega}$ $c\Delta(1/N)$, 279

$`\tilde{\omega}$ $(s-m)^{-2}\tilde{\omega}$, 299

Grad the vector operator $(\partial/\partial\alpha, \partial/\partial\beta, \partial/\partial\gamma)$, 261

*Affixes*†

$X'$ relates to the image space (of $K$ or a particular $\mathscr{S}$), if $X$ relates to its object space, 3

$X_k$ if $X_j$ relates to $\mathscr{S}_j$, the value $k$ of $j$ relates to the last surface of $K$, i.e. the surface adjacent to the image space, 269

$X_n$ $n$th-order part of $X$, or a coefficient of this, 31

$X^{(n)}$ $(2n-1)$th-order part of $X$, or a coefficient of this, where $K$ is at least doubly plane-symmetric, 50

$\underset{m}{X}$ part of $X$ of chromatic order $m$, or a coefficient of this, 224

$\underset{0}{X}$ value of $X$ associated with the base-colour $\lambda$, 223

$\bar{X}$ only when $X = \xi, \eta, \zeta$: the particular rotational invariants $\bar{\xi} = \beta'^2 + \gamma'^2$, $\bar{\eta} = \beta'\beta + \gamma'\gamma$, $\bar{\zeta} = \beta^2 + \gamma^2$, 48

$X^{\dagger}$ relates to the spherical point characteristic, 100

$\hat{X}$ refers to non-standard positions of reference planes, 95

$X^*$, $\bar{X}^*$ jointly distinguish effective from characteristic aberrations coefficients, 74

$X^{\ddagger}$, $\bar{X}^{\ddagger}$ analogous to $X^*$, $\bar{X}^*$ when considering a curved posterior base-surface, 101

$X_y^{\#}$, $\bar{X}_z^{\#}$ dual of $X_y, X_z (= -X_z, X_y)$, 172

$\tilde{X}$ distinguishes coefficients relating to skew aberration function, 175

$X^{\#}$, $\tilde{X}^{\#}$ jointly distinguish quantities associated with parts of the aberration function of semi-symmetric or $c$-symmetric systems, 174, 194

† $X$ generally denotes any quantity appropriate to the context, and it functions merely as a carrier of the various affixes.

*Miscellaneous conventions*

$\langle X_j \rangle$ denotes the pseudo-expansion of $X_j$, 297

**X** denotes any pair of quantities $X_y$, $X_z$ which transform as the components of a two-vector under rotations about $\mathscr{A}$, 39

X denotes ordinary three-vectors, 3

**A.B** $A_y B_y + A_z B_z$, 39

# BIBLIOGRAPHY

Born, M. and Wolf, E. *Principles of Optics*. Pergamon Press, Oxford, 1959.

Buchdahl, H. A. *Optical Aberration Coefficients*. Dover Publications, Inc., New York, 1968.

Conrady, A. E. *Applied Optics and Optical Design*. Dover Publications, Inc., New York, 1957 (Part I), 1960 (Part II).

Hamilton, W. R. *Mathematical Papers, Vol. I*. Cambridge University Press, 1931.

Herzberger, M. *Strahlenoptik*. Springer, Berlin, 1931.

Kline, M. and Kay, I. W. *Electromagnetic Theory and Geometrical Optics*. Interscience, New York, 1965.

Luneburg, R. K. *Mathematical Theory of Optics*. University of California Press, Berkeley, 1964.

Picht, J. *Einführung in die Theorie der Elektronenoptik*. 2nd edition. J. A. Barth, Leipzig, 1957.

Steward, G. C. *The Symmetrical Optical System*. Cambridge University Press, 1928; also, *Acta Mathematica*, vol. 67, p. 213.

# INDEX